Nanoscale Science and Engineering Forum 2016

Core Programming Area at the 2016 AIChE Annual Meeting

San Francisco, California, USA
13-18 November 2016

ISBN: 978-1-5108-3441-5

Printed from e-media with permission by:

Curran Associates, Inc.
57 Morehouse Lane
Red Hook, NY 12571

Some format issues inherent in the e-media version may also appear in this print version.

Printed by Curran Associates, Inc. (2017)

For permission requests, please contact AIChE
at the address below.

AIChE
120 Wall Street, FL 23
New York, NY 10005-4020

Phone: (800) 242-4363
Fax: (203) 775-5177

www.aiche.org

Additional copies of this publication are available from:

Curran Associates, Inc.
57 Morehouse Lane
Red Hook, NY 12571 USA
Phone: 845-758-0400
Fax: 845-758-2633
Email: curran@proceedings.com
Web: www.proceedings.com

TABLE OF CONTENTS

454451 Bio-Electronic Devices for Healthcare: From Wearable Biosensors to Nanorobots

Abstract:

Research Interests: Wearable Devices, Flexible Electronics, Nanomaterials, Nanomachines, Nanorobotics, Biosensors, Nanomedicine, Nanomotors.

Wearable biosensors are essential to the realization of personalized medicine through continuously monitoring an individualâ€™s state of health. Human sweat, which is rich in physiological information, is an excellent candidate for non-invasive monitoring. Given the complexity of sweat secretion, simultaneous and multiplexed screening of sweat biomarkers is critical. I have developed a mechanically flexible and fully-integrated sensor array for multiplexedÂ *in-situ*Â perspiration analysis, which simultaneously and selectively measures sweat metabolites and electrolytes, as well as the skin temperature (to compensate other sensorsâ€™ readings) (Gao et al. Nature, 2016, 529, 509-514). The platform is the first ever demonstration of a complete system for accurate sweat analysis which merged skin-conforming plastic-based sensors and conventional integrated-circuit components at an unprecedented level of integration, not only to measure the output of multiplexed and selective sensors, but also to obtain an accurate assessment *via* signal processing of the physiological state of the human subjects. This wearable system has been used to measure the detailed sweat profile of human subjects engaged in prolonged indoor and outdoor physical activities. Currently I have greatly expanded the scope of platform to monitor a wide panel of analytes including heavy metals, peptides and proteins. I envision that wearable sweat biosensors can be used for diverse clinical and physiological applications (*e.g.* monitoring pre-diabetes, dehydration, heavy metal poisoning, and major depressive disorder). While wearable devices play an important role in health monitoring on the skin, nanorobots offer great promise to efficiently deliver drug payloads and perform precision diagnosis/surgery inside human body. The nanoscale propulsion is challenging due to low Reynolds fluids. I have invented the first fully biocompatible and biodegradable nanorobots that can move autonomously in human biofluids based on Zn or Mg micromotors. I have also demonstrated the first *in vivo* study of synthetic nanorobots in living mice. The acid-driven propulsion effectively enhances the delivery of cargo payloads onto the stomach wall. In addition, I pioneered a variety of artificial nanomachines that can achieve nanoscale propulsion. After modification with different functional groups or bio-receptors, these nanorobots are able to selectively transport and isolate specific targets (such as protein, DNA, bacteria and cancer cells). These works are anticipated to significantly advance the emerging field of nanomachines and to open the door to *in vivo*evaluation and clinical applications of the synthetic nanorobots.

Teaching Interests: I am interested in teaching most of the the major courses for Chemical Engineering, especially Heat and Mass Transfer, Fluids Dynamics, and Thermodynamics, Separation Process. I am also interested in other courses including Electrochemistry, Nanotechnology and Nanoengineering, Nanomedicine.

Tailoring Inorganic Materials with High Surface Area for Electronic Applications

Wanmei Sun[1], Richard B. Kaner[2], Micah Green[1]

[1] Artie McFerrin Department of Chemical Engineering, Texas A&M Univeristy
[2]Department of Chemistry & Biochemistry, University of California Los Angeles

wanmei@tamu.edu

Synthesis of 2D Materials

Graphene Oxide

--From modified Hummers' method

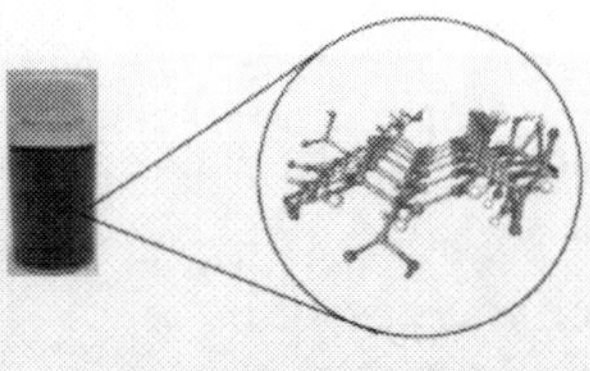

Advantages of Hummers' method

- Easy dispensability in water and other organic solvents
- Sheet size ~ 50 μm
- 1-5 layers

Graphene

--From electrochemical exfoliation

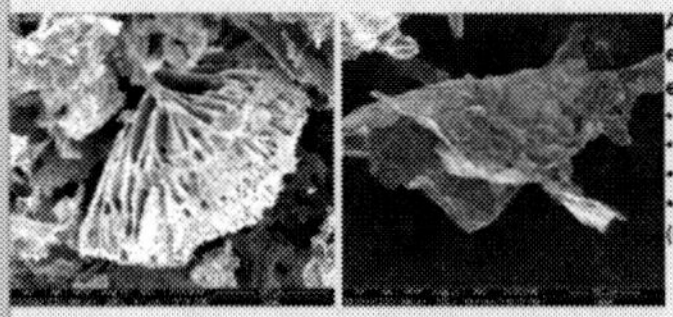

Advantages of electrochemical exfoliation

- Up scalability
- Solution processability
- Eco-friendliness
- Large sheet size (up to 200 μm)

Black Phosphorene (in progress)

--From high energy mechanical milling and solvent exfoliation

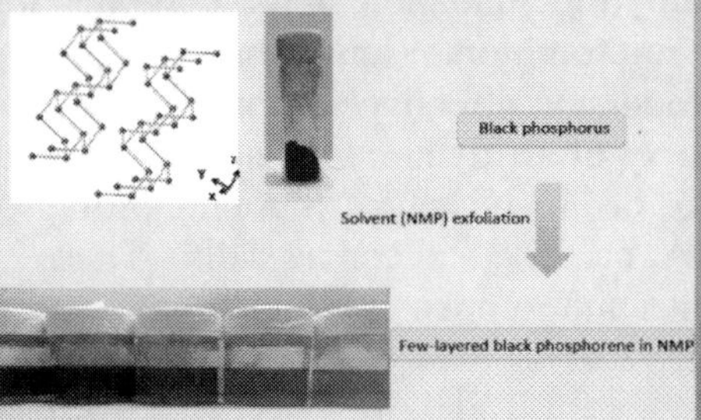

Advantages of black phosphorene

- Tunable direct band gap (0.3 – 2.0 eV for bulk to monolayer)
- High hole mobility (up to 286 $cm^2V^{-1}S^{-1}$)
- Anisotropic electronic, transport, optoelectronic and mechanical properties

Mxene (in progress)

--Selective etching of Al from Ti_2AlC without HF

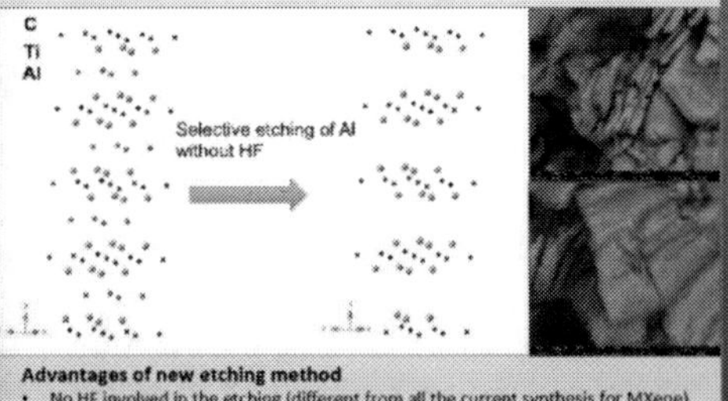

Advantages of new etching method

- No HF involved in the etching (different from all the current synthesis for MXene)
- Tunable functional groups on the surface
- Mxene is hydrophilic and highly conductive

Synthesis of 3D Materials

Hexagonal Tungsten Trioxide

--From hydrothermal reaction

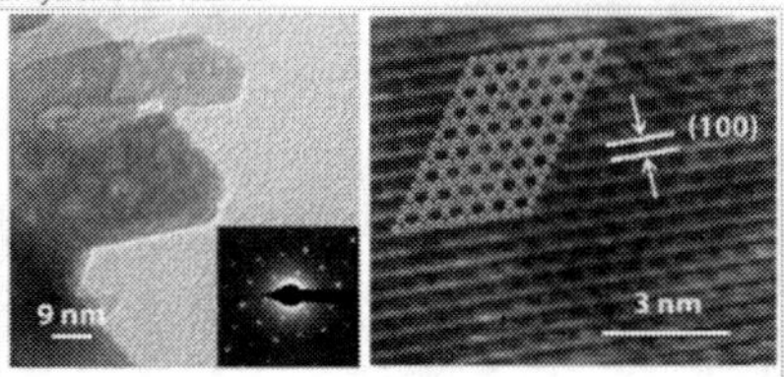

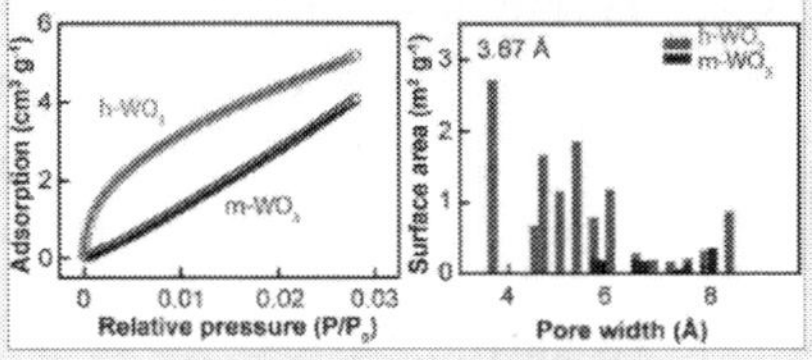

Transmission electron microscopy (TEM) and selected area electron diffraction (SAED) images show high crystallinity. Low pressure CO_2 adsorption isotherms and nonlocal density functional theory fitting shows the existing of intralcrystalline tunnels and the tunnel size is 3.67 Å. Thermal gravimetric analysis(TGA), inductively coupled plasma atomic emission spectroscopy(ICP-OES) show the open and rigid tunnels adsorb H^+ and Li^+, but not Na^+ in aqueous solution.

Synthesis of 0D & 2D Materials

Transition Metal Oxides on Reduced Graphene Oxide

--From *in situ* reduction (in microwave, flash *et al*)

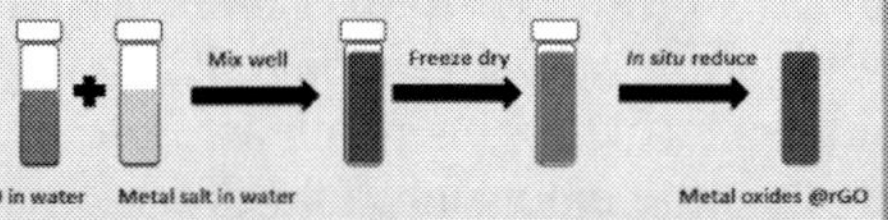

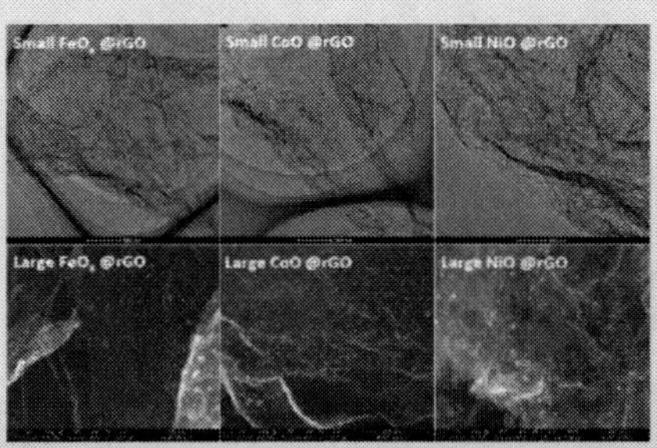

Synthesis of 1D Materials

Tungsten Trioxide, Molybdenum Trioxide *et al*

--From physical vapor deposition

Electrochemical Energy Storage

High Surface Area Improves Electrochemical Prosperities

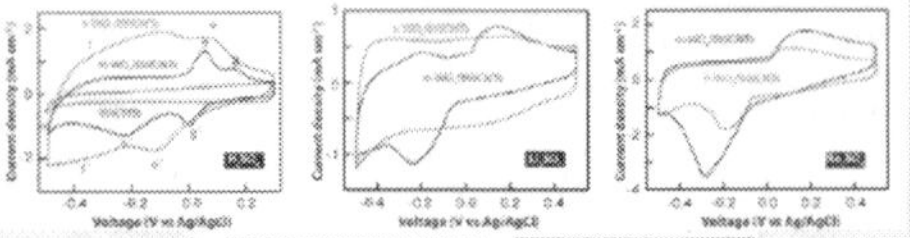

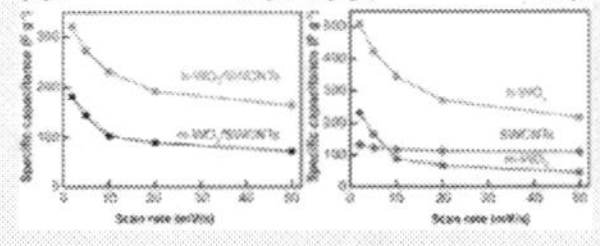

With increased surface area(from intracrystalline tunnels), hexagonal tungsten trioxide processes high specific pesudocapacitance in small ion electrolytes(such as H^+ and Li^+), not for big ion electrolytes(such as Na^+)

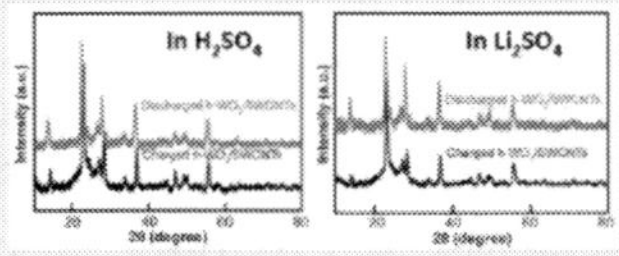

No phase changes for hexagonal tungsten trioxide in small ion electrolytes (such as H^+ and Li^+)

Research Interesting

- Rational design and synthesis of novel 2D materials
- Tailoring the surface of 2D materials for related applications
- 2D materials for electromagnetic interference shielding
- 2D materials and their composites for energy storage
- 0D and 2D materials for biosensor and bio images

Resume

Working experience

- Postdoctoral Research associate — Mar. 2016-Present

Education

- Joint Ph.D in Chemistry — Oct. 2012-Oct.2014
  Department of Chemistry and Biochemistry,
  University of California, Los Angeles
- Ph.D in Physical Electronic — Sept. 2010-Mar. 2016
  Wuhan National Laboratory Optoelectronics,
  School of Optical and Electronic Information,
  Huazhong University of Science and Technology, China
- B.A. in Applied Chemistry — Sept. 2006-Jun. 2010
  School of Chemistry and Chemical Engineering,
  Huazhong University of Science and Technology, China

Publications

- Wanmei Sun, Michael T. Yeung, Andrew T. Lech, Cheng-Wei Lin, Chain Lee, Tianqi Li, Xiangfeng Duan, Jun Zhou, Richard B. Kaner "High Surface Area Tunnels in Hexagonal WO_3" *Nano Letters* 2015, 15 (7), 4834-4838
- Xianghui Zhang, Li Gong, Kang Liu, Yuanzhi Cao, Xu Xiao, Wanmei Sun, Xuejiao Hu, Yihua Gao, Jian Chen, Jun Zhou, Zhong Lin Wang " Tungsten Oxide Nanowires Grown on Carbon Cloth as a Flexible Cold Cathode" *Advanced Materials* 2010, 46 (22), 5292-5296
- Tianqi Li, Majid Beidaghi, Xu Xiao, Liang Huang, Zhimi Hu, Wanmei Sun, Xun Chen, Yury Gogotsi, Jun Zhou "Ethanol Reduced Molybdenum Trioxide for Li-ion Capacitors" *Nano Energy* 2016, 26, 100-107
- Jia Li, Kang Liu, Xiang Gao, Bin Yao, Kaifu Huo, Yongliang Cheng, Xiaofeng Cheng, Dongcheng Chen, Bo Wang, Wanmei Sun, Dong Ding, Meilin Liu, Liang Huang " Oxygen- and Nitrogen-Enriched 3D Porous Carbon for Supercapacitor of High Volumetric Capacity" *ACS Applied Materials & Interfaces* 2015, 7 (44), 24622-24628

459786 Fluorescent Nanosensors for Biomolecular Targets

Abstract:

Research Interests: Single walled carbon nanotubes (SWCNTs) have remarkable optical properties, and they benefit from the ease of surface functionalization and biocompatibility, which render them favorable agents for numerous biomedical applications including imaging, sensing, targeting, delivery, and therapy. Semiconducting SWCNTs have intrinsic bandgap fluorescence of excitonic nature in a wide range of the near infrared (nIR) part of the spectrum. In addition to their high photo-stability, lack of photobleaching, and narrow spectral bandwidth, the SWCNTs nIR fluorescence emission overlaps with the tissue transparency window and benefits from reduced tissue absorption, scattering, and autofluorescence, making them favorable candidates for in vivo imaging and biomedical sensing applications. Upon tailored surface functionalization, SWCNTs can spectrally response to their surroundings, exhibiting a modulation of their fluorescent intensity or a shift of the peak emission wavelength, enabling real-time optical detection. I will present a recent discovery of a SWCNT-heteropolymer complex that recognizes the protein fibrinogen with high selectivity, using the concept of Corona Phase Molecular Recognition (CoPhMoRe). In this approach, a heteropolymer is adsorbed onto and templated by a nanoparticle surface to recognize a specific target analyte. This is the first demonstration of CoPhMoRe for macromolecular targets, opening new avenues for synthetic, non-biological antibody analogues with great promise for medical and clinical applications.

Teaching Interests: I believe that as an educator my role is to catalyze the learning process of the students. I wish for my students to be deep learners and to gain genuine understanding of the material, granting them tools to creatively solve problems they had never encountered before. Since I consider teaching to be an essential aspect of the academic career, I am fully committed to facilitating these goals by setting up a fruitful and inclusive teaching environment, while motivating and engaging the students in class using active learning techniques and educational technologies.

473675 Chemical Engineering Faculty Candidate with Specialization in Nanoscale Science and Engineering

Abstract:

Research Interests:

Research should not always be constrained to theoretical treatise of some interesting phenomenon performed, but should strive toward finding practical solutions for day-to-day problems. My doctoral and postdoctoral research has focused on synthesizing and characterizing organic-inorganic hybrid materials for practical applications. I have developed several multifunctional hybrid-materials and made significant progress toward meeting the necessary requirements for industrial and defense applications. I have broad research interests in materials synthesis and nanocrystal growth, functional materials for energy conversion and antimicrobial activity, high surface area materials for molecular and energy storage and biocompatible materials for controlled drug release. I am especially interested in surface interaction and mechanisms of chemical reaction of interfacial molecules in the broad fields of energy conversion, catalysis and nanotechnology. My research covers wide ranging questions with applications of emerging functional materials from the molecular and nanostructural levels to the applied environment. As a PI and Co-PI on two research grants, I was recently awarded $583k from the Department of Defense to investigate the multifunctionality of quantum-dot incorporated nanostructured functional materials and their optoelectronic sensitivity and photocatalytic activity. I am looking for a tenure-track faculty position in a research oriented university where I can explore my research and teaching potential in chemical engineering.

Teaching Interests:

Proposed Courses
In addition to the basic undergraduate courses required for chemical engineering students, I would like to propose one or more undergraduate/graduate level courses under the following topics: Mass Balances, Fluid Mechanics, Heat Transfer, Kinetics and Reactor Design, and Process Control.

475507 Taking the Temperature of the Interiors of Magnetically Heated Nanoparticles and Optical Biomolecular Chemical Sensing Using Single Wall Carbon Nanotubes

Abstract:

2nd Year Postdoctoral Associate

Research Interests:

My research has been focusing on using nanomaterials for controlled drug delivery and for optical biomolecular sensing. For my Ph. D work, I applied surface modified mesoporous silica nanoparticles for anti-cancer drug delivery, and studied the particle interior magnetic heating effect using temperature-dependent luminescence of nanocrystals. The knowledge on the physical and chemical properties of silica nanoparticles facilitated their biomedical use for in vitro and in vivo studies. During my postdoctoral training, I investigated on using near IR fluorescent carbon nanotubes to detect protein content as well as to examine their binding kinetics. I have developed a miniature microarray platform based on optical signals for label-free multiple protein molecule sensing.

Postdoctoral Project: "***Multiplexed protein detection using single wall carbon nanotube microarray***"

PI: Michael S. Strano, Department of Chemical Engineering, MIT

PhD Dissertation: "Physical Properties of Mesoporous Silica Nanoparticles for Stimuli-Responsive ***Drug***

Delivery"

PI: Jeffrey I. Zink, Department of Chemistry and Biochemistry, UCLA

Research Experience:

- Label-free Single Wall Carbon Nanotube Microarray for Protein Profiling
    - Surface modification of single wall carbon nanotubes with antibody and dsDNA
    - Protein binding kinetic analysis and batch processing
    - Carbon nanotube near IR fluorescence spectroscopy and microscopic imaging
    - Near IR microarray detector construction

- ***In Situ*** Optical Measurement of Nanoparticle Interior Temperature
    - Synthesis of thermally sensitive upconversion lanthanide nanocrystals and dual core nanoparticle
    - ***In situ*** interior temperature study of magnetically heated silica nanoparticles
    - Mechanism investment of nanomachine operation in response to continuous near-IR irradiation
    - Simultaneous spectral detection of nanoparticle interior temperatures and the corresponding cargo release

- Mesoporous Silica Nanoparticles for Controlled Drug Delivery
    - Synthesis of polymer coated silica particles with nanovalves for enhanced biomedical properties
    - Construction of Gd incorporated nanomachines for a multi-functional delivery vehicle
    - Synthesis of functionalized mesoporous silica nanoparticle for ***in vivo*** and ***in vitro*** drug delivery studies

Teaching Experience:

- Teaching certificate, MIT Kaufman Teaching Certificate Program
- Chemistry content expert, MIT BLOSSOMS (outreach program)
- Teaching assistance for general chemistry discussion sessions, UCLA (five quarters)
- Teaching assistance for chemical laboratory sessions, UCLA (two quarters)
- Mentor for two graduate students and one undergraduate student

Teaching Interests:

I believe the goal of a college education is not only to teach the knowledge and skill sets that are needed for work but also to prepare students with the curiosity and methods to continue the learning throughout their lives, for career and personal development. To meet this goal, I will try to improve the learning efficiency by integrating more critical thinking, active learning and connection-building components into our teaching process. Especially for upper level classes, I will introduce advanced projects, in which students are asked to pick research papers that they are interested in, orally present the work to the class and articulate their own opinions about it in the format of a short study report.

Future Direction:

I aim at using carbon nanotube sensing platform to perform early cancer diagnosis by detecting the trace amount of circulating tumor biomarkers. Compared to current immunoassays, nanotube molecular sensors are highly sensitive (down to single molecule detection sensitivity), require no fluorescent labeling and could quantify multiple protein components for correlation studies. Hence, they are especially advantageous in examining the small amount of signature biomolecules for early tumor diagnosis.

In addition, I want to develop a controlled therapeutic delivery platform based on biofunctional molecules. With my experience in using antibody as the recognition site for biomolecular sensing, I am planning on integrating the antibodies with the mesoporous silica nanoparticles. Antibodies offer the flexibility of targeting the receptors of tumor tissues, and hence improve the nanoparticle localization and uptake. In addition, antibody could potentially be the substrate for enzymatic activities and therefore, introduce a new series of nanomachine triggering mechanism. Besides antibody, other biofunctional molecules, such as DNA aptamars and proteins could also be integrated with the silica particles as the next generation of functional delivery vehicles.

Selected Publications:

- **J. Dong** and J. I. Zink. "Simultaneous Spectroscopic Measurements of the Interior Temperature and Induced Cargo Release from Pore-restricted Mesoporous Silica Nanoparticles." ***Nanoscale* 8**, 10558-10563 (2016).
- **J. Dong** and J. I. Zink. "Light or Heat? The Origin of Cargo Release from Nanoimpeller Particles Containing Upconversion Nanocrystals under IR Irradiation." ***Small 11***, 4165-4172 (2015).
- **J. Dong** and J. I. Zink. "Taking the Temperature of the Interiors of Magnetically Heated Nanoparticles." ***ACS Nano 8***, 5199-5207 (2014). **Highlighted** in ***Nature Physics***.
- **J. Dong**, M. Xue and J. I. Zink. "Functioning of Nanovalves on Polymer Coated Mesoporous Silica Nanoparticles." ***Nanoscale 5***, 10300-10306 (2013).

476010 Nanoscale Engineering and Model-Guided Design of Advanced Energy Storage and Conversion Technologies Utilizing Ultrathin Polymer Films

Abstract:

Research Interests:

Due to CO_2-induced climate change, there is currently great interest in using sustainable energy resources such as sunlight. The major challenge is that the collection and storage of solar energy should be economical. A promising approach to address this grand challenge is the application of inexpensive novel polymers in solar power conversion and storage systems. Therefore, my research focuses on scientific exploration of innovative material development in polymers for energy storage and harvesting systems. By developing new chemical vapor deposition (CVD) processes, I have successfully integrated nanoscale polymer thin films (<30nm) into the nanostructured electrodes of dye sensitized solar cells (DSSCs) and supercapacitors, and have improved device performance and stability. Concurrently, much of the current material research in the areas of energy storage and conversion has been based on scientific intuition and trial and error experimentation—a slow and inefficient process with many of the directions taken to address device challenges being without theoretical guidance. In fact, many of the recent advances in the DSSC and supercapacitor fields have relied on the computational design and screening of materials. Therefore, a major focus of my research has been to use first principles modeling to gain insight into the key physical and electrochemical processes occurring in these energy systems, and then apply that knowledge towards enhancing device performance by optimizing material properties and architecture. This holistic synergetic approach between experiment and modeling has allowed me to gain a more complete understanding into the fundamental processes within these devices along with rapidly optimizing material selection, device architecture, and performance.

In the area of energy conversion, DSSCs represent one of the newer generations of solar technologies. Current DSSC technology utilizes an organic liquid electrolyte; however, liquid electrolytes suffer from practical issues of leakage, evaporation, and stability as well as fundamental issues of rapid charge recombination loss at the photoanode-electrolyte interface that limits photocurrent and photovoltage. One promising solution is to replace the liquid with a solid polymer electrolyte material. Polymer electrolytes are an attractive alternative because they do not suffer from the practical disadvantages of liquid electrolytes and could potentially enhance the cell's I-V behavior. However, one of the major challenges with using polymers is the difficulty in integrating them within the tortuous mesoporous TiO_2 photoanode of the DSSC. The photoanode consists of a sintered network of TiO_2 nanoparticles, which results in an electrode layer of at least 10–15 µm thick with 10–25 nm interconnected pores. Conventional liquid processing techniques like spin coating or solvent casting lead to ineffective filling of the pores due to wettability and surface tension issues, resulting in premature pore blockage in this high aspect ratio (>1000) material. My work has addressed these challenges by utilizing a novel initiated chemical vapor deposition (iCVD) technology to directly synthesize and grow polymer electrolytes within the mesoporous TiO_2 photoanode[1]. In this way, there is tight integration of the polymer electrolyte with the photoanode with good contact at the electrode-electrolyte interface. In tandem with these experiments, I've formulated first-principles mathematical models composed of coupled partial differential equations to accurately describe the dynamic internal and interfacial processes during DSSC operation to understand how properties were affected by polymer chemistry and to determine how chemistry affects overall performance. The models accurately predict the J-V behavior of polymer electrolyte DSSCs and has been used for parametric sensitivity studies and parameter estimation to derive cell parameters that can be directly related to polymer properties. The work discovered that the polymer's Lewis acid/base character has a significant impact on the interaction of the polymer with the TiO_2 at the interface, which significantly influences charge conduction and electrode potential. These insights offered for the first time a way to combine modeling and experiments to design polymer electrolytes that would lead to more enhanced DSSCs with much higher power conversion efficiencies, greater than 50% improvement in conversion efficiency when compared to a liquid electrolyte DSSC and this work was published in the ***Journal of Power Sources***[2]. In fact, very recently, I've successfully utilized these insights to carefully select a new polymer electrolyte that enabled an increase in both photocurrent and photovoltage, a breakthrough that was previously not possible with other polymer electrolytes studied that showed only an increase in either the current or voltage but not both. This work has been formally published as an invited paper in a Special Issue on Recent Advances in Energy Conversion and Storage Devices in ***Chemical Engineering Science***[3]. Extending from this work, I've formed successful collaborations that take advantage of my strengths and expertise in DSSC materials, fabrication, and analysis. This has led to another contribution in a Special Issue on Solar Cells in ***AIMS Materials Science***[4] and a submitted paper to the Journal of Physical Chemistry C[5].

In the area of energy storage, there is currently much research interest in developing new, inexpensive, and lightweight energy storage devices that are environmentally friendly. Supercapacitors represent a viable storage solution that provides high power density. However, current supercapacitors suffer from lower energy density compared to other storage technologies like batteries because they only use physical, electrostatic interactions to store charge. A promising approach to enhancing the energy density in supercapacitors is by adding a redox polymer film on the porous carbon electrodes of these devices. These polymers, like polypyrrole, polythiophene and polyaniline, can participate in Faradaic redox reactions that provide further charge storage through more energy-dense chemical reactions, which can double charge storage of these devices. However, integrating polymer coating into these porous electrodes is non-trivial and challenging, but worthwhile because of the possible 2x performance improvement. Current methods for coating polymers onto carbon substrates such as chemical bath deposition, electrodeposition, and casting from suspension are unable to uniformly and conformally coat the microporous (< 2 nm) carbon electrodes throughout the entire electrode thickness, typically 100 µm, without blocking pore accessibility to ions. The high aspect ratios (>10,000) and tortuous pore structure along with liquid surface tension, solution viscosity, poor wettability, and solute steric hindrance make conformal coatings very challenging. Furthermore, poor solubility of conducting polymers in common solvents due to their rigid backbone structure limits processibility, and for electrochemical deposition, the need for a conductive substrate further limits broad utility. Overcoming these challenges is worthwhile because a highly porous structure with an ultrathin uniform conformal polymer coating would maintain the rapid movement of ions through the pores during charge/discharge reactions, preserving the high electrochemical high double layer capacitance (EDLC) while simultaneously dramatically improving the capacitance along with power and energy densities due to the polymer's redox reactions. By developing novel oxidative chemical vapor deposition (oCVD) processes, I was able to synthesize redox (conjugated) polymers and form highly conformal, ultrathin (< 10 nm) coatings around the topologically complex surfaces of porous nanostructured carbons. This initial effort has led to a publication in ***ACS Nano*** in which I was a co-author[6]. Very recently, I have successfully showed, for the first-time, that oCVD can produce polyaniline, which has the highest theoretical capacitance of all redox polymers, and significantly improve the energy capacity and stability of porous carbons derived from carbide materials through a collaborative effort with Prof. Yury Gogotsi and his group, recognized as one of the world's experts in carbon nanomaterials and energy storage. This work is submitted to ***Advanced Materials***[7]. The findings indicate that the performance of the polymer can be tuned based on the reaction conditions and optimal conditions enable oCVD to produce ultrathin (<30 nm) conducting polymer films that conformally coat pores as small as 1.7 nm with the resulting supercapacitors showing >100% increase in specific capacitance compared with bare supercapacitors and exhibiting excellent cyclability, lowering to only 90% of the initially stabilized value (~100 F/g) after 10,000 cycles—which is unprecedented for conducting polymers.

Looking forward, there is a wide range of research areas that could benefit from the integration of ultrathin polymer films using a holistic, synergetic modeling and experimental approach that I have developed and used throughout my research. These areas include functionalizing nanostructured surfaces, energy applications such as solar cells, batteries, fuel cells, and supercapacitors, along with research in separation and membranes. The combined mathematical modeling and iCVD & oCVD expertise is a unique attribute that will facilitate positive research contributions and developing new processing insight particularly with the design of polymeric electronic materials for nanoscale science and engineering.

Teaching Interests:

I received my B.S. in Chemical Engineering at the University of Virginia where I did research for Prof. Teresa B. Culver in sustainable stormwater management. I am currently co-advised by Prof. Soroush and Prof. Lau at Drexel University. Academically, I enjoy learning and have maintaining a perfect 4.0 GPA (grade point average, equivalent to an overall A grade) in all my Ph.D. graduate courses at Drexel, including all the required core graduate chemical engineering courses (Mathematical Methods, Thermodynamics, Transport Phenomena, and Kinetics). I have greatly enjoyed helping teach undergraduate chemical engineering course while I was a TA, some of my favorite course to teach would be Kinetics and Thermodynamics.

I believe that both teaching and research are greatly enhanced when they are integrated together as an educational program. Throughout my Ph.D., I have mentored one high school and nine undergraduate students on research projects related to my research and their educational curriculum. I have trained and guided these students to perform polymer thin film synthesis and processing using CVD approaches as well as materials/device characterization under the broad theme of energy. I have also given back to the Drexel and Philadelphia community by helping to initiate research projects with the Drexel Smart House, a physical building and living environment run by Drexel undergraduate students with the goal of exploring and understanding different sustainable technologies for the future and showcasing those technologies to the larger Philadelphia community. I have also been the President of our department's Graduate Student Association as well as Vice President of the College of Engineering Graduate Association and helped give back to the graduate college.

I am enthusiastic and passionate about research, which is shown tangibly by the number of publications and by the wide range of major professional conferences I have presented at, such as the American Institute of Chemical Engineers (AIChE), American Vacuum Society (AVS), American Chemical Society (ACS), and the Electrochemical Society (ECS)--including one invited talk at the ECS conference in Cancun, Mexico that acknowledged the importance of my DSSC work. I have presented 11 oral presentations and 5 poster sessions. Furthermore, I have formed collaborations with W.L. Gore, Prof. Gogotsi in Drexel's Material Science department, Prof. Ji in Drexel's Chemistry Department, Prof. Li in Penn's Chemical Engineering Department, and Prof. Kalra in the Chemical and Biological Engineering Department at Drexel.

As a testament to my work ethic, I have been the recipient of numerous honors, including the Department of Chemical and Biological Engineering's 2016 Outstanding Ph.D. Student Award, the College of Engineering's 2016 Outstanding Graduate Student Award, and most recently Drexel University's highest honor with the 2016 Research Excellence Award for Outstanding Research given to only four students out of the entire university's graduate population. I have also been the recipient of numerous merit fellowships, including the 2016 Leroy L. Resser Endowed Fellowship, the 2015 George Hill, Jr. Endowed Fellowship, and the 2015 Koerner Family Award. In addition, I have been a GAANN fellow for the Department of Education's National Academy of Engineering Grand Challenges from 2014–2016. I was also one of the six national finalists for the highly competitive W.L. Gore Fellow Award in 2015.

(1) Janakiraman, S., Farrell, S. L., Hsieh, C.-Y., Smolin, Y. Y., Soroush, M. & Lau, K. K. S. Kinetic analysis of the initiated chemical vapor deposition of poly(vinylpyrrolidone) and poly(4-vinylpyridine). ***Thin Solid Films* 595, Part B**, 244-250, (2015). http://www.sciencedirect.com/science/article/pii/S0040609015005088

(2) Smolin, Y. Y., Nejati, S., Bavarian, M., Lee, D., Lau, K. K. S. & Soroush, M. Effects of polymer chemistry on polymer-electrolyte dye sensitized solar cell performance: A theoretical and experimental investigation. ***Journal of Power Sources* 274**, 156-164, (2015). http://www.sciencedirect.com/science/article/pii/S0378775314016371

(3) Kuba, A. G., Smolin, Y. Y., Soroush, M. & Lau, K. K. S. Synthesis and integration of poly(1-vinylimidazole) polymer electrolyte in dye sensitized solar cells by initiated chemical vapor deposition. ***Chemical Engineering Science***. http://www.sciencedirect.com/science/article/pii/S000925091630238X

(4) Johnson, N. M., Smolin, Y. Y., Shindler, C., Hagaman, D., Soroush, M., Lau, K. K. S. & Ji, H.-F. Photochromic Dye-Sensitized Solar Cells. ***AIMS Materials Science* 2**, 503-509, (2015)

(5) Johnson, N. M., Smolin, Y. Y., Hagaman, D., Soroush, M., Lau, K. K. S. & Ji, H.-F. Suitability of N-Propanoic Acid Spiropyrans and Spirooxazines for Use as Sensitizing Dyes in Dye-Sensitized Solar Cells. ***Journal of Physical Chemistry C*, Submitted**, (2016)

(6) Nejati, S., Minford, T. E., Smolin, Y. Y. & Lau, K. K. S. Enhanced Charge Storage of Ultrathin Polythiophene Films within Porous Nanostructures. ***ACS Nano* 8**, 5413-5422, (2014). http://dx.doi.org/10.1021/nn500007c

(7) Smolin, Y. Y., Van Aken, K, Boota, M., Soroush, M., Gogotsi, Y.,Lau, K. K. S. Engineering Ultrathin Polyaniline in Carbide Derived Carbon Supercapacitors using Oxidiative Chemical Vapor Deposition. ***Advanced Materials*, Submitted**, (2016)

476050 Elucidation of Atomic-Scale Structure/Function Relationships: Toward Predictive and Rational Design of Nanoscale Materials

Abstract:

Research Interests: Nanoscale materials, atomic-scale structure/function relationships, synchrotron radiation characterization methods, catalysis, biomimetics

Teaching Interests: With a background in physics and chemistry, and well suited to teach a wide-range of class work. I would be interested in teaching materials related courses, thermodynamics, and/or transport courses.

Elucidation of Atomic-Scale Structure/Function Relationships: Toward Predictive and Rational Design of Nanoscale Materials

Nicholas M. Bedford, National Institute of Standards and Technology, Boulder, CO 80305

Postdoctoral Fellow

Research Interests: The emergence of nanotechnology has produced advances in various application areas of technological importance. Unfortunately, the creation of nanoscale materials is too often achieved serendipitously or via Edisonian approaches, impeding rapid technological development. As the arrangement of atoms within a nanoscale material directly influences the materials properties, it is imperative to have a complete atomic-scale structural understanding to elucidate, predict, and ultimately design/tune materials properties for a given material. As such, my research interests are centered on understanding synthesis-driven structure/function relationships using atomic-scale synchrotron radiation characterization methods. *In-situ* characterization methodologies are also of interest to understand atomic-scale structural changes during relevant chemical processes, *e.g.* charge/discharge cycles of a battery materials or during catalytic turnovers occurring at a catalysts surface.

Successful Proposals: NRC Research Associateship Program Postdoctoral Fellowship (AFRL); DAGSI Graduate Student Research Fellowship Program; 20+ successful synchrotron general user facility proposals

Additional Proposal Experience: Panel reviewer for the Catalysis and Biocatalysis program at CBET-NSF; reviewer for SSRL general user beam time requests; Submission of multiple peer-review proposals to DOE, AFOSR, and ARO.

Postdoctoral Projects: "*Development and Characterization of Peptide-Assembled Nanoparticles*"

Under supervision of Dr. Rajesh Naik, Air Force Research Laboratory

"Elucidation of Structure/Function Relationships for Engineering Nanomaterials"

Under supervision of Dr. Tim Quinn, National Institute of Standards and Technology

Ph.D. Dissertation: *"Electrospun Fibers for Energy, Electrical, and Environmental Applications"*

Under the supervision of Prof. Andrew J. Steckl and Donglu Shi at the University of Cincinnati Department of Material Science and Engineering

Research Experience: My research experience is highly interdisciplinary in nature with a common underlying theme of understanding nano- and atomic-scale phenomena. My formal Ph.D. training was in synthesis and characterization of polymeric and inorganic nanomaterials, while during my postdoctoral appointments I was immersed into the area of synchrotron radiation characterization, including high-energy X-ray diffraction (HE-XRD) coupled to atomic pair distribution function (PDF) analysis, X-ray absorption spectroscopy (XAS), and small angle X-ray scattering (SAXS). Through intensive training at several national laboratories and DOE user facilities, I have become proficient in all aspects of the aforementioned structural characterization methods, which can yield nanomaterial structural models with atomic-scale precision. Using such models, structure/function relationships can be readily examined, providing the needed understanding to create new nanomaterials with improved properties.

Teaching Experience: During my career, I have been involved in various avenues of teaching and mentorship. While at AFRL, I held a dual appointment in the Department of Chemistry at the University of Miami, where I was an instructor for the summer section of Principles of Chemistry I. As the instructor, I was responsible for all aspects of the class. While in graduate school, I was a TA for many engineering courses, including Basic Heat Transfer and Intro to Thermodynamics. During my undergraduate studies, I was laboratory assistant for both organic and general chemistry labs. Additionally, I have mentored several graduate, undergraduate, and high school students during my postdoctoral tenures at both NIST and AFRL.

Future Directions: As faculty I plan to continue applying my expertise in nanomaterials synthesis and synchrotron radiation characterization to establish synthesis-based nanomaterials design rules for a variety of materials systems. My initial focus will involve nanomaterials with emergent catalytic, electrocatalytic, photocatalytic, energy harvesting and/or energy storage properties. Using XAS, HE-XRD coupled to atomic PDF, and SAXS, structural information can be accessed over various length scales, providing the needed data to build full nanomaterial structural models with atom-scale resolution. These models will then be used to assess synthesis-driven structure/function relationships to be implemented in future nanomaterial design. In-situ methodologies will be explored as well, wherein structural transitions occurring during relevant chemical events can provide additional insights in assessing and predicting materials properties.

Synthetically, I plan to explore multiple avenues for the creation of nanomaterials synthesis, where the initial focus will be on biomimetic approaches. Using the specificity of biological molecules, nanomaterials can be synthesized and assembled with unpredicted precision. Additionally, the complexity and modularity of chemical functional groups within biomolecules provide a large basis set of possible biotic/abiotic interactions to manipulate materials properties. In tandem with the aforementioned synchrotron radiation characterization expertise, I fully expect the capability to use biomimetic approaches to achieve nanomaterials with enhanced properties through the development of structure/function relationships.

Selected Publications (*denotes corresponding author), current H-index=11 (Google Scholar):

N. M. Bedford,* Z. Hughes, Z. Tang, B. D. Briggs, Y. Ren, V. G. Petkov, R. R. Naik,* M. R. Knecht,* T. R. Walsh,* "Probing the Sequence-Dependent Structure/Function Relationships of Catalytic Peptide-Enabled Au Nanoparticles", *J. Amer. Chem. Soc.* **2016**, 138, 540.

N. A. Merrill, E. M. McKee, K. C. Merino, L. F. Drummy, S. Lee, B. Reinhart, Y. Ren, A. I. Frenkel, R. R. Naik, **N. M. Bedford**,* M. R. Knecht* "Identifying the Effects of Metal Composition on the Structure and Catalytic Activity of Peptide-Templated Materials", *ACS Nano*, **2015**, 9, 11968.

B. D. Briggs†, **N. M. Bedford**†, S. Seifert, H. Koerner, H. Ramezani-Dakhel, H. Heinz, A. I. Frenkel, R. R. Naik,* M. R. Knecht,* "C-C Coupling of Peptide-Capped Pd Nanoparticle Progresses Through an Atom Leeching Mechanism", *Chem. Sci.* **2015**, 6, 6413. **†Equal author contribution**

N. M. Bedford,* H. Ramezani-Dakhel, J. M. Slocik, B. D. Briggs, Y. Ren, A. I. Frenkel, V. G. Petkov, H. Heinz,* R. R. Naik,* M. R. Knecht,* "Elucidation of Biologically Programmed Atomic-Scale Structure of Nanoparticle Interfaces that Modulates Catalytic Activity", *ACS Nano* **2015**, 9, 5082. **Featured on the front coverN. M. Bedford**, R. Bhandari, J. M. Slocik, S. Seifert, R. R. Naik, M. R. Knecht,* "Peptide-Modified Dendrimers as Templates for the Production of Highly Reactive Catalytic Nanomaterials", *Chem. Mater.* **2014**, 26, 4082.

N. M. Bedford, M. Pelaez, C. Han, D. D. Dionysiou, A. J. Steckl,* "Photocatalytic cellulosic electrospun fibers for the degradation of potent cyanobacteria toxin microcystin-LR", *J. Mater. Chem.* **2012**, 22, 12666. **Published as HOT Article by RSC**

N. M. Bedford, M. B. Dickerson, L. F. Drummy, H. Koerner, K. M. Singh, M. C. Vasudev, M. F. Durstock, R. R. Naik,* A. J. Steckl,* "Nanofiber-Based Bulk-Heterojunction Organic Solar Cells Using Coaxial Electrospinning", *Adv. Ener. Mater.* **2012**, 2, 1136.

476080 Fast Modeling Protein Corona on Nanoparticle Based Biosensors in Complex Solvent Environments/ Cell Membrane By a Coarse Grained Simulation System

Abstract:

Research Interests:

Understand protein behavior in a complex environment is the key in rationalizing the design of many novel techniques such as protein biosensors based on novel nanoparticle (NP) materials. Protein folding and aggregation properties can be largely affected by the environments such as solid abiotic interfaces and complex co-solvents, which would directly affect the expected functionality or performance of the biosensors. Difficulties remain in applying current experimental methods to obtain a detailed picture of protein structures on a solid surface, so our ability is limited by these factors in the design of biosensor surface chemistry a priori for desired protein behavior. To address this point, in this work we present a coarse-grained simulation system that is able to model protein folding, misfolding, and aggregation on various NP surfaces and in the presence of co-solvents based on the Karanicolas and Brooks (KB) Go-like protein model. A flat-surface potential was first included and carefully parameterized based on a large experimental data set. This surface model has been shown in many studies to be able to reproduce and predict protein behavior at different types of surfaces as verified by experimental tests. Further effort in developing the surface potential was applied in describing surface curvature effects on protein behavior as a means of understanding nano-sized biosensor materials. Tanford's transfer free energy model was recently added to this simulation system as a descriptor of co-solvent effects or cell membrane environments. The energy of each part of the underlying folding/surface/solvation potential has been well balanced so that each element works collaboratively as a single complex environment for the protein. Together, this coarse-grained modeling system is expected to accurately predict the protein adsorption, folding/unfolding, aggregation, and protein-protein interactions in a complex environment as it would experience in differing cellular or solvent environments as well as in the context of a NP-based biosensor.

Teaching Interests:

I would like to practice my teaching under the guidance of the student-centered learning philosophy, which emphasizes personal reading, active peer discussing, and collaboration. The eventual goal of my teaching is to help them find their way to acquire knowledge and tools independently but not simply from the class. As a potential new faculty member, I am very interested in keeping optimizing my teaching techniques through my career as an educator.

Specifically, I would like to teach engineering classes that are currently given by the department on undergraduate or graduate levels, such as: Engineering Mathematics, Thermodynamics, Kinetics, and Transport Phenomena. Also, I want to bring classes such as molecular simulation and statistical mechanics for biological engineers, which are directly related to my research experiences and interests. A variety of simulation methods including quantum chemistry calculation, multi-levels molecular dynamics/ Monte Carlo simulation, and quantitative structure-activity relationship (QSAR) modeling will be taught in the classes.

475972 Designing Functional Self-Assembled Structures Via Complex Colloidal Interactions

Abstract:

Research Interests:

Colloidal materials and soft matter in general exist in the forefront, background, and future of our everyday lives. They compose many of the products we interact with and rely on to maintain (or repair) our health, facilitate the harvesting of resources (water, food, energy, etc.), and overall enhance our quality of life. The beauty of soft materials lies in the hierarchical structures that lead to their beneficial properties, which can be tuned by chemically modifying the colloidal building blocks of which they are composed or the medium in which they are suspended. *This chemistry-structure-property nexus has been and will continue to be the crux of my research interests.* In particular, the fundamental relationships between surface chemistry and colloidal interactions and subsequently between solution structure and colloidal dynamics are challenging and intriguing issues I have explored during my PhD. For example, I have identified protein surface charge heterogeneity as the origin of a hierarchical solution structure resulting in a unique gelation hysteresis behavior. Given the implications of the structure-dynamics-properties relationship for designing materials, I am excited to continue exploring similar physical phenomena. Future applications I am interested in include designing soft materials ranging from biocompatible materials for pharmaceutical formulation and delivery, to functional proppants for enhanced oil recovery, to self-assembled networks for separations or electronic applications. Specifically, I intend to exploit the versatility of colloidal chemistry and shape to study the effect of anisotropy and competing interactions on their microscopic behavior and macroscopic properties. While I have previously used proteins as colloids with anisotropic structure and interactions, recent advances in colloid synthesis now provides boundless opportunities to explore their effects on material properties using a range of chemistries that can be tailored for specific applications.

The methodology I have developed to explore my research interests encompasses modelling, simulation, and a range of experimental techniques. During both my PhD and my current post-doctoral position I have developed viscosity and mass transport models, respectively, and utilized Monte Carlo (MC) and Brownian dynamics (BD) simulations, respectively, in conjunction with rheology, small angle scattering, and dynamics measurements. I believe these three techniques are individually important, yet complimentary, and in my experience yield even more powerful insight if implemented in conjunction. In particular, combining simulation and experiment is necessary to understand the dynamics and material properties of anisotropic colloids, for which no models currently exist. I plan to employ this toolbox of research techniques to explore the self-association behavior of anisotropic colloidal building blocks and identify new material properties to engineer more effective soft matter materials relevant to the pharmaceutical and energy industries.

Teaching Interests:

I have enjoyed tutoring and mentoring others during each phase of my education and believe it has enhanced my own understanding of various subjects. My teaching experience is primarily from tutoring students as a TA, first for engineering math courses while at the University of Virginia and then for a pilot scale distillation experiment for the senior lab course in chemical engineering at the University of Delaware, for which I received very good reviews. As a post-doc, I have mentored graduate students in our lab and an undergraduate student on a side project I developed. Also while at MIT, I participated in the Kauffman Teaching Certificate Program, which teaches current best practices in designing courses and implementing active learning techniques in the classroom.

475989 Synthesis and Optimization of Nanomaterials for Sustainable Energy Generation and Catalysis

Abstract:

Research Interests:

The effectiveness of a supramolecular substrate, a solar cell, or a catalyst, heavily depends on the successful manipulation of its constituent components, at the nanoscale. Such manipulation begins at early-stage materials development, whether it is extraction of a protein, synthesis of a new photo-active material, fabrication of a composite energy-generating device, or development of novel catalysts. Understanding the key chemical/physical characteristics of these materials which render them their unique activity is the foundation leading to their successful isolation/fabrication and consequent optimization.

The highly robust, Mycobacterial channel protein â€œMspAâ€, has been isolated in high yield and purity, which served as a building block for synthesis of supramolecular aggregates of protein-organic dye and protein-metal/metal oxide nanoparticle clusters. The specific extraction and purification techniques employed for MspA, had a remarkable impact on its self-organization behavior in aqueous media, which in turn, had a significant impact on its aggregation behavior with external nanomaterials. The optimal aggregate size for various applications (such as drug delivery, hyperthermia and channel activity) of these clusters was studied extensively under various mediums and temperatures. The protein-organic dye clusters in particular, displayed remarkable thermo-stability and showed dynamic surface charge interactions with a buffer medium. These were used successfully in a dye-sensitized TiO_2 based photo-voltaic device, to conduct electricity under simulated sunlight, thus forming the first ever, hybrid, nano-protein solar cell prototype.

Iron/iron(III) oxide core-shell nanoparticles and micro-mesoporous titanosilicates are two types of highly valuable catalysts, with industrial appeal. The core-shell nanoparticles were functionalized with sulfonic acid and used as a highly selective catalyst to hydrolyze cellulose into glucose. A statistical technique was used to systematically optimize the reaction conditions of the above. The statistically-guided optimization led to high selectivity toward glucose, while production of 5-HMF and other monosaccharides were decreased. Reduction of 5-HMF is particularly significant as it is toxic to yeast, which are used industrially, to convert cellulose degradation products into ethanol. The catalyst was found to be highly stable and reusable, due to its magnetic property, thus, leading to a potential sustainable, technology for conversion of plant matter into ethanol. The same statistical approach mentioned above, was utilized to simultaneously probe synthetic parameters of a novel, hierarchically ordered, micro-mesoporous titanosilicate (MMTS), with the aim of achieving enhanced catalytic performance. MMTS was used to catalyze the epoxidation of cyclohexene with *tert*-butylhydroperoxide, and demonstrated unprecedented catalytic activity and high product selectivity. Thus, systematic nanoscale manipulation of reaction conditions and material synthesis, drastically improves the efficiency of catalysis.

The above is a summary of work from my PhD, followed by a short post-doctoral study, conducted at Kansas State University (advisor Prof Stefan Bossmann), and of my current post-doctoral research, conducted at University College London (advisor Prof Marc-Olivier Coppens). My future research plans include development of novel nanomaterials *via* sustainable routes for energy generation and catalysis.

Teaching Interests: I am experienced in conducting lectures and lab classes in areas of physical organic chemistry, synthetic organic chemistry and organometallic chemistry. I am interested in further expanding my scope of teaching into thermodynamics and nanomaterials engineering.

476194 Microfluidic Platform Technologies for Detection of Biochemical Markers

Abstract:

Research Interests: Microfluidics, Bio/chemical sensors, Micro/nanofabrication, Functional sensing materials, Reaction and transport phenomena.

Teaching Interests: Microfluidics, bio/chemical sensors, Interfacial science, Reaction and transport phenomena.

There is a growing interest in microfluidic based technologies for detection of various (bio-)chemical markers in wide variety of applications ranging from healthcare to environment. In microfluidic sensors the microchannel surfaces are functionalized with stack of intermediate molecular layers to the specific probe molecules. The efficiencies of these microfluidic sensors depends on the effective capture of analytes flowing in the carrier fluid by the surface immobilized receptors. The effective diffusion of analytes to the surface bound molecules primarily depends on two important category of parameters i.e. surface engineering and fluid flow characteristics. The surface engineering parameters such as number densities, orientations, surface free energies and lengths of molecular chains/stacks, and interfacial phenomena at fluid-wall interfaces along with the flow characteristics such as nature of liquids, frictional resistance, and flow profiles plays a significant role in defining the performance of the microfluidic sensor. Hence, the design and development of an effective microfluidic sensor requires a rational approach, based on the precise engineering of reactions and transport phenomena which are essential to ensure the optimum performance. Therefore, my research would focus on fundamental understanding of the interfacial phenomena, reaction and transport phenomena along with the biological aspects to engineer the microfluidic platform technologies for detection of various (bio-)chemical markers for applications ranging from healthcare diagnostics to environmental sensors. Being a chemical engineer and having worked in various industries along with semiconductor Fab, I have research interests in microfluidics, bio/chemical sensors, micro/nanofabrication, functional sensing materials, and reaction and transport phenomena. As a future faculty member my goal is to design and develop microfluidic platform technologies by using various smart and soft materials, nanomaterials, electronic materials along with flexible substrates from optical to electrochemical detection methods to meet the demanding needs in the domain of healthcare and environment. Previously, I worked on developing silicon based electrolyte insulator semiconductor based microfluidic immunosensors for early detection of cancer biomarkers and presently my research is focused on developing microfluidic based point of care platform technologies on cellulose materials for detection of various (bio-)chemical markers. These (bio-)chemical markers ranges from proteins, miRNA to DNA for various diseases such as Osteoporosis, Cervical Cancer, Tuberculosis, bacterial and viral infections, and Kidney rejection markers. In my poster, I would be presenting the work on silicon based microfluidic immunosensors, cellulose based point of care devices, and some key highlights from my past and current research projects along with the future projects in the microfluidic platform technologies.

476081 Understanding Structure-Property Relationships for Complex Fluid-Fluid Interfaces

Abstract:

Javen S. Weston

*Postdoctoral Researcher:*Georgetown University Department of Physics and National Institute of Standards and Technology

Research Interests: Material-material interfaces are everywhere around us and understanding how to control and modify them is at the center of many current research problems, from formulating laundry detergents to creating highly efficient photovoltaic solar cells. My research interests primarily revolve around studying how to create, stabilize, and modify these interfaces; with particular interests in surfactant adsorption, emulsion and foam stabilization, nanoparticle dispersions/composites, and nanoparticle assembly at fluid-fluid interfaces. Throughout my career I have both gained experience using a variety of characterization techniques and made valuable contacts at national labs and industrial R&D departments which, when combined, provide me with an excellent foundation to build a research career upon.

Successful Proposals: ADNOC-OSC 2-year Research and Development Grant

Postdoctoral Project: *"Developing Measurements of Structure-Property Relationships for Complex Fluids in Extreme Environments."*

Under supervision of Steve Hudson (NIST, Mat. Sci. and Eng. Division), Daniel Blair (Georgetown, Dept. of Physics), and Katie Weigandt (NIST, Center for Neutron Research)

PhD Dissertation: *"Metal Oxide Nanoparticles: wettability modification for emulsion stabilization and electrostatically induced gelation"*

Under supervision of Jeffrey H. Harwell, Chemical, Biological and Material Engineering, University of Oklahoma

Research Experience: My scientific research journey has introduced me to a wide variety of different topics within the realm of colloid and interface science. During my undergraduate career, I worked as a research assistant in OU's Applied Surfactant Lab (ASL), which balanced industrially-sponsored research projects with more academic grant-funded research. Over my time in the ASL, I worked on a variety of different projects including: formulating a high performance 'green' hard surface cleaner with SC Johnson, investigating the competitive adsorption of surfactants and polyelectrolytes on metal oxide nanoparticles, optimizing nanoparticle propagation in oil reservoir rock with the Advanced Energy Consortium, developing ultra-low interfacial tension surfactant blends for use in a variety of environmental remediation applications, improving performance of erasable inks with Papermate, and incorporating polyelectrolytes into detergent formulations with Clorox. My graduate research was completed as a member of the Center for Interfacial Reaction Engineering with funding from a DoE-EPSCoR Grant. I modified the surfaces of nanoparticles and used them to stabilize Pickering emulsions; investigating how the surface modification alters the emulsion properties. Additional work studying interactions between oppositely charged nanoparticles dispersed in water using microscopy, small-angle scattering, and rheology rounded out my graduate studies. This wide range of experience has left me with a broad understanding of the current state of academic and industrial research in the field of colloids and surfactants.

Teaching Interests : Throughout the course of my graduate and post-graduate work I have had several opportunities to gain hands-on teaching experience. During my first year of graduate school I was given opportunities to act as a guest-lecturer and TA in undergraduate-level laboratory courses. Then, during fall semesters throughout the remainder of my graduate studies, I was asked to act as the Instructor of Record for a section of a First-Year Engineering Experience course in OU's College of Engineering, for which I was awarded an Outstanding Graduate Assistant Teaching Award. As faculty I would be interested in teaching a wide variety of courses and would be excited to teach any course. Specific interests would include materials, fluid mechanics, thermodyanamics, and colloids/surfaces focused courses.

Proposed Future Research: As a faculty member, my research philosophy would rest on the belief that the best way to maintain a well-funded, relevant set of research goals is to strike a balance between basic scientific research that may or may not be of immediate commercial relevance and industry-funded research/consulting projects. Combining these two goals can be difficult, but I believe it is the best way to stay up to date with the cutting edge of academic research while simultaneously anticipating industry needs, which are often an excellent indicator of future research directions in engineering fields. I also believe in using a diverse group of researchers in terms of experience levels; using undergraduates, graduate students, and postdoctoral researchers to maximize the productivity of the research group, as a whole. In particular, I believe bringing in researchers from other STEM majors or creating opportunities for interdepartmental collaboration can help solve problems far more synergistically than can occur when all of the students and faculty involved are approaching the research problem from a similar background.

Scientifically, I plan to focus my research on studying how interfacial properties alter bulk performance in three different applications:

- Nanoparticle stabilized foams - How do interparticle forces, particle morphology, and surface modifications affect foam properties like stability, bubble size, and rheology
- Structure-property relationships in bicontinuous microemulsions - linker molecules are used to modify the interfacial rigidity of microemulsions and alter coalescence time, neutron scattering techniques and rheological measurements can be used to determine how changes in interfacial morphology and dynamics affect kinetics of emulsion coalescence
- Membranes via emulsion polymerization - Developing a technique to produce nanoporous membranes via polymerization of bicontinuous emulsions where the monomer is used as the organic phase could allow for greater control and tunability of pore size distributions and pore densities

Selected Publications:

- **JS Weston**, RE Jentoft, BP Grady, DE Resasco, and JH Harwell. "Silica Nanoparticle Wettability: Characterization and Effects on Emulsion Properties," *Industrial & Engineering Chemistry Research*, DOI: 10.1021/ie504311p, 2015.
- **JS Weston**, D Venkataramani, CP Aichele, BP Grady, JH Harwell, and DE Resasco. "Pseudosolid, Shear-Thinning Gel Formation in Binary Dispersions of Metal Oxide Nanoparticles at Low Volume Fractions," *Langmuir*, 30 (49), 14982, 2014.
- **JS Weston**, JH Harwell, B-J Shiau, and M Kabir, "Disrupting admicelle formation and preventing surfactant adsorption on metal oxide surfaces using sacrificial polyelectrolytes," *Langmuir*, 30 (22), 6384-6388, 2014.
- N Briggs, **JS Weston**, B Li, D Venkataramani, CP Aichele, JH Harwell, and S Crossley. "Multiwalled Carbon Nanotubes at the Interface of Pickering Emulsions," *Langmuir,* 31, 13077, 2015.

476298 Protein Self-Assembly Toward Engineering of Biofunctional Nanomaterials

Abstract:

Research Interests: Making nanomaterials directly from proteins can enable spatial and quantitative controls of protein molecules at the nanoscale. Such controls are expected to extend the utility of proteins through manipulation of physical properties such as stability, molecular transport, and specific binding interactions. Using fundamentals of protein assembly, I plan to run research projects that focus on (1) design of self-assembling protein modules, (2) fundamental study of protein assembly processes, and (3) application of biofunctional protein nanomaterials for nanomedicine, biosensors, and nano-biocatalysis.

Naturally-derived or artificially designed protein motifs present characteristic structural and physical properties, which can be harnessed to design self-assembling modules. For examples, modular combination of protein motifs results in self-assembly of protein suprastructures with various morphologies which include spheres [1], vesicles [2], sheets, porous structures [3], and artificial nano-shapes that are unprecedented in nature. I previously demonstrated protein self-assembly approaches using simple protein motifs such as coiled coils, elastin-like polypeptides, and artificial random coils. I will present protein assembly strategies to retain both self-assembled morphology and biological functionality of protein nanomaterials. To incorporate biological functionality, I will develop functional protein modules. In combination with computational design of protein structures, experimental characterization to engineer self-assembly processes and biological functionality will be discussed. In addition, I will present design strategies that relate nanostructural characteristics and specific applications.

Teaching Interests: At the graduate level, I plan to design a class for protein nanomaterials, which is at the interface between protein engineering and nanotechnology. While I provide knowledge about fundamentals of protein engineering, biochemistry, and nanotechnology, I will focus on teaching students to develop their ability to solve engineering problems in the interdisciplinary area. Also, I am interested in teaching undergraduate chemical engineering classes, using examples of biological systems at the nanoscale. My research background on protein self-assembly covers topics in chemical engineering, such as thermodynamics and kinetics of biomolecules. I would like to introduce examples of engineering problems in bio- and nanotechnology.

[1] W. M. Park and J. A. Champion, *Angew. Chem. Int. Ed.*, **2013**, 52, 8098-8101.
[2] W. M. Park and J. A. Champion, *J. Am. Chem. Soc.*, **2014**, 136, 17906-17909.
[3] W. M. Park, C. M. Yee, J. A. Champion, *J. Mater. Chem. B*, **2016**, 4, 1633-1639.

476317 Graphene and Other Nanosheets: Exfoliation and Processing for Nanocomposites and 3D Macrostructures

Abstract:

Research Background:

The emergence of graphene as a single layer nanosheet with unique properties has paved the way to exploit the optical, electrical, thermal and mechanical properties of other nanosheets such as MoS_2, WS_2, and boron nitride. Many of the existent layered materials have been exfoliated to produce unprecedented nanosheets including Mxenes and black phosphorous. The dependence of the nanosheets properties and potential applications on their structure and morphology, particularly on the number of layers in an individual nanosheet, necessitates a precise control over their production techniques. Current nanosheets exfoliation techniques including the liquid-phase exfoliation lack such a control at atomic level and are challenged by the instability and reaggregation of nanosheets in liquid media. On the other hand, the yield of these techniques is not sufficient for bulk usage of nanosheets in applications such as composites or battery electrodes. Hence, strategies for scalable, high-yield production of nanosheets that offer atomic-level control over their structure and properties are in high demand.

To this point, in Prof. Micah Green's laboratory, I have tried to address the low yield of liquid phase exfoliation by using dispersant molecules for stabilization of colloidal graphene at higher concentrations. I have also tailored the polymer molecules to synthesize polymeric dispersant compatible with pristine graphene, a product with highest quality in graphene family. This allows for the usage of minimal graphene content for mechanical and electrical properties enhancement in polymer nanocomposites. The change of morphology of graphene from 2D nanosheets to 3D crumpled particles is another strategy that I have explored to prevent aggregation during the processing of graphene for other applications. Recently, I have focused on understanding the colloidal interactions in the graphene oxide dispersions in order to control their assembly into 3D macrostructures; ideally, these porous conductive 3D networks could be the optimum electrodes for lithium ion batteries.

Research Interests:

Being aware of the similarities and differences of graphene and other nanosheets, as a faculty member, I will expand my research to new materials and technologies. The research carried in my future laboratory will focus on (i) developing alternative synthesis and/or exfoliation methods for existing and novel 2D nanomaterials, (ii) fundamental study of the processing-structure-property relationships in each nanosheet family, (iii) relate the nanoscale properties of nanosheets to their bulk performance in macrostructures, and (iv) scalable manufacturing strategies to enable usage of these materials in industrial end products in response to real-life needs.

Teaching Interests:

In my laboratory and future classes, I would apply the methodology that I have developed during my research career and my teaching experience as a teaching fellow and teaching assistant (Numerical Analysis for Chemical Engineers & Reaction Design courses) to create a vivid learning environment for my future students. While they will obtain a fundamental understanding of the physics behind their research topic, they will get the opportunity to use their engineering background to design and build products with industrial applications. Obviously, collaboration with both industrial organizations and academic scientists with different computational and experimental skills is essential to accomplish these goals.

478651 Engineered Nanostructured Materials for Efficient Separation and Storage

Abstract:

Postdoctoral Projects:

Georgia Institute of Technology, Chemical & Biomolecular Engineering, 08/2015 – Now

Advisors: Prof. Krista S. Walton, Prof. David Sholl

(1) Synthesis and Characterization of Water-Stable Metal-Organic Frameworks (MOFs)

(2) Structural Stability of MOFs in Hot Liquid Water

(3) Multi-component Gas Adsorption on MOF materials.

(4) MOFs for Adsorption-Based Hydrocarbon Separations and High-Performance Oil/Water Separation

University of South Carolina, Chemical Engineering, 04/2013 –04/2015

Advisor: Prof. Miao Yu

(1) Thin, High-Quality Membranes for Separation. Membrane Materials: Zeolites (NaY, SAPO-34/11, AlPOs, and NaA-Polyamide composite) and Graphene Oxide. Relevant applications: H_2 Separation, N_2/CH_4, CO_2/CH_4, Propane/Propylene Separation, Oil/Water, and Desalination.

(2) Nano-valved Adsorbents prepared by Molecular Layer Deposition (MLD) for Efficient Separation and storage (e.g., CH_4 and H_2 Storage, Shape & Size Selective Separation of Organic Molecules (e.g., Ethanol/1-Butanol))

(3) Study the Growth Mechanisms of High-Aspect-Ratio Zeolites (e.g., SAPOs and AlPOs)

University of Oklahoma, Chemical Engineering, 08/2011 – 04/2013

Advisor: Prof. Daniel E. Resasco

(1) Study of The Hydrothermal Stability of FAU Zeolite – A Primary Component in FCC Catalysts

(2) Silylated Hydrophobic Zeolites with Enhanced Tolerance to Hot Liquid Water and Improved Thermal Stability under Reaction Conditions in Water/oil Emulsions

PhD Dissertation: "Zeolite Nanocrystals and Hierarchical Porous Structures: Synthesis, Characterization and Formation Mechanism"

Advisors: *Prof. Huanting Wang* (Monash University, Australia) and *Dr. Anita J. Hill* (Chief of Process Science and Engineering, The Commonwealth Scientific and Industrial Research Organization)

Background and Research Interests:

All across the world, people are facing a wealth of new and challenging problems, particularly the energy and environmental issues. For example, billions of tons of annual CO_2 emissions are the direct result of fossil fuel combustion to generate electricity. To minimize the impact of CO_2 on the environment, design of high-performance separation materials and technologies for efficient carbon capture and sequestration (CCS) is urgent and essential. *My research in this area is creating novel nanostructured (membrane) materials with enhanced transport properties by ordering their nano-architectures via different methods and meanwhile exploring their novel and energy-sustainable scale up.*

On the other hand, enhanced demand for fuels worldwide not only decreased world oil reserves but also increased climate concerns about the use of fossil-based fuel. To address these energy and environmental problems, efforts have been made towards improved utilization of fossil fuel and development of renewable energy production. With the abundant availability and carbon-neutral nature, biomass is recognized as one of the most promising renewable energy resource. A number of transportation fuels can be produced from biomass, helping to alleviate demand for petroleum products and improve the greenhouse gas emissions profile of the transportation sector. Traditional catalysts suffer from many undesirable properties, such as small accessible pore size, low hydrothermal stability, and less controllable active sites. Among these, low hydrothermal stability at upgrading temperatures greatly hinders conversion of lignocellulosic biomass to biofuel. *My proposed research will focus on synthesizing a new class of ultra-stable catalysts with tunable nanostructure and functionalities for efficient bio oil upgrading, with special emphasis on study of their hydrothermal stability.*

Oil pollution is an another serious global issue because of the large amounts of oily wastewater produced by petrochemical and other industries, as well as by frequent off-shore oil-spill accidents. Therefore, it is in great need to develop effective techniques to treat oil-polluted wastewater at such low oil/grease concentrations in order to satisfy the stringent governmental limitations and preserve the environment. Membrane techniques have been widely employed for water purification and are very effective in separating stabilized oil emulsions-especially for removing oil droplets. However, current membranes suffer from membrane fouling both on surface and in internal structures, which significantly limits their service time and degrades separation performance in practical operations. *My proposed research in this field attempts to adopt the concept of biomimetic hierarchical roughness in membrane design for creating superoleophobic membrane surfaces from a vast pool of candidate materials, such as zeolites, metal-organic frameworks (MOFs), and single-layered graphene oxide.* My research also focuses on development of facile, low-cost preparation technique which would open a completely new direction for the membrane society. If the proposed membrane could be successfully fabricated, further investigation on scaling-up will be pursued.

The practice of drug delivery has changed dramatically in the last few decades and even greater changes are anticipated in the near future. It is because controlled delivery systems (CDSs) are one of the promising applications for human healthcare. In pharmaceutical market, the CDS is growing fast with approximate 10% annual increase. However, an important challenge in this area is the efficient delivery of drugs in the body using non-toxic nanocarriers. Most of the existing carrier materials show poor drug loading (usually less than 5 wt% of the transported drug versus the carrier material) and/or rapid release of the proportion of the drug that is simply adsorbed (or anchored) at the external surface of the nanocarrier. Zeolite having a pore size from 0.5 nm to 2 nm and more than 100 different pore structures is a promising candidate. *I aim to produce uniform zeolite nanocrystals and nanostructures for the adsorption of drug molecules with sizes below and above 2 nm, respectively. I will employ an experimental approach to fundamentally study the interaction of drug molecules with microporous zeolite matrices, and realize a controllable releasing of drug molecules to target sites.*

Teaching Interests: I am very comfortable in teaching core Chemical Engineering courses, such as Thermodynamics, Heat Transfer, Transport Phenomena, Reaction Engineering, and Mass Transfer. My teaching interests are also expanded many specialized Chemical Engineering courses, such as Phase Equilibrium and Stage-based Separations, Surfaces and Adsorption, Adsorption and Nanomaterials Characterization, and Process Simulation. I also plan to develop courses in the areas of solid & polymer surface chemistry, hierarchical porous materials design, and membrane separation technologies.

Selected Publications:

[1] K.H. Chu, **Y. Huang**, M. Yu, N. Her, J. Flora, C.M. Park, S. Kim, J. Cho, Evaluation of Humic Acid and Tannic Acid Fouling in Graphene Oxide-coated Ultrafiltration Membranes. *ACS Applied Materials & Interfaces*, 8, 22270–22279 (2016).

[2] **Y. Huang**, L. Wang, Z.N. Song, S.G. Li, M. Yu, Growth of High-Quality, Thickness-Reduced Zeolite Membranes towards N_2/CH_4 Separation Using High-Aspect-Ratio Seeds. *Angewandte Chemie International Edition*, 127, 10993–10997 (2015). *Featured as Back Cover Article.*

[3] J.J. Song, **Y. Huang**, S.W. Nam, M. Yu, J. Heo, N. Her, J.R.V. Flora, Y. Yoon, Ultrathin Graphene Oxide Membranes for the Removal of Humic Acid. *Separation and Purification Technology*, 144, 162–167 (2015).

[4] **Y. Huang**, H. Li, L. Wang, Y.L. Qiao, C.B. Tang, C. Jung, Y.M. Yoon, S.G. Li, M. Yu, Ultrafiltration Membranes with Structure-Optimized Graphene Oxide Coatings for Anti-Fouling Oil/Water Separation. *Advanced Materials Interfaces*, 2, 1400443–1400450 (2015). *Highlighted by the Journal as Top 1 (2014~2015) and Top 4 (2016) Most-Read Full Research Papers.*

[5] Z.N. Song, **Y. Huang**, W.W. Xu, L. Wang, Y. Bao, S.G. Li, M. Yu, Continuously Adjustable, Molecular-Sieving "Gate" on 5A Zeolite for Distinguishing Small Organic Molecules by Size. *Scientific Reports*, 5, Article number: 13981 (2015).

[6] Z.N. Song, **Y. Huang**, W.W. Xu, M. Yu, Composite 5A Zeolite with Ultrathin Porous TiO_2 Coating for Selective Gas Adsorption. *Chemical Communications*, 51, 373–375 (2014).

[7] H. Li, **Y. Huang**, Y.T. Mao, W.W. Xu, H. Ploehn, M. Yu, Tuning Underwater Oleophobicity of Graphene Oxide Coatings via UV Irradiation. *Chemical Communications*, 50, 9849–9851 (2014)

[8] H. Li, Z.N. Song, X.J. Zhang, **Y. Huang**, S.G. Li, Y.T. Mao, H.J. Ploehn, Y. Bao, M. Yu, Ultrathin, Molecular-Sieving Graphene Oxide Membranes for Selective Hydrogen Separation. *Science*, 342, 95–98 (2013).

[9] P.A. Zapata, **Y. Huang**, M.A. Gonzalez-Borja, D.E. Resasco, Sylilated Hydrophobic Zeolites with Enhanced Tolerance to Hot Liquid Water. *Journal of Catalysis*, 308, 82–97 (2013). *Invited Paper for the*

[10] **Y. Huang**, J.F. Yao, X.Y. Zhang, C.H. (Charlie) Kong, H.Y. Chen, D. X. Liu, M. Tsapatsis, M.R. Hill, A.J. Hill, H.T. Wang, Role of Ethanol in Sodalite Crystallization in an Ethanol-Na_2O-Al_2O_3-SiO_2-H_2O System. *CrystEngComm*, 13, 4714–4722 (2011).

[11] **Y. Huang**, D.H. Dong, J.F. Yao, L. He, J. Ho, C.H. Kong, A. J. Hill, H.T. Wang, In Situ Crystallization of Macroporous Monoliths with Hollow NaP Zeolite Structure. *Chemistry of Materials*, 22, 5271–5278 (2010).

[12] **Y. Huang**, K. Wang, D.H. Dong, D. Li, M.R. Hill, A.J. Hill, H.T. Wang, Synthesis of Hierarchical Porous Zeolite NaY Particles with Controllable Particle Sizes. *Microporous and Mesoporous Materials*, 127, 167–175 (2010).

[13] J.F. Yao, **Y. Huang**, H.T. Wang, Controlling Zeolite Structures and Morphologies Using Polymer Networks (Review paper). *Journal of Materials Chemistry*, 20, 9827–9831 (2010).

[14] **Y. Huang**, J. Ho, Z. Wang, P. Nakashima, A.J. Hill, H.T. Wang, Mesoporous Carbon Confined Conversion of Silica Nanoparticles into Zeolite Nanocrystals. *Microporous and Mesoporous Materials*, 117, 490–496 (2009)

[15] D.H. Dong, **Y. Huang**, X.Y. Zhang, L. He, C.Z. Li, H.T. Wang, Shape Forming of Ceramics with Controllable Microstructure by Drying-Free Colloidal Casting. *Journal of Materials Chemistry*, 19, 7070–7074 (2009).

476265 Programmable Peptide-DNA Hybrid Nanomaterials

Abstract:

Research Interests: Natural biomaterials systems evolved over billions of years to solve some of the most challenging engineering problems. These optimized properties result from hierarchical designs exhibiting structure and function on multiple scales. The structure and function of natural materials have inspired scientists in attempts to mimic their designs and construction principles with the goal of mimicking the native qualities using synthetic materials. Moreover, by employing novel chemistries, processing and fabrication tools now available, we might be able to improve on Nature's designs and functions. Although tremendous advance was achieved in mimicking the structural and functional properties of native biomaterials, some key features are still very challenging to reproduce. These include (i) development of matter that is programmable on both temporal and spatial scales (ii) Controlling the Hierarchical Structure (iii) Integrating Sense and Response to the structural design, and more. Specifically, creation of regenerative material and tissue engineering systems with this intricate functionality necessitates new composites that are highly dynamic and can coordinate multiple interacting components. Merging different types of molecules such as peptides and nucleic acids into hybrid materials will generate a new class of materials that are amenable to structural and functional programming. **Postdoctoral Project:** "*Instructing Cells With Programmable Peptide-DNA Hybrid Nanomaterials*."Under supervision of Samuel I. Stupp, Simpson Querrey Institute for BioNanotechnology, Feinberg School of Medicine, Northwestern University, Chicago, IL 60611, United States **PhD Dissertation:** "*Functionalized Semiconductor Quantum Dots for Optical Sensing*" Under supervision of Itamar Willner, The Institute of Chemistry, The Hebrew University of Jerusalem, Jerusalem, Israel. **Research Experience:** I have had highly multidiciplinary training. I obtained my undergraduate degree in chemistry and computer science at Bar-Ilan University in Israel, followed by a Master's degree in chemistry at The Hebrew University of Jerusalem. I then began doctoral studies in chemistry and nanobiotechnology at The Hebrew University of Jerusalem, Israel, in the lab of Prof. Itamar Willner.

My PhD focused on the synthesis of functionalized semiconductor and metallic nanoparticles as well as DNA nanostructures for the multiplexed analysis of biomarkers and probing intracellular processes. During my postdoctoral training at Northwestern University I studied self-assembling nanomaterials for regenerative medicine in the lab of Prof. Samuel Stupp and introduced a new area into the lab, focusing on DNA-peptide hybrids for constructing cell scaffolds with programmable control over structure and function. I also collaborated extensively with medical doctors and biologists to create nanomaterials of biomedical interest in areas like neural stem cell differentiation, regeneration of neurons, and siRNA delivery. As a result of my versatile training and extensive collaborations I have gained experience in many fields such as synthetic chemistry, material science, biology, medicine, optical imaging, and biomedical engineering. **Teaching Interests:** Aside from my research career, I also have extensive teaching experience. At the Hebrew University of Jerusalem I was a teaching assistant of a variety of courses including Organic Chemistry, Biological Chemistry of Materials, Computational Quantum Chemistry and several introductory labs in chemistry and nanotechonolgy. At Northwestern, I served as a co-instructor with my postdoctoral advisor teaching a course on Materials and Nanochemistry intended for undergraduate (senior year) and graduate students. Lastly, I also have had the privilege of mentoring several graduate and undergraduate students. **Fellowships:** 1. European Molecular Biology Organization Long-term Postdoctoral Fellowship. (2013-2015) 2. The Shlomiuk award for outstanding P.hD. research study. (2013) 3. The Charles Clore fellowship for outstanding graduate students in natural sciences. (2010-2012) 4. Converging Technologies fellowship for excellent graduate students. (2008-2009) **Future Direction:** As faculty I would like to continue developing hybrid materials based on nucleic acids and peptides for different applications in nanomedicine. In particular I would like to integrate my PhD experience in developing nanoparticles-based sensing devices, DNA switches and machines with my postdoctoral experience in peptide self-assembly to create biomaterials, or systems of materials, that simultaneously provide structure, sensing, and response - all integrated into function. The oligonucleotide-peptide hybrids will merge the programmability and nanoscale control afforded by DNA nanotechnology with the biological function of peptides and proteins. Applications that will drive and inspire the research include systems that can regulate cell behavior (stem cell; immune cells), injectable materials that "know" to travel to an injury site and trigger tissue regeneration by harnessing the body's own power to heal, and new generations of wearable materials (electronic and others) as “Sense and Treat” systems that could be used as scaffolds to follow certain biomarkers and promote wound healing. **Selected Publications: H-Index: 33; Citations – 4016** 1. **Freeman R.**, Stephanopoulos N., Sur S., Boekhoven J., Lee S.S., Stupp S.I., Instructing cells with programmable DNA peptide hybrids, *Nature*, 2016 (in revisions). 2. **Freeman R.**, Boekhoven J., Dickerson M.B., Naik R.R., Stupp S.I., Biopolymers and supramolecular polymers as biomaterials for biomedical applications, *MRS Bulletin*, 2015, 40, 1089-1101. 3. N. Stephanopoulos, **R. Freeman**, H.N. Scheler, S. Sur, S. Jeong, F. Tantakitti, J.A. Kessler, S.I. Stupp, Bioactive DNA-Peptide Nanotubes Enhance the Differentiation of Neural Stem Cells Into Neurons, *Nano Letters,* 2015, 15, 603-609. 4. **Freeman R.**, Girsh J., Fang-ju Jou A., Annie Ho J., Hug T., Dernedde J, Willner I. Optical Aptasensors for the Analysis of the Vascular Endothelial Growth Factor (VEGF), *Anal. Chem.*, 2012, 84, 6192-6198. 5. **Freeman R.**, Liu X.Q., Willner I. Chemiluminescent and Chemiluminescence Resonance Energy Transfer (CRET) Detection of DNA, Metal Ions and Aptamer-Substrate Complexes Using Hemin / G-Quadruplexes and CdSe/ZnS Quantum Dots, *J. Am. Chem. Soc.*, 2011, 133, 11597-11604. **(cited 287 times)** 6. **Freeman R.**, Finder T., Gill R. Willner I. Probing Protein Kinase (CK2) and Alkaline Phosphatase with CdSe/ZnS Quantum Dots. *Nano Lett.*,2010, 10, 2192-2196. 7. **Freeman R.**, Finder T., Willner I., Multiplexed analysis of Hg2+ and Ag+ ions by nucleic acids-functionalized CdSe/ZnS quantum dots and their use for logic gate operations, *Angew. Chem. Int. Ed.*, 2009, 121, 7958 - 7961. **(cited 344 times)**

476267 Point-of-Care Molecular Detection with Surface Engineering of Nanomaterials for Diagnostic Platforms

Abstract:

Research Interests:

A fundamental axiom of medicine is that faster diagnosis yields earlier intervention, reducing the spread of infection and improving patient outcomes. The development of effective, point-of-care (POC) methods for the detection of disease biomarkers would thus increase the frequency with which many diseases are detected and identified, and decrease the delay between diagnosis and treatment. In order to be efficient, POC diagnostic devices must put clinically actionable information in a doctor's hands during a patient's first visit rather than during a follow-up. Here we proposed a new signaling mechanism for detection of a variety of molecules that displays clinically relevant sensitivity and specificity for variety of targets, and ideally is: 1-rapid relative to the 15 minutes of a typical patient/provider interaction, 2-selective enough to work in unprocessed bodily fluids, 3-amenable to small volumes rather than venous blood draws, 4-quantitative, and 5-able to diagnose multiple analytes simultaneously. This electrochemical DNA-based sensing platform is designed to detect real antibodies in human blood based on steric hindrance effects of the big molecule to influence the DNA-DNA hybridization. We hypothesized that when a large macromolecule binds to a DNA strand, this should, in principle, reduce the number of DNA strands that can hybridize to their complementary strand attached to a surface. We set out to demonstrate this steric hindrance mechanism using electrochemistry by employing a redox-labeled signaling DNA strand and a gold electrode that contains the complementary capturing DNA sequence at high surface density. Upon binding to the capturing strands, the redox-labeled signaling strands generate a large electrochemical signal by bringing the redox labels close to the electrode surface. In the presence of a large target protein that binds a recognition element on the signaling strand, we hypothesize that fewer copies of this strand will be able to reach the surface-bound capturing strand due to steric hindrance, therefore generating lower electrochemical currents. We eventually demonstrate the multiplexing detection of multiple antibodies within less than 10 minutes in whole blood having this detection platform. Additionally, the platform can be broadly adapted to selectively and specifically detect many other proteins having well-designed recognition ligand to that protein. Thus, this detection platform holds great promise to be well positioned for adaptation as, for example, point-of-care diagnostics.

As a faculty candidate, I envision three potential multidisciplinary projects, which are in line with the interdisciplinary and collaborative research environment: (1) Point-Of-Care diagnosis: Single-step methods for rapid and quantitative diagnosis of disease biomarkers. (2) Electrochemical 2D and 3D -sensors for direct and/or indirect molecular detection. Improvement of the catalytic activity of nanostructured Pt (Pd, or Au) by electrochemical alloying, (3) Development of Giant MagnetoResistnace in nanoscale magnetic Multilayer for biosensing application.

Teaching Interests:

I am eager to continue to the undergraduate and graduate education. I would be exited to teach courses across the curriculum of Materials Engineering, Chemical Engineering, Nanotechnology Engineering and Biomedical Engineering, as well as advanced course subjects in the field of nanobiotechnology, surface engineering, electrochemistry and analytical chemistry.

470102 Regulation of Intracellular Delivery through Peptide-Based Nanocarrier Design (Invited Talk)

Abstract:

Nucleic acid cargoes offer unmatched diversity in gene regulatory potential and therapeutics, and understanding of nucleic acid functionality continues to expand rapidly and dramatically through seminal discoveries including RNA interference approaches and gene editing technologies. In nature, the basis for gene regulation is ultimately encoded by the exquisite specificity with which cells are able to control both the location and accessibility of nucleic acid constructs to govern their activation states. My research program seeks to understand and control gene activation using synthetic constructs through nature-inspired approaches to control and quantify cell binding interactions and stability in polymer and peptide nanocarriers. The basis of our approaches is the design of stimuli-responsive polymers and peptides whose interactions with nucleic acids and cells can be controlled dynamically by specific intracellular or external triggers. We exploit our ability to control nucleic acid binding/release and cellular processing to gain new mechanistic insights over nucleic acid delivery, leading to design advances including histone-inspired DNA targeting, light-responsive gene silencing, and collagen turnover-stimulated gene expression. This talk will highlight two ways we have used nature-inspired peptides to control gene transfer in regenerative medicine.

Our approaches are exemplified by our work in histone-targeted nanocarrier design. Histones have received great interest as potential gene carriers for several decades due to their seminal role in chromatin packaging and gene transfer, yet therapeutic efforts with histones have lacked both a well-controlled materials approach and a deeper knowledge of cellular processing mechanisms. Hence, histone-based carriers have failed to reach clinical efficacy. We have capitalized on newly recognized and highly pivotal roles for histone tails in native gene regulatory control to develop a gene transfer method that utilizes native, histone-based processing pathways *via* incorporation of post-translationally modified (PTM) histone tails within controllably-assembled DNA vehicles (polyplexes). Our efforts proved that polyplexes displaying PTM-modified histone tails promote nuclear accumulation, DNA release, transcription, and enhanced transfection. Moreover, our group has combined detailed nanostructure engineering with sophisticated cellular imaging to identify novel aspects in the cell biology framework regulating polyplex transport to the nucleus.

We have also focused on novel mechanisms to exploit natureâ€™s ability to harness extracellular matrix (ECM) proteins such as collagens to sequester and control delivery of bioactive nanostructures. Our specific approaches have capitalized on a class of peptides known as collagen-mimetic peptides, or CMPs, that have been recognized for their unique affinity for native collagen, which can be tailored through alterations in CMP amino acid sequence and molecular weight. CMPs incorporate themselves into the natural collagen triple helical structure via strand invasion, in a reversible process previously that has been used to modify extracted collagen *in vitro* and exclusively target remodeling collagen *in vivo*. In our studies, we employed a proline-rich CMP designed to act not only as an adjustable tether to regulate collagen-polyplex affinity, but also as an adhesive/endocytic ligand for polyplexes. The use of a collagen scaffold afforded our system structural support and innate bioactivity to encourage cellular ingrowth and proliferation, whereas altering the extent of the modification of our vector provided additional tunability to allow tailorable release for prolonged time periods. This CMP-based approach also consistently and fully maintained polyplex activity in the presence of serum for at least a week, whereas most bolus and substrate-mediated gene delivery approaches report rapid reductions within hours or a few days, and the level of transgene expression directly correlated with MMP-concentrations and the extent of collagen remodeling, demonstrating â€œon demandâ€ release. The ability to tailor release over extended periods via physical attachments, combined with the ability to provide cell-trigger release and collagen-mediated uptake, make this approach very attractive for many applications in regenerative medicine.

470495 Role of SNP Characteristics on the Endocytosis and Intracellular Trafficking of siRNA

Abstract:

Short interfering RNA (siRNA) therapeutics, have garnered considerable attention for their potential to treat a wide variety of disease-associated proteins through the use of RNA interference (RNAi). siRNA therapeutics have been developed for cancers, genetic disorders, and infectious diseases, but their progress through clinical pipelines has been limited by delivery vehicles that are toxic, immunogenic, or ineffective. Despite the variety of delivery vehicles that have been developed, the essential chemical and physical characteristics of effective delivery vehicles remain unknown. It is also unclear if RNAi is initiated more potently following endocytosis by particular mechanisms and if this variability is specific to a certain cell type or delivery vehicle chemistry.

Our goal is to define nanoparticle characteristics that are essential for potent siRNA delivery. Using silica nanoparticles (SNPs) for their tunable synthesis, we are investigating delivery criteria in four main categories: sequence specific mRNA degradation (silencing), membrane translocation, endosomal escape, and association with RNAi proteins. Using SNPs, vehicle size, structure, charge, and surface functionalization can be varied to determine optimal vehicle design criteria in a variety of cell types.

The presentation will discuss our novel live cell assay that allows independent intracellular trafficking of delivery vehicle, antisense, and sense siRNA. Using confocal microscopy, we track vehicle-siRNA complexes during endocytosis (whether by Clathrin, Caveolin, ARF6, GRAF1, Flotillin, and Macropinocytosis), endosomal trafficking (Early, Late, Recycling, and Lysosome), trafficking to the endoplasmic reticulum or golgi, and exocytosis. Our results to date demonstrate that multiple vehicle characteristics affect siRNA delivery and demonstrate the critical role of intracellular trafficking in maximizing siRNA potency.

462951 Engineering Periodic shRNA Delivery Systems with High Silencing Efficacy

Abstract:

Challenges in the delivery of small interfering RNAs (siRNAs) to disease sites have hampered clinical translation of RNA interference, particularly in cancer therapeutics. Due to its low valency and high rigidity, siRNA is not readily condensed into stable nanoparticles that can successfully protect it from nuclease degradation and deliver it into target cells. Periodic short hairpin RNAs (p-shRNAs), consisting of repeating RNAi sequences linked together, can potentially address these delivery barriers by improving complexation with cationic delivery materials. Ranging from under 200 to over 5000 nucleotides in length, p-shRNA can be generated in large quantities through rolling circle transcription, in which an RNA polymerase repeatedly traverses a small circular DNA template. We have engineered these p-shRNA molecules via template design and selective enzymatic digestion of their hairpin loops to generate an open-ended p-shRNA (op-shRNA) that induces greater gene silencing than p-shRNA and siRNA in multiple cancer cell lines up to nine days.

The high valency and flexibility of op-shRNA dramatically improves complexation with low molecular weight polycations compared to siRNA. To develop a delivery vehicle optimized for the biophysical properties of op-shRNA, we have synthesized a library of biodegradable poly(beta-amino ester)s (PBAEs) via factorial design. Our designed PBAEs condense op-shRNA into compact nanoparticles below 100 nm in diameter, and demonstrate high transfection efficiencies at polymer and RNA concentrations much lower than typically required for PBAE-mediated gene delivery. By varying the molecular weight and incorporation of alkyl chains in our PBAE library, we were able to develop correlations between these parameters and multiple parameters influencing op-shRNA delivery, including complex assembly, stability, intracellular disassembly, cellular uptake, and cytotoxicity. In particular, PBAEs of low to intermediate molecular weight and increasing alkyl chain content delivered op-shRNA *in vitro* most efficiently, generating levels of gene knockdown comparable to those by commercial transfection reagents. Through nucleic acid engineering and rational carrier design, we have successfully developed a stable, potent RNAi delivery platform that can trigger significant gene silencing at low doses, and potentially enable higher therapeutic efficacy *in vivo*.

460542 Polypeptide/Nucleic Acid Complexes As Delivery Vehicles

Abstract:

Electrostatically driven polymer self-assembly mechanisms, such as polyelectrolyte complexation, are vastly underexplored compared to amphiphilic-based assemblies, and yet offer unique opportunities for both the encapsulation of charged therapeutics and controlled delivery, via the tailoring of intermolecular interactions using pH and salt. Nanoscale polyelectrolyte complexes micelles can be created using block-copolymers that incorporate a neutral, yet hydrophilic block to stabilize the polyelectrolyte complex on the nanoscale. The modular nature of these assemblies enables the addition of a targeting ligand conjugated to the hydrophilic block to increase the efficacy of delivering nucleic acids, creating polyelectrolyte complex micelles with a targeting ligand outside the corona of the micelle. Ultimately, we demonstrate the use of targeted polyelectrolyte complex micelles formed using therapeutic nucleic acids that inhibit microRNAs involved in atherogenesis, with the goal of creating a treatment for atherosclerosis. In doing so, we create trifunctional polymers that contain a targeting peptide (that target fibrin or endothelial cells), a polyethylene glycol domain to facilitate microphase separation, and a polylysine domain used to complex the microRNA inhibitors. We characterize these nanoscale assemblies and proceed to explore the ability of these micelles to inhibit microRNA-92a, which regulates the expression of endothelial transcription factors involved in the progression of atherosclerosis. Also we explore the ability to inhibit microRNA-33a in macrophages, which is involved in the expression of ATP binding cassette transporters involved in the trafficking of cellular cholesterol to the liver. Ultimately, we show the effectiveness and biodistribution of our targeted micelle assemblies using a murine model.

462124 Highly Potent mRNA Delivery In Vivo with Intravenously-Administered Ionizable Lipid Nanoparticles

Abstract:

Introduction: Messenger RNA (mRNA) has tremendous therapeutic applications, including protein replacement therapy, cancer immunotherapy, and genomic engineering. However, successful intracellular delivery of mRNA remains challenging, as nucleic acids cannot readily cross cellular membranes alone due to their relatively large size, negative charge, and hydrophilicity. Over the past decade, researchers have developed lipid nanoparticles (LNPs) to encapsulate and deliver small interfering RNAs (siRNAs) to silence protein expression in the liver, but these LNPs have only recently been reported to encapsulate and deliver mRNA to produce protein *in vivo*. Because of the considerable size, stability, and charge differences between siRNA and mRNA, we hypothesized that LNP formulation parameters of historically-used siRNA-LNPs could be better optimized to more efficiently deliver larger mRNA payloads. To this end, we developed a generalized approach for optimizing LNP formulations for mRNA delivery *in vivo* using Design of Experiment (DOE) methodologies. After using this strategy to significantly improve mRNA-LNP potency *in vivo*, we then applied this mRNA-optimized formulation with new classes of ionizable lipid materials and discovered some of the most efficacious intravenously-administered mRNA delivery vehicles reported to date in the literature.

Materials / Methods: The most common siRNA-LNP formulations contain four components: an amine-containing ionizable lipid, phospholipid, cholesterol, and lipid-anchored polyethylene glycol, the relative ratios of which can have profound effects on the formulation potency. We first encapsulated mRNA encoding for Erythropoietin (EPO), a secreted serum protein, into the original siRNA-based LNP formulation (which employed the ionizable lipid C12-200) to establish a baseline from which to improve. LNPs were synthesized via microfluidic mixing of an ethanolic lipid phase with an aqueous RNA phase to form particles approximately 100 nm in diameter. Then, we employed DOE to modulate several formulation parameters simultaneously (e.g. lipid:mRNA weight ratio, phospholipid structure, molar composition of excipients). DOE allowed for the reduction of the total number of formulations required to deduce statistically significant trends in our large, multi-dimensional design space, which is important for the economical screening of LNP formulations due to poor *in vitro*/*in vivo* translatability and costs associated with screening large libraries *in vivo*. EPO mRNA-loaded C12-200 LNPs were injected intravenously at 15 μg total mRNA per mouse, and blood was collected 6 hours post-injection; an ELISA assay was then used to quantify serum EPO levels. Using Definitive Screening and Fractional Factorial DOE Designs, we screened a total of 38 distinct LNPs *in vivo*and characterized top-performing LNPs via their mRNA encapsulation efficiency, diameter, polydispersity, pKa, zeta potential, and morphology. Then, we applied the formulation parameters from the most efficacious C12-200 LNP formulation to a new class of ionizable lipid materials formed through a ring opening reaction between alkenyl epoxides and a polyamine core; unlike C12-200, which has fully saturated hydrocarbon tails, this new library of alkenyl amino alcohol (AAA) ionizable lipids contained degrees of unsaturation in its tails.

Results: After employing this DOE-based optimization process, we significantly ($p < 0.05$) increased the efficacy of intravenously-delivered liver-targeting C12-200 LNPs seven-fold for EPO mRNA delivery relative to the formulation previously used for siRNA delivery. At a 15 μg (0.75 mg/kg) mRNA dose, we achieved serum EPO protein concentrations in mice of approximately 7,000 ng/mL, which is 5-6 orders of magnitude greater than physiological levels. We found that increased lipid:RNA weight ratios and the use of the phospholipid 1,2-dioleoyl-sn-gylcero-3-phosphoethanolamine (DOPE) were the most crucial factors in determining LNP efficacy. Interestingly, when our mRNA-optimized C12-200 LNP formulation was loaded with siRNA, it silenced protein expression in hepatocytes nearly identically to the original siRNA formulation, suggesting that siRNA-loaded LNPs may be more tolerant than mRNA-loaded LNP of formulation design space differences. Lastly, when C12-200 was substituted with a library of novel alkenyl amino alcohol (AAA) lipids and formulated into LNPs with EPO mRNA, lead lipid OF-02 was discovered to afford EPO serum concentrations of approximately 14,000 ng/mL, the highest efficacy yet reported in the literature. Furthermore, mRNA delivered with OF-02 LNPs was expressed predominately in the liver and had a linear dose-response curve ranging from 0.75 to 2.25 mg/kg RNA dose.

Conclusions: In conclusion, we significantly increased the potency of liver-targeting C12-200 LNPs for mRNA delivery using Design of Experiment methodologies, while showing that the same mRNA-optimized LNPs did not confer additional potency when delivering siRNA. The optimized LNP formulation showed applicability to other types of ionizable lipids, such as the novel lipid OF-02, which made highly potent intravenously-delivered mRNA-LNPs. This work provides evidence that some technologies previously used for siRNA delivery may require further optimization for better mRNA delivery. Additionally, the general strategy presented here could be used to screen multi-dimensional design spaces in vivo for optimizing other biomaterials and payloads.

467122 Folate Receptor-Targeted Aminoglycoside-Derived Polymers for Transgene Expression in Cancer Cells

Abstract:

Targeted delivery of anticancer drugs can overcome the limitations of poor drug bioavailability, drug resistance, and unintended side effects associated with current chemotherapeutic regimens. Folate-receptors are overexpressed in several epithelial malignancies, including ovarian, renal, lung, triple-negative breast and bladder cancers, making these attractive targets for the targeted delivery of small molecule and/or nucleic acid therapeutics to cancer cells. Aminoglycosides are hydrophilic molecules that contain multiple sugars, amines and hydroxyls, and are attractive as monomers for generating diverse libraries of macromolecules. We first cross-linked aminoglycosides with diglycidyl ethers and subsequently derivatized these polymers with folic acid resulting in the formation of folate-targeted polymers. These folic acid-conjugated aminoglycoside-derived polymers were investigated for transgene delivery and expression in triple negative breast and bladder cancer cells at different polymer: plasmid DNA weight ratios. The targeted folic acid-conjugated polymers demonstrated significantly higher levels of luciferase expression in cancer cells compared to their parental polymers. Competitive inhibition of the folate receptor using free folic acid significantly inhibited transgene expression activity, indicating a role for the receptor in transgene expression. The targeted polymers were employed for the delivery of a newly constructed pEF-TRAIL plasmid, which expresses the TNF-alpha Related Apoptosis Inducing Ligand (TRAIL) protein, which is known to iduce death selectively in cancer cells. Cancer cell death following delivery of the pEF-TRAIL plasmid was investigated as a single-agent treatment as well as with other chemotherapeutics resulting in a combination treatment approach. Our results indicate that folate-targeted, aminoglycoside-derived polymers may be promising new materials for delivering nucleic acids to cancer cells in several malignancies.

460866 Sustained Transgene Expression Via Substrate-Mediated Gene Transfer Results from Multiple Transfection Events

Abstract:

Statement of purpose: Sustained delivery of therapeutic genes in vitro and in vivo has a wide range of applications in studying biology and in developing therapies to treat disease. Non-viral vectors such as cationic polymers still present promising approaches; however, bolus transfection methods with polyethyleneimine (PEI)-based DNA polyplexes suffer from considerable levels of cytotoxicity and short-lived transgene expression levels. Here, we designed and characterized a hyaluronic acid (HA)-based porous hydrogel system for non-viral gene delivery by loading with surface-associated DNA polyplexes (Figure 1A). With this, we observed enhanced and sustained transgene expression over 30 days of cell culture, with better cell viability, both marked improvements over comparable bolus transfection techniques. Finally, we investigated mechanisms thought to be responsible for the sustained expression profile.

Methods: HA hydrogels were formed via a Michael-addition reaction between acrylated HA and thiolated peptide crosslinkers with the addition of RGD peptide for cell adhesion. The sphere templating method was used to introduce 60-µm diameter pores. Plasmid DNA was complexed with polyethyleneimine to form DNA polyplexes, which were surface coated by electrostatic association to the pore surfaces of the hydrogel by incubating the formed hydrogel in the polyplex solution. DNA loading was assessed with ^{32}P-labelled DNA and scintillation counting. Transfection profiles were obtained by loading DNA encoding for the Gaussia luciferase gene, seeding D1 mouse mesenchymal stem cells in the hydrogels, and detecting the expressed protein with the associated assay.

Results: ^{32}P analysis of DNA-loaded porous hydrogels showed that up to 4 µg of DNA could be loaded into the scaffold. Although the DNA was electrostatically immobilized to the scaffold, minimal polyplex release was observed with only 5% of the initial loaded DNA released over 7 days. Cell culture within this scaffold resulted in sustained transgene expression over a period of more than 30 days at levels higher than those seen using bolus transfection techniques (Fig. 1B). In addition, adjusting the DNA loading concentration resulted in modulation of transgene expression (Fig. 1B). While repeated bolus transfections in 2D culture result in marked toxicity (Fig. 1C), culture in surface-coated hydrogels was significantly less severe (Fig. 1D). To determine if sustained expression was due to re-transfection events, it was observed that levels of internalized plasmid DNA in cells cultured in surface-coated hydrogels was sustained at higher levels over time than in cells cultured in non-coated hydrogels but administered a bolus polyplex transfection (Fig. 1E).

Conclusions: After DNA loading by surface coating, there was minimal basal DNA release over time, suggesting that this may serve as a robust system for long-term DNA availability and expression. Indeed, sustained expression was seen over thirty days of culture and could be enhanced by increasing DNA loading. Although repeated bolus transfections result in severe toxicity, repeated internalization events are possible in surface-coated scaffolds without the same toxic effects and are responsible for the sustained expression observed. This system is a promising means of sustained gene delivery for various biomedical applications without the need for genomic integration via viral methods, which have innate safety concerns.

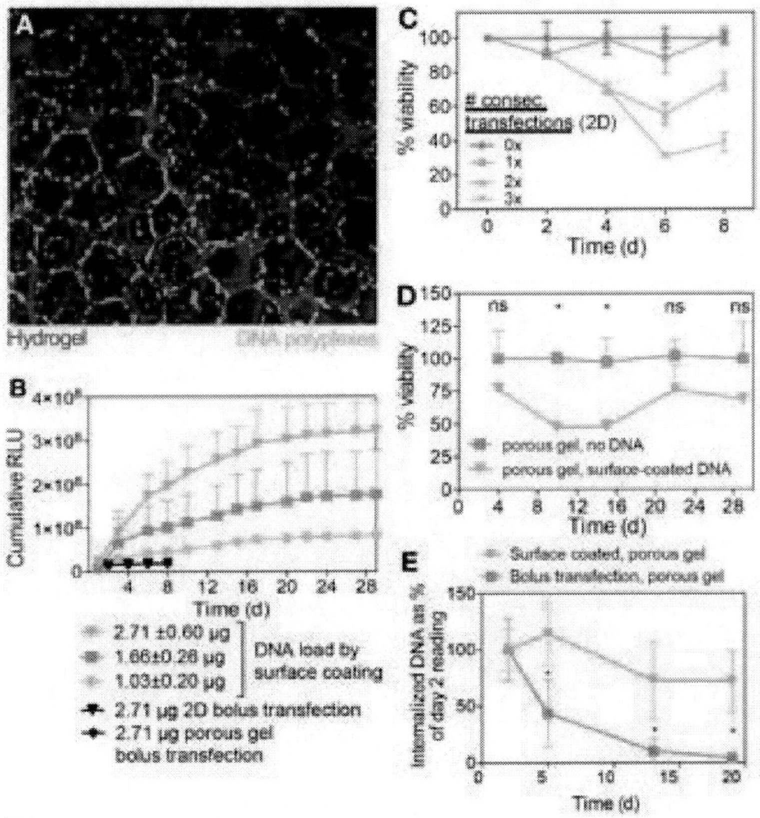

467956 Injectable Hydrogel Beads for Delivery of High Concentration Mab Formulations

Abstract:

Monoclonal antibody (mAb) based therapeutics are now the largest and fastest growing sector of the biopharmaceutical market. These products have shown excellent results in clinical settings as a result of the high binding specificity and minimal side effects of mAbs. However, our inability to predict protein-protein interactions poses a challenge to formulation efforts for these products. As an example, several recent studies have demonstrated that concentrated mAb formulations necessary for subcutaneous (SC) injection are plagued by high viscosities that are detrimental to this form of drug delivery. To mitigate these hindrances, we have developed a hydrogel delivery vehicle for the encapsulation of mAbs at very high concentrations while maintaining formulations with low viscosity suitable for SC injection. Monoclonal antibody encapsulated beads are produced using a thermally responsive nanoemulsion that can be cross-linked by UV curing. The temperature dependence can be utilized to tune the internal structure of the hydrogel bead and subsequently, control the local concentration and release of mAb. Additionally, the concentrated local environment within the hydrogel matrix is expected to help not only to stabilize the protein for long-term storage and but also to retain its native structure and function upon release.

471510 Development and Physicochemical Characterization of Tacrolimus-Loaded Nanocomposite Microparticles for the Treatment of Pulmonary Hypertension

Abstract:

Pulmonary arterial hypertension (PAH) is a progressive cardiovascular disease that leads to limited exercise capacity, right heart failure, and ultimately death. The primary treatment options currently available for patients with PAH are based on drugs with vasodilatory properties that improve cardiopulmonary function. Unfortunately, these drugs show no effect on prohibiting the progress of obliterative vascular pathology. Because of this, the only option for many patients to improve their health and quality of life is heart-lung transplantation. Therefore, new approaches to utilize compounds that focus on activating cellular mechanisms to reverse vascular remodeling are under investigation. Improving the function of the bone morphogenetic protein receptor-2 (BMPR2) signaling pathway is a potential direction in the development of improved PAH therapies, as the mutations of BMPR2 have been identified as the main cause of inherited PAH, accounting for 60 to 80% of familial cases. Since challenges exist in clinical application of BMPR2 gene therapy, tacrolimus (TAC) can be applied instead, as it increases signaling through the BMPR2 pathway. However, the application of TAC is hindered by its poor solubility, instability, and poor bioavailability. In addition to these limiting pharmacokinetic properties, systemic side effects such as neurotoxicity and nephrotoxicity also complicate the clinical use of TAC.

To overcome this problem, nanocomposite microparticles (nCmP) in the form of dry powder aerosols can be applied to deliver TAC more safely and effectively. This system is comprised of TAC-loaded acetalated dextran (Ac-Dex) nanoparticles (NP) entrapped in microparticle carriers using the excipient mannitol to allow for the delivery of TAC NP to the lungs. Upon pulmonary administration, the nCmP will deposit on the mucus in the respiratory tract, decompose into free NP and mannitol, and allow the nanoparticles to penetrate the mucus barrier and then release drug to the targeted site in a sustained rate. The goal of the described research was the initial development and physicochemical characterization of the nCmP systems with ability of targeted delivery, improved drug solubility, and potential of efficient penetration of mucus barrier and of controlled release via particle engineering.

The NP were approximately 200 nm in diameter with narrow size distribution both before loading into and after redispersion from nCmP. The NP exhibited smooth, spherical morphology and the nCmP were raisin-like spheres. High encapsulation efficacy was achieved both in the encapsulation of tacrolimus in NP and that of NP in nCmP. nCmP exhibited desirable aerosol dispersion properties allowing them to deposit into the deep lung regions for effective drug delivery. Overall, the designed nCmP system is a promising application in pulmonary TAC delivery for the treatment of PAH due to its novel features including targeted pulmonary delivery, improved solubility of tacrolimus, potential of permeation through mucus barrier, and of controlled drug release

468314 Nanoparticle-Mediated Inhibition of DNA Repair Sensitizes Brain Tumors to Radiotherapy

Abstract:

Radiotherapy (RT) is an integral component of the treatment for pediatric and adult brain tumors. However, survival is frequently accompanied by one or more radiation-induced adverse developmental and psychosocial sequelae, especially in children. Therefore, strategies that enhance RT in brain tumors while sparing adjacent normal brain are expected to improve life-long outcomes. The multifunctional DNA repair protein Ape1/Ref-1 has been implicated in conferring radiation resistance in brain tumors as Ape1 activity is inversely correlated with survival in brain cancer patients. Therefore, inhibiting Ape1 specifically in brain cancer cells should greatly enhance the effects of RT. We have developed a nanoparticle (NP) delivery vehicle that can stably bind, protect, and deliver nucleic acids to brain cancer cells and tumors. Here, we used our NPs to deliver siRNA against Ape1 (siApe1). We found that NP-mediated siApe1 delivery reduced Ape1 expression and activity, which was accompanied by a greater abundance of DNA damage. This diminished the resistance of brain cancer cells to RT. Therefore, NP-mediated inhibition of Ape1 may help enhance the therapeutic effect of RT in brain tumor patients and reduce treatment-induced morbidity.

461925 Theranostic Nanoparticles for Traumatic Brain Injury

Abstract:

Traumatic brain injury (TBI) is the leading cause of disability and death in people under 45. Currently, there are no effective treatment options available for the approximately 10 million new cases each year worldwide. Damage caused by TBI occurs in two phases: (1) the primary insult, and (2) the secondary phase where damage can spread beyond the primary injury leading to severe neurocognitive, physical, and psychosocial impairment. We have developed image-guided, Gd-conjugated, oxygen reactive polymer (ORP) nanoparticles to reduce the reactive oxygen species (ROS)-dependent secondary phase of the injury. ORP has a high ROS sponge capacity and is able to protect cell cultures from ROS-mediated damage. Magnetic resonance imaging (MRI) revealed high uptake and retention or ORP in the damaged region of brain in controlled cortical impact mouse models of TBI, which was attributed to a passive targeting effect similar to the enhanced permeability and retention effect observed in tumors. Importantly, ORP significantly reduced neurodegeneration and gliosis surrounding the primary injury in these mouse models. ORP shows initial promise as an effective therapy for TBI and will help provide a better understanding of nanomaterial interaction with damaged brain.

471209 Design, Synthesis, and Biological Evaluation of Novel Lipid Nanoparticle Materials for the In Vivo Delivery of Messenger RNA

Abstract:

The delivery of messenger RNA (mRNA) cargoes entrapped in lipid nanoparticles (LNPs) represents one potential strategy to replace deficient or non-existent proteins within target cell populations *in vivo*. Recent efforts have demonstrated that systemic administration of mRNA LNPs can induce functional protein expression in specific tissues in the body, but lead materials remain limited by their efficacy, biodistribution, and toxicity. Here, we employ rational design principles and synthesize novel LNP materials that can either *i.* induce higher levels of protein expression at a given dose or *ii.* modify the site of protein induction. Additionally, we establish critical structure-function parameters within our new mRNA LNP materials, demonstrating that *cis* alkenes play a pivotal role in improving the efficacy of both degradable and non-degradable mRNA LNP materials. *In vivo* experiments demonstrate batch-to-batch consistency, dose-response curves, and time course data for our lead mRNA LNPs. Toxicity studies also demonstrate that these materials are well tolerated *in vivo*. In sum, these studies not only establish highly potent materials for mRNA delivery, but also highlight the importance of employing rational synthetic design principles to better understand and improve our ability to deliver nucleic acids *in vivo*.

468226 Engineering Polymer Drug Conjugates to Synergistically Schedule Chemotherapeutics

Abstract:

Combination chemotherapy is commonly used to treat metastatic breast cancer; however, the success is limited due to systemic toxicity. To improve therapeutic efficacy, polymer drug conjugates carrying synergistic ratios of chemotherapy drugs can be used to selectively target cancer cells. While the importance of drug ratio on synergy has been previously investigated, the effect of temporal scheduling is not well understood. Here, we systematically evaluated the effect of temporal scheduling of doxorubicin (DOX) and gemcitabine (GEM) to determine both synergistic drug ratios and drug schedules. To engineer the temporal scheduling of DOX and GEM, hyaluronic acid drug conjugates with distinct linkers conjugating both DOX and GEM were optimized to control: (i) their individual release kinetics so as to match the optimized schedule and (ii) their molar ratios so as to optimize the synergistic dose. We show that polymer conjugates that deliver DOX and GEM in synergistic schedules and ratios are more effective at killing breast cancer cells compared to healthy epithelial cells. These results emphasize the importance in understanding the effect release rates have on the efficacy of synergistic drugs and provide the groundwork for the design of novel delivery systems capable of delivering exact molar ratios of synergistic drugs with temporal control.

462228 Hybrid Nanoparticles for Sequential and Controlled Delivery

Abstract:

Despite the great advance of nano-medicine in cancer treatment, it was reported that less than 10% of injected dosage could be delivered to the tumor sites. Co-delivery of drugs exploiting the blockage of multiple survival pathways in cancerous cells to achieve synergistic effect offers an alternative solution to this issue. In addition, recent growing knowledge of cellular networking points out that the release order of drugs plays a critical role in optimal therapeutic outcomes, which raises an another challenge in designing drug carrier. In this presentation, novel hybrid nanoparticles co-encapsulating hydrophilic and hydrophobic agents will be introduced. In this nanoparticle design, the release order of the agents can be programmed by loading them onto the surface or inside the nanoparticle. To validate the design of the nanoparticles, the size, structure, and drug release kinetics were assessed along with cytotoxicity tests.

462908 Solvent-Dependent Conductivity Enhancement of Carbon Nanotube Structures through Iodine Monobromide (IBr) Doping

Abstract:

Bulk carbon nanotube (CNT) electrical conductors are of interest for aerospace and terrestrial electrical transmission applications due to their low density, flexure tolerance, chemical stability, and high electrical and thermal conductivities. One research area of opportunity is the improvement of the electrical conductivity of bulk CNT structures towards the exceptionally high intrinsic conductivity of individual CNTs. To this end, the electrical conductivity of CNT structures can be modified by the incorporation of chemical dopants. Several methods of dopant delivery such as gaseous deposition and aqueous soaking have demonstrated success depending on the specific dopant species of interest. Recent work with interhalogen compounds, such as IBr and ICl, has shown the strong potential of these species to increase the conductivity of CNT thin films and wires, motivating the study and refinement of the delivery methodology for these dopants.

In this work, various CNT structures and types, including single- and multi-wall CNTs (SWCNTs and MWCNTs) in sheet and yarn formats, were doped with IBr, and the solvent-dependent conductivity enhancement was quantified. The solvatochromic shift of IBr in solutions of acetone, chloroform, dimethylacetamide (DMA), dimethyl sulfoxide (DMSO), ethanol, hexanes, isopropanol, methanol, or water was quantified using optical absorption spectroscopy. When CNT samples were exposed to equivalent concentrations of IBr in various solvents, differing levels of conductivity enhancement were observed. Top performing organic solvent systems included hexane and chloroform solutions of IBr, which caused a 5.7x increase over the purified CNT sheet conductivity, compared to an only 1.7x increase from doping solutions of IBr in acetone or DMSO. Conductivity enhancement was correlated with the solvatochromic shift, with a 6.7x increase for water being a notable outlier, indicating that dopant interaction with the CNTs in a bulk structure is dependent on the dopant-solvent and solvent-CNT interactions.

IBr adsorption onto the CNTs was also studied using time-dependent optical absorbance spectroscopy to gain further understanding of the dopant-CNT-solvent system, such as determining how much dopant was adsorbed by the CNTs as a function of time, and at what point the CNTs were saturated. This method allowed for the relative quantification of dopant interaction with CNTs and demonstrated that the doping solutions which showed greater increase in CNT conductivity are those that promoted greater dopant adsorption. Finally, differences in doping efficacy across CNT sample types is demonstrated to be an effect of varying dopant adsorption limitations, such as greater dopant adsorption on SWCNTs than on MWCNTs with a correspondingly larger increase in SWCNT conductivity than MWCNT conductivity.

These observations highlight the importance of dopant delivery methodology when increasing the electrical conductivity of bulk CNT structures. The trends depicted enable the selection of solvents that promote increased dopant interaction with CNTs and the determination of effective dopant exposure times on the order of half an hour at room temperature. The resulting increase in electrical conductivity for bulk CNT conductors can spur the adoption of multifunctional CNT materials for advanced wiring architectures.

466726 Surface Hydrogen Enables Sub-Eutectic Vapor-Liquid-Solid Semiconductor Nanowire Growth

Abstract:

Semiconductor nanowires are emerging as indispensable nanoscale building blocks for next generation energy conversion, electronic, and photonic devices. The bottom-up vapor-liquid-solid (VLS) mechanism â€“ whereby a liquid eutectic "catalyst" droplet collects precursor molecules (or atoms) from the vapor and directs crystallization of the solid nanowire â€“ is a nearly ubiquitous method for nanowire synthesis. While VLS growth below the bulk metal-semiconductor eutectic temperature has long been known, the fundamental processes that govern this behavior are poorly understood. Here, we show that hydrogen atoms adsorbed on the Ge nanowire sidewall enable AuGe catalyst supercooling and control Au transport. Our experimental approach combines in situ infrared spectroscopy to directly and quantitatively determine hydrogen atom coverage with a "regrowth" step that allows catalyst phase to be determined with ex situ electron microscopy. Maintenance of a supercooled catalyst with only hydrogen radical delivery confirms the centrality of sidewall chemistry. This work underscores the importance of the nanowire sidewall and its chemistry on catalyst state, identifies new methods to regulate catalyst composition, and provides synthetic strategies for sub-eutectic growth in other nanowire systems. We leverage this newfound understanding of nanowire growth chemistry to fabricate large-area arrays of high quality axial Si/Ge heterostructures.

469942 Bsmv As a Novel Biotemplate for Palladium Nanomaterial Synthesis and Mechanistic Comparisons

Abstract:

A new virus was applied to the synthesis of palladium nanorods using the hydrothermal synthesis. The Barley stripes mosaic virus (BSMV) was successfully used to produce metal nanorods of similar quality to those produced with the Wild type Tobacco mosaic virus (WtTMV). Biomineralization was characterized in terms of the adsorption, reduction (after decoupling), and nanocrystal formation processes. The BSMV surface-mediated reduction of $Pd^{(2+)}$ proceeded via 1st order kinetics in both $Pd^{(2+)}$ and BSMV. The adsorption equilibrium relationship of $PdCl_3H_2O^-$ on the BSMV surface was described by a multi-step Langmuir isotherm suggesting alternative adsorbate-adsorbent interactions when compared to those on WtTMV. The total adsorption capacity of palladium species on BSMV is more than double of that on WtTMV. Finally, all length scales of the as-synthesized materials could be characterized in a statistically relevant way by combining ultra-small angle X-ray scattering (USAXS) with small angle X-ray scattering (SAXS) experiments and particle scattering models. Results confirm the presence of core-shell cylindrical particles with 1-2 nm grains. The materials synthesized were uniform and monodisperse, with controllable diameters and therefore, of comparable quality to those synthesized with WtTMV. Overall, BSMV has been confirmed as a viable and promising alternate biotemplate for biomineralization of inorganic materials. With this new virus, the biotemplating toolbox has been expanded for synthesis of new materials and comparative study of biomineralization processes.

469387 A Computational Investigation of the Surfactant-Mediated Carbon Nanotube Stabilization in a Liquid Epoxy Resin

Abstract:

We have investigated the carbon nanotube (CNT) stabilization in a liquid epoxy resin using a set of cationic and anionic surfactants with varying concentrations at room and elevated temperatures. Our main objective has been to gain a fundamental molecular understanding of the effects of surfactant type, concentration, and temperature on the mitigation of CNT â€œbundleâ€ formation (aggregation) in the epoxy resin, which arises due to strong van der Waals interactions between the nanotubes. Epoxy, being one of the widely used resins in aerospace-grade carbon fiber-based composites, has been enhanced by CNTs to provide, for example, lighting strike protection. However, the uniform dispersion of the CNTs in the resin has to date remained challenging. In this work, we have performed a series of molecular dynamics (MD) simulations of single-CNT and dual-CNT systems containing three different concentrations of 1) cationic surfactants, i.e., cetyltrimethylammonium bromide (CTAB) and sodium cholate (SC), 2) anionic surfactants, i.e., sodium dodecyl sulfate (SDS) and sodium dodecylbenzenesulfonate (NaDDBS), and 3) a 1:1 mixture of CTAB+SDS. We have performed the simulations at room (298 K) and elevated temperature (398 K). All simulations have been performed in the LAMMPS simulation package. We have quantified the CNT-surfactant interfacial interactions, and the surfactant geometries on the CNT surfaces to elucidate their CNT stabilizing mechanism. We have also investigated the effect of temperature increase on the CNT-surfactant interactions in the single-CNT systems. As for the dual-CNT systems, we have calculated the potential of mean force (PMF) and compared the stabilizing effects of the different surfactants at different concentrations.

472454 Nanoporous Graphene for Energy Storage Applications

Abstract:

Reduced graphene oxide (RGO) with modified morphologies has shown potential application as an electrode material for supercapacitors and batteries. By creating nanopores within RGO sheets, the specific surface area of electrode and diffusion of electroactive species can be improved. Here, we present a simple, solution based process to make nanoporous graphene from graphene oxide (GO) via microwave (MW) radiations. GO aqueous dispersion prepared by modified Hummerâ€™s method can be converted to RGO via thermal, chemical or photochemical processes. Further, RGO in presence of hydrogen peroxide is irradiated with microwaves. The MW generated OHâ€¢ radicals oxidize the carbon atoms around active defective sites in RGO, leaving behind carbon vacancy, which extends to form the nanopores. This nanoporous conductive material is further investigated for electrochemical and diffusional properties.

459738 Novel Energy Sources Based on Excess Thermopower and Carbon Nanotube Fibers

Abstract:

As cylindrical forms of graphene sheets, carbon nanotubes are unique in their high surface area for molecular adsorption and exceptional electrical conductivity. The theory describing so called thermopower waves, self-propagating reaction waves that generate electrical current from a thermally driven wavefront, explains the resulting power as arising from two additive components: the thermoelectric effect and the chemical potential gradient across the reaction front or excess thermopower. In this work, we validate this theory and generate unique power sources by creating conduits driven by only excess thermopower. Electrical current (11.9 μA/mg) is reversibly produced across a carbon nanotube fiber by localized chemical doping with acetonitrile. We show that the voltage is driven by a spatial gradient in the Fermi energy across the carbon conduit and can reach an unprecedented 525 mV in magnitude. Chemical modification of the nanotube surface, specifically, oxidation of the conduit is shown to enhance the effect. Closed circuit experiments further reveal an inverse length-scaling of the maximum power as $L^{-1.03}$. The resulting specific powers as large as 30.0 kW/kg for oxidized nanotubes highlight the potential for microscale energy generation and harvesting. These results constitute a strong validation of Excess Thermopower theory and may enable unique power sources capable of continual operation based on repeated solvent exposure.

474979 NSEF Forum Award Lecture - Toward a Distributed Renewable Electrochemical Energy and Mobility System (DREEMS): Electrocatalysis for Automotive Fuel Cells

Abstract:

One of the grand challenges facing humanity today is the development of an alternative energy system that is safe, clean, and sustainable and where combustion of fossil fuels no longer dominates. A distributed renewable electrochemical energy and mobility system (DREEMS) could meet this challenge. At the foundation of this new energy system, we have chosen to study a number of electrochemical devices including fuel cells, electrolyzers, and flow batteries. For all these devices electrocatalysis plays a critical role in controlling their performance, cost, and durability, and thus their economic viability. In this presentation, I will focus on our recent work on hydroxide exchange membrane fuel cells (HEMFCs) which can work with non-precious metal catalysts and inexpensive hydrocarbon polymer membranes. More specifically I will show why hydrogen oxidation reactions are slower in base than in acid for precious metal catalysts, and how we have developed the most active non-precious metal catalysts. I will also outline the remaining challenges for automotive HEMFCs.

474981 NSEF Young Investigator Award Lecture - Nanostructured Composite Materials for Thermoelectric Applications

Abstract:

The rapid development of nanomaterials in the past decade has provided a possibility of directly converting waste heat back to electricity based on the thermoelectric Seebeck effect. In the past few years, we have developed a transformative approach to pioneered low cost and scalable solution-phase growth methods to mass produce thermoelectric nanowires and nanowire heterostructures to match the physical and economic magnitudes of energy use and economical entertainment in the manufacture/recycling. These nanostructured thermoelectric materials show a significantly enhanced performance compared to the bulk crystals based on the quantum confinement, energy filtering, and minority carrier blocking. In addition, our research in flexible nanostructure-based thermoelectric systems further enabled the waste heat harvesting from virtually any shaped surface.

467643 Water-in-Water Emulsion Based Synthesis of Hydrogel Nanospheres with Tunable Release Properties

Abstract:

Poly(ethylene glycol) (PEG) micro/nanospheres have several unique advantages as polymer based drug delivery systems (DDS) such as tunable size, large surface area to volume ratio, and colloidal stability. Emulsification is one of the widely used methods for facile synthesis of micro/nanospheres. Two-phase aqueous system based on polymer-polymer immiscibility is a novel approach for preparation of water-in-water (w/w) emulsions. This method is promising for the synthesis of PEG micro/nanospheres for biological systems, since the emulsion is aqueous and do not require organic solvents or surfactants. Here, we report the synthesis of nano-scale PEG hydrogel particles using w/w emulsions using phase separation of dextran and PEG prepolymer. Dynamic light scattering (DLS) and scaning electron microscopy (SEM) results demonstrated that nano-scale hydrogel spheres could be obtained with this approach (**Figure 1**). We investigated the release kinetics of a model drug, pregabalin (PGB) from PEG nanospheres and demonstrated the influence of polymerization conditions on loading and release of the drug as well as the morphology and size distribution of PEG nanospheres. The experimental drug release data was fitted to a stretched exponential function which suggested high correlation with experimental results to predict half-time and drug release rates from the model equation. The biocompatibility of nanospheres on human dermal fibroblasts using cell-survival assay suggested that PEG nanospheres with altered concentrations are non-toxic, and can be considered for controlled drug/molecule delivery.

SHAPE * MERGEFORMAT

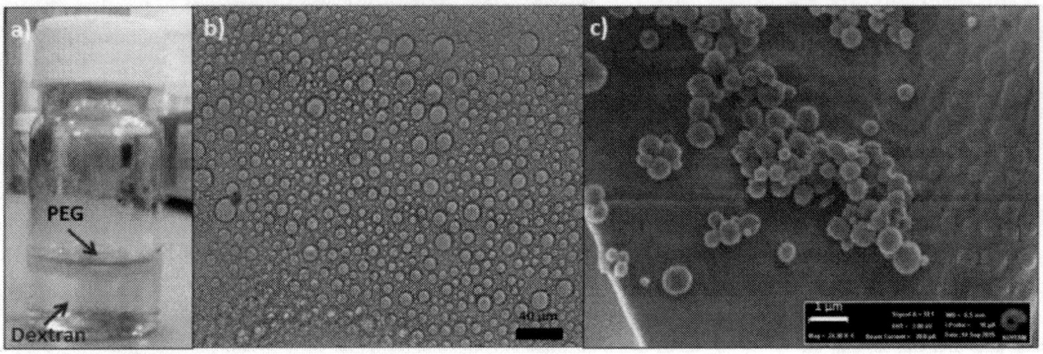

Fig. 1. a) Two-phase separation of dextran-PEG polymer b) Light microscope image of dextran-PEG w/w emulsion (scale bar: 40 μm) c) SEM image of dried PEG (scale bar: 1 μm)

471831 A Pharmacokinetic Model of a Tissue Implantable Insulin Sensor

Abstract:

While implantable sensors such as the continuous glucose monitoring system have been widely studied, both experimentally and mathematically, relatively little attention has been applied to the potential of insulin sensors. Such sensors can provide feedback control for insulin infusion systems and pumps and provide platforms for the monitoring of other biomarkers in vivo. In this work, the first pharmacokinetic model of an affinity nanosensor is developed for insulin operating subcutaneously in the limit of where mass transfer across biological membranes reaches a steady state. Using a physiological, compartmental model for glucose, insulin, and glucagon metabolism, the maximum sensor response and its delay time relative to plasma insulin concentration, are calculated based on sensor geometry, placement, and insulin binding parameters for a sensor localized within adipose tissue. A design relation is derived linking sensor dynamics to insulin time lag and placement within human tissue. The model should find utility in understanding dynamic insulin responses and forms the basis of model predictive control algorithms that incorporate sensor dynamics.

472055 Liposome-Encapsulated Synergistic Drug Combinations for Low Dose Chemotherapy

Abstract:

PEGylated liposomes revolutionized drug delivery for toxic chemotherapy drug doxorubicin, tokened as "red death", by reducing cardiotoxicity in patients. However, it remains unclear whether liposomal doxorubicin is therapeutically superior to free doxorubicin. We developed a novel PEGylated liposomal system, named DAFODIL (Doxorubicin And 5-Fluorouracil Optimally Delivered In a Liposome) which surpasses therapeutic effects of free drug administrations. DAFODIL delivers synergistic ratios of doxorubicin and 5-fluorouracil to tumors, and, with remarkably low doses, achieves greater than 90% reduction in tumor growth of a murine 4T1 mammary carcinoma model. Co-delivery of optimal, synergistic ratios ensured therapeutic effects at drug concentrations far below their maximum tolerable drug doses. Therefore, our approach improves both the therapeutic efficacy and toxicity of liposomal drug delivery systems, and provides a safe and effective means to deliver combination chemotherapies. This methodology is made feasible by extending the well-established ammonium-sulfate gradient encapsulation method to nucleobase analogues, a liposomal entrapment method initially only useful for anthracyclines. Therefore, our strategy can be utilized to efficiently evaluate various chemotherapy combinations in an effort to translate more effective combinations clinically.

469014 Photoexcited Quantum Dots Potentiate Antibiotic Activity in Multidrug-Resistant Bacteria

Abstract:

The rise of multidrug-resistant (MDR) bacteria is a growing concern to global health and is exacerbated by the lack of new antibiotics being developed. In order to treat MDR infections, new classes of antibiotics must be developed that are effective against already pervasive MDR infections. While current antibiotics are derived from natural products, investigating true synthetic antibiotics provides opportunity to design therapeutics for antimicrobial targets of interest. In this study we present a novel, nanoparticle based therapeutic that is able to inhibit MDR clinical isolates of *Escherichia coli*, *Salmonella typhimurium*, and *Klebsiella pneumoniae* both as a monotherapy as well as combination therapy with antibiotics that the isolates were previously resistant against. The nanoparticles used in this study generate specific, light activated reactive species (LARS) which alter the drug resistance phenotype and lower the level of resistance of the clinical strains to traditional antibiotics. We show that the redox homeostasis perturbing nanoparticles in combination with protein synthesis inhibiting antibiotics, namely chloramphenicol, streptomycin, and clindamycin, lower the IC_{50} of resistant MDR clinical isolates below the Clinical Laboratory Standards Institute (CLSI) breakpoint. We find that this phenomenon is not only limited to protein synthesis inhibiting antibiotics, but extends to other classes of antibiotics such as fluoroquinolones. Using a multiplicative model, we show that this combination therapeutic effect is additive and the effect is concentration dependent on both the antibiotic as well as the nanoparticle concentration. Interestingly the nanoparticle is able to have this drastic effect on antibiotic resistance at nanomolar concentrations highlighting the strong effect of the LARS produced. The bactericidal effect of the combination therapy is greater than that of each mono-therapy alone highlighting the potential utility of this combination therapy for treating MDR infections.

468253 Evaluation of the Cancer-Preventive Effect of Resveratrol-Loaded Nanoparticles on the Formation of Tumor Spheroids

Abstract:

Despite the recent advancements in the treatment of cancer, designing advanced therapeutics remains an enormous challenge for scientists. It is estimated in the U.S. that 1,685,210 new cases of cancer will be diagnosed in 2016 and 595,600 people will die from cancer, corresponding to approximately 1,600 deaths per day[1].

Chemotherapy has been extensively used to combat cancer, however its overall success is rather limited for lung cancer. An emerging alternative over the past few decades has been the use of chemopreventive phytochemicals, which are constituents of fruits and vegetables with potential pharmacological and/or nutritional functions. These agents may reverse or inhibit the carcinogenic process resulting in a decreased cancer risk [2]. The FDA has approved several agents for cancer prevention, including Tamoxifen for breast cancer and the Human Papilloma Virus (HPV) vaccine for cervical intraepithelial neoplasia. More than 200 other chemopreventive agents are being evaluated in clinical trials [3].

Resveratrol (3,5,4â€²-trihydroxystilbene, RSV) is a phytoalexin, a substance synthesized by plants to counteract pathogens [4]. It is abundantly found in red grapes and in lower amounts in peanuts and blueberries. In 1997, interest in RSV peaked when Jung *et al.* reported its chemopreventive effects on all stages of neoplastic activity (initiation, promotion and progression) [5]. *In vitro* studies show the promising properties of resveratrol; however, its limited bioavailability in the bloodstream makes it difficult to translate into clinical studies. A study by Dolfini *et al.* [6] showed that RSV induces drastic growth inhibition on a human breast cancer cell line in *in vivo* and *in vitro* assays. Also, RSV-loaded gel nanoparticles enhanced intracellular uptake and cytotoxicity in a NCI-H460 lung cancer cell line [7] and RSV-loaded nanostructure lipid carriers exhibited higher penetration into the skin in comparison to solid lipid nanoparticles [8], demonstrating improvement with RSV in nanoparticle systems.

The aim of this project was to evaluate the chemopreventive effect of RSV and RSV-loaded nanoparticles (RSV-NP) on the formation of tumor spheroids. We synthesized biodegradable nanoparticles comprised of the biodegradable polymer acetalated dextran loaded with RSV and we evaluated the efficacy of pure RSV and RSV-NP using A549 adenocarcinoma lung cancer cell line in 2D and 3D cell culture studies. 3D multicellular spheroids (MCS) reflect many of the properties of solid tumors and better mimic the barriers of drug diffusion, transport, and distribution within tumors [9, 10]. RSV and RSV-NP exerted an inhibitory effect on the growth of MCS, showing a promising alternative as a chemopreventive drug, which could palliate the incidence or mortality from lung cancer.

References

[1] Siegel, R.L., K.D. Miller, and A. Jemal, *Cancer statistics, 2016.* CA Cancer J Clin, 2016. 66(1): p. 7-30.

[2] Sporn, M.B., et al., *Prevention of chemical carcinogenesis by vitamin A and its synthetic analogs (retinoids).* Fed Proc, 1976. 35(6): p. 1332-8.

[3] Davis, J.S. and X. Wu, *Current state and future challenges of chemoprevention.* Discov Med, 2012. 13(72): p. 385-90.

[4] Singh, C.K., M.A. Ndiaye, and N. Ahmad, *Resveratrol and cancer: Challenges for clinical translation.* Biochim Biophys Acta, 2015. 1852(6): p. 1178-85.

[5] Jang, M., et al., *Cancer chemopreventive activity of resveratrol, a natural product derived from grapes.* Science (New York, N.Y.), 1997. 275(5297): p. 218-20.

[6] Dolfini, E., et al., *Resveratrol impairs the formation of MDA-MB-231 multicellular tumor spheroids concomitant with ceramide accumulation.* Cancer Lett, 2007. 249(2): p. 143-7.

[7] Karthikeyan, S., et al., *Anticancer activity of resveratrol-loaded gelatin nanoparticles on NCI-H460 non-small cell lung cancer cells.* Biomedicine & Preventive Nutrition, 2013. 3(1): p. 64-73.

[8] Gokce, E.H., et al., *Resveratrol-loaded solid lipid nanoparticles versus nanostructured lipid carriers: evaluation of antioxidant potential for dermal applications.* Int J Nanomedicine, 2012. 7: p. 1841-50.

[9] Sutherland, R.M., *Cell and environment interactions in tumor microregions: the multicell spheroid model.* Science, 1988. 240(4849): p. 177-84.

[10] Minchinton, A.I. and I.F. Tannock, *Drug penetration in solid tumours.* Nat Rev Cancer, 2006. 6(8): p. 583-92.

468082 Preparation, Characterization and *in Vitro* Validation of a Novel Paclitaxel Transport System to Target HER2-Positive Breast Cancer

Abstract:

Breast cancer is the second kind of cancer with the highest frequency worldwide, and the leading cause of oncological death among women in both, developed and developing countries [1]. Among the different clinical subtypes, the HER2-positive one, associated with a worse prognosis for patients, represents the 15-30% of all of the breast tumor cases [2].

One of the most efficient drugs used in the management of this subtype of cancer is paclitaxel, a natural terpenoid compound [3]. However, it has a poor aqueous solubility and an elevated toxicity, and these facts limit its clinical application.

With the aim of sorting out both disadvantages, a novel transport system for paclitaxel has been designed: polymeric nanoparticles of sodium alginate and piperazine, developed thanks to the electrostatic interactions that take place between both compounds at slightly acid pH values [4].

In such transport system, paclitaxel has been included in Î²-cyclodextrins in order to improve its aqueous solubility and in addition, trastuzumab, an antibody that binds specifically to the HER2 receptor, has been attached for avoiding the severe side effects that the terpenoid drug has. Both, the paclitaxel-Î²-cyclodextrins complexes and the antibody, have been attached to nanoparticlesâ€™ surface by a diimide-activate amidation process, and the conjugation efficiency of both of them to the nanoparticles has been analyzed. Later, the characterization of the transport system has been carried out by determining the average size and zeta potential of the nanoparticles, their stability over time under different conditions and the paclitaxel release kinetics. Finally, the effectiveness of this novel paclitaxel transport vehicle has been validated *in vitro*, by MTT assays performed with the BT474 cell line, with which positive results have been obtained. In this manner, a new door could be opened to improve the specific treatment of the HER2 positive breast cancer.

References

[1] Amaro, J., Severo, M., Vilela, S., Fonseca, S., Fontes, F., La Vecchia, C. et al. 2013. Patterns of breast cancer mortality trends in Europe. The Breast 22, 244-253.

[2] Figueroa-Magalhaes, M.C., Jelovac, D., Connolly, R.M., Wolff, A.C. 2014. Treatment of HER2-positive breast cancer. The Breast 235, 128-136.

[3] Singla, A.K., Garg, A., Agarwal, D. 2002. Paclitaxel and its formulations. Int. J. Pharm. 235, 179-192.

[4] Román, J.V., Rodríguez-Rodríguez, J.A., Martín del Valle, E.M., Galán, M.A. 2016. Synthesis of a new nanoparticle system based on electrostatic alginate-piperazine interactions. Polymer Adv. Tech. 27(5), 623-629.

472116 Adsorption, Stabilization and Recovery of Polyphenolic Flavonoids By TiO_2 Functionalized Mesoporous Silica Nanoparticles

Abstract:

Flavonoids such as quercetin and its derivatives, which are normally isolated from plants, are antioxidants with medicinal and therapeutic applications for the treatment of cancer, cardiovascular disease and other conditions because they act against tumors, inflammation, and reactive oxygen species. At the same time, because they are photoactive compounds, flavonoids are highly susceptible to degradation and activity loss after extraction from biological sources. Therefore, methods are needed to retain the activity of flavonoids during physicochemical processes and storage. A possible material of interest for binding, protection and release of a wide variety of compounds are engineered silica particles. Porous silica nanoparticles are of interest for the recovery, and more recently, the delivery of therapeutics. Recent progress in metal oxide chemistry allows the synthesis of silica nanoparticles of controlled size, pore properties, and surface functionalization. Titania (TiO_2) dispersed in mesoporous silica nanoparticles (MSNP) is hypothesized to have the potential to act as a high surface area adsorbent for quercetin due to its ability to bind dienols through chelation.

Here, the adsorption of quercetin and rutin (0.05-10 mg/ml in ethanol) is investigated on MSNPs (165 nm diameter) that are functionalized with submonolayer coverage (7-85%) of TiO_2. The amount of flavonoid adsorbed onto the particles was found to be a strong function of titania grafting density on the silica surface. As a function of quercetin concentration, strong binding occurs at low solution concentration followed by a gradual approach to saturation at higher concentration. Nanoparticles with approximately 60% of monolayer coverage of titania adsorb 230 mg quercetin/gram silica, which is more than 100 times the capacity of nonporous particles with similar titania coverage. The flavonoid recovery from nanoparticles was also investigated using biocompatible ethanolic citric acid solution (20% w/v) and a five step multistage desorption process were able to recover more than 90% of the bound flavonoids cumulatively. The radical scavenging activity (RSA) of recovered and particle bound quercetin were measured and most of the RSA was retained after binding quercetin, and after recovery of bound quercetin into solution. The materials developed here have the potential to be used to strongly and directly bind flavonoids and related secondary metabolites from plant cells, and will serve as a synthetic platform to isolate, recover, and potentially deliver these compounds in biological systems.

473620 Degradation Kinetics of PLGA and PLGA Conjugated with Alendronate Nanoparticles

Abstract:

A drug carrier has to be able to target, transport and delivery drug to a specific place in the body. This carrier should be degraded by the body as quick as possible in order to avoid accumulation. Also the carrier have to be made from a biodegradable material which could be degraded by the body into nontoxic products. PLGA or poly(lactic-co-glycolic acid) is a biodegradable material used for drug delivery system. This material can be modified with different moieties like alendronate, in order to target to a specific place. Alendronate is a molecule with a higher affinity to bone tissue so, with this material could be possible to target an active compound like N-acetylcysteine to bone tissue. Although, a lot studies has been done about drug delivery systems, only few has been done about degradation of PLGA nanoparticles modified with a particular moiety, and we could not find reports about degradation kinetics of PLGA nanoparticles modified by alendronate (PLGA-ALE). The aim of this research was to compare the degradation kinetic of PLGA, PLGA-ALE and PLGA-ALE-NAC (nanoparticles made with PLGA-ALE and loaded with N-acetylcysteine) nanoparticles, the first one has been reported for many authors, the second and three one has not been reported.

Conjugation of PLGA with alendronate was made by carbodiimide chemistry, joining the acid terminated of PLGA with the amine of alendronate. The reaction was verified by NMR and FTIR. Nanoparticles were produced by nanoprecipitation method. Degradation kinetics was done suspended nanoparticles in PBS at 37 °C for 33 days, molecular weight of the polymer was measured by GPC.

Polymer degradation exhibited a first order kinetics, with values for the specific kinetics constant of 0.082, 0.0159 y 0.0163 $days^{-1}$ for PLGA, PLGA-ALE and PLGA-ALE-NAC nanoparticles respectively. The first value was the highest of the three, which means that alendronate protect polymer chain from a fast degradation. Values of PLGA-ALE and PLGA-ALE-NAC are similar with a difference of 2.5% in favor to PLGA-ALE-NAC indicating a small effect of NAC in favor to polymer degradation. As a conclusion modification of PLGA with alendronate reduce the polymer degradation rate, but, NAC increase 2.5% the degradation rate.

463064 Confinement Facilitated Protein-Protein Stacking: As Investigated By Neutron Scattering

Abstract:

Confinement of biomolecules in structured mesoporous materials can offer many advantages such as chemical and thermal stability, and resistance to degradation in both biological and synthetic systems. This has led to the application of protein confinement in implementation of this platform to applications as diverse as pharmaceutical formulation, bio-catalysis and sensing. A fundamental understanding of protein behaviour under confinement is paramount to design the best mesoporous host for a particular application. While many studies have linked changes in confined secondary structure and enzymatic activity, the understanding of overall confined protein tertiary structure and orientation under confinement is only partially understood. Here, we discuss recent advances elucidated via small-angle neutron scattering (SANS), using protein-loaded mesoporous silica materials known to exhibit a link between a better preserved confined secondary structure and higher enzymatic activity in narrow pores in which the curved surface interacts electrostatically with the enzyme[1]. SANS enabled the general behaviour of two model proteins, representing distinct electrostatic conditions, to be observed over three confinement geometries in both the Henry and high concentration adsorption regimes. This analysis results in the observation of protein-protein stacking, which is driven by protein coordination that is facilitated by the geometry of the confining mesopores and the protein loading concentration.

References: [1] Sang L-C, Coppens M-O. Effects of surface curvature and surface chemistry on the structure and activity of proteins adsorbed in nanopores. Phys Chem Chem Phys. 2011;13(14):6689

455709 Energy Innovation - Challenges to Commercialization of New Energy Technology

Abstract:

Meeting growing global demand for energy in the 21st century in an environmentally responsible manner will require a significant escalation in innovation and the development of new, clean energy sources. Innovation is defined as invention plus commercialization, so that the innovation process isn't complete until the invention has been turned into a useful product that provides real world benefits. Innovation in energy is complex, has a high rate of failure, takes a long time. Commercializing new energy technology requires numerous technical and business disciplines, partnerships with large energy entities, support from service organizations and most importantly, significant investment.

This presentation will explore the energy commercialization process and how a technology accelerator, such as the Houston Technology Center, can reduce the risk of failure and speed up the innovation process by helping energy entrepreneurs navigate the commercialization process, develop support networks, acquire skills and resources, develop strategic relationships and attract funding.

463641 Energy and Nanomaterials: Interfacial Intersection

Abstract:

With energy at the forefront of the national economy and security energy-directed applications are of particular interest. Not surprisingly, energy and materials are intimately related. Many forms of energy utilization, conversion and storage and generation are dominated by interfacial chemistry. Therein nanomaterials as an interfacial modifier can play a critical role in these processes.

Specific energy applications we have explored include the following:

1) Storage: Increased energy density in Li ion batteries and supercapacitors using carbon nanotubes

2) Efficiency: Reduced friction using nanolubricants between moving parts

3) Transfer: Improved thermal management using nanofluids in heat transfer applications

4) Conservation: Lightweight polymeric composites incorporating nanotubes, nanoclays and graphene oxide for vehicle composites

5) Control: Gas sensors based on nanoscale metal oxide semi-conductors for process control and monitoring

6) Generation: Catalysts and photocatalysts using nanostructured oxides for accelerated charge transfer and minimal recombination losses

Highlights in each application will be presented.

Though synthesis of a host of carbon and metal oxide nanomaterials has been demonstrated, their integration into practical applications remains highly challenging. This talk will provide an overview of their synthesis, integration and value in energy storage, conservation, transfer, efficiency, control and generation. Alternative, scalable synthesis approaches such as flame synthesis and associated laser-based optical diagnostics will be touched upon.

470168 Carbon Nanotubes in Real World Applications – A Perspective on Translational Challenges and Industry Progress

Abstract:

It's been 25 years since the discovery of the carbon nanotube (CNT). CNTs were lauded as a breakthrough materials, with potential to revolutionize virtually every industry from energy to health. Early research indicated that many of the challenges of working with these materials were related to scaling of high purity, high perfection materials and directing their organization into controlled morphologies. Over the past decade, there has been a decline in carbon nanotube suppliers and decrease in press releases, accompanied by a shift in attention to newer materials such as graphene and other two-dimensional materials. Yet, many companies have been quietly incorporating CNTs into applications. This talk will focus on the state of the art for CNTs in applications, including their use in functional coatings, energy storage, and light-weighting. Our group's efforts at Battelle are in translational science – using new science discoveries to solve a broad range of industrial problems. The dominant theme in this type of research is how to control structures at interfaces to enable functionality, durability, and stability, while minding aspects of scalability and manufacturability. The strategic fit of a new material into an existing device is often a major hurdle to be managed. A case study from our group will be used to highlight translation of a CNT coating into a vehicle application -- a low-power, lightweight ice protection system for unmanned aerial vehicles (UAVs). Icing is a serious problem of UAVs, due to their flight characteristics, limited power budget, and restrictions on added weight and size. Our approach is to integrate an electrically-powered heater coating between the standard coating system and selectively power to keep flight surfaces clean. Due to their high conductivity and ability to be spray-applied, CNTs are an excellent candidate for creating heaters that can be retrofitted onto existing aircraft. Successful deployment of these systems required balancing several other requirements, primarily related to power, weight, and lifetime. Through fundamental study of conduction mechanisms in CNT networks, their environmental response, and fatigue behavior, approaches for creating stable, functional systems were identified. We will discuss our path to managing power budget, stability, and durability to create a system that has demonstrated successful performance in a simulated field environment. The key lessons learned from this case study will be summarized and generalized for others going through the product development process. We expect to see growth in the number of applications of CNTs, as the utility of CNTs are demonstrated in real-world environments.

473378 Solution Processed Inorganic Solar Cells

Abstract:

SOLUTION PROCESSED INORGANIC SOLAR CELLS

Caleb Miskin and Rakesh Agrawal
School of Chemical Engineering
Purdue University, West Lafayette, Indiana, USA

With the abundant availability of renewable energy from the sun, one can easily obtain all the power needed for society to continue its growth. What is holding the progress from moving forward is not the capability, but rather the economics. Thin-film solar cells create one step towards the reduction of costs by utilizing less material, but they have difficulty in scalability. To assist in scalability and further reduce costs, we have turned to the development of solution processed solar cells.
In our solution-based approach, we have taken a two prong approach. In the first approach, we synthesize nanoparticles of candidate compound semiconductors and disperse them in solvents to formulate inks. In the second approach, we are developing a new chemistry based on amine-thiol mixtures to dissolve precursor compounds that could not be dissolved earlier in common solvents to prepare molecular precursor inks. These inks are then used to print micron thick films on suitable substrates and heat treated to grow large grains suitable for good optoelectronic properties. Our group is studying the dissolution mechanism for molecular precursors, mechanism for grain growth in nanoparticle films during annealing under different environments and is characterizing final devices for their optoelectronic performance. We are also using the solution chemistry to synthesize nanoparticles of some new and promising semiconducting materials and their thin films. Some examples of the semiconducting materials being explored in our lab are: Cu(In,Ga)(S,Se)2, Cu2Zn(Sn,Ge)(S,Se)2, Cu3P(S,Se)4, Cu3As(S,Se)4, ZnSnP2 , SnS etc.
In this talk, we will highlight some of the important findings of our groupâ€™s research including fabrication of

Cu(In,Ga)(S,Se)2 and Cu2Zn(Sn,Ge)(S,Se)2 solar cells with power conversion efficiencies of 15% and 9.4% respectively. The excitement of this routeâ€™s science and technology along with its challenges will be presented.

471833 Processing as Viable Strategy for Forming High Performance Lithium Ion Battery Electrodes

Abstract:

Future applications of energy storage devices such as lithium ion batteries will require substantial improvement in electrochemical performance. Key metrics will be higher specific capacity leading to high energy and power densities, high Coulombic efficiency, and low capacity fade over several hundred to several thousand charge-discharge cycles.

Developments in this area can be divided into two distinct parts (i) efforts to discover new materials, or (ii) devise novel strategies for placing already known materials into optimal configurations. The latter strategy will be reviewed in this presentation. Vapor phase and colloidal processing techniques to create nanostructured anodes will be described, and the performance of half cells created by these methods will be analyzed. Several challenges still remain, and these will be discussed.

461202 Nanostructured Block Copolymers for Lithium Batteries and Biofuels Purification

Abstract:

Nanostructured block copolymers enable the design of membranes with optimized transporting channels for a variety of applications. This presentation addresses the use of these membranes to transport lithium ions in polymeric lithium battery electrolytes, and to separate biofuels such as ethanol and butanol from dilute fermentation broths where they are produced. Previous studies of electrolytes showed an inverse correlation between ionic conductivity and shear modulus: increasing conductivity invariably resulted in a decrease in the shear modulus. Similarly, previous studies of alcohol-selective membranes showed an inverse correlation between alcohol separation factor (ratio of alcohol concentration in the exiting stream to that of the incoming stream) and total flux: increasing the separation factor invariably resulted in a decrease in total flux. Block copolymer membranes do not suffer from these inverse correlations. Practical implications of our discoveries for enabling safe rechargeable batteries and second generation biofuels will be discussed. A large fraction of energy and materials that we use today are derived from industries that consume non-renewable resources. The two projects described here may enable a gradual transition toward to a more sustainable approach.

474230 Bone Target N Acetylcysteine Loaded in PLGA-ALE Nanoparticle to Osteoporosis Treatment. an *in Vitro* Test

Abstract:

N acetylcysteine (NAC) is a derivative of cysteine which is a precursor of well known antioxidant glutathione. This kind of components have shown the potential to attenuate osteoclasts activity *in-vitro* experiments. Osteoporosis is an illness characterized by an imbalance between osteoclast and osteoblast activity, so a component like NAC could be used as a therapeutic agent for its treatment. However, studies on the osteoporosis treatment with this agent have not yet shown beneficial effects. Despite there is no a conclusive explanation for this lack of activity, a possible cause is the low bioavailability of NAC (6% oral), preventing the required NAC doses to be delivered in the bone tissue to get a therapeutic effect.

In order to increase the bioavailability of NAC, and to selectively target the bone tissue, conjugated alendronate PLGA nanoparticles (PLGA-ALE) were used as carrier for NAC. This carrier should not react or destabilize the active compound, particularly as NAC denature by internally disulfide bond formation. Because of that, it is necessary to verify if the active compound retain its biological action after the loading process.

To confirm how suitable carrier are the PLGA-ALE nanoparticles to maintain NAC stability, this work developed an *in vitro* test to evaluate the NAC activity on osteoclast-cells like, previously generated by polyethylene glycol (PEG) induced macrophage fusion.

Synthesis of PLGA-ALE was done through carbodiimide chemistry. Nanoparticles were produced by the nanoprecipitation method and the *in vitro* test was done by fusion of U937 cells in the presence of PEG, following a fusion hybridoma protocol. Fused Cells were seeded on sterile bone slices for 4 to 6 hours, to attach the cells to the support. After that, a fetal bovine serum and free NAC supplemented culture medium was added and loaded within nanoparticles. Finally cells were incubated during 4 days, and subjected to reactive oxygen species (ROS), bone resorption pits and immunohistochemistry analysis. As osteoclastic activity surrogates, presence of different proteins were evaluated: matrix metalloproteinase (MMP), integrin aVb3 receptor (IB3), subunit of vacuolar ATPase (ATPase), DC-STAMP, Carbonic Anhydrase II (CAII), Cathepsin K, ADAM-12 and Calcitonin Receptor.

According with results both, free or polymer-loaded NAC, caused a decrease expression of the different osteoclast proteins (free NAC 74% and loaded NAC 71%). This experiments indicated that the polymer helps maintaining the stability of the active molecule up to 4 days. Taking into account that previous experiments demonstrated a high affinity of the polymer nanoparticles for the bone tissue, it is expected that a loaded NAC osteoporosis treatment to be more active than a free NAC one. Still *in-vivo* tests are necessary to confirm our findings.

As a conclusion, NAC loaded in PLGA-ALE nanoparticles retain the antioxidants properties of free NAC. The carrier did not have cytotoxic properties because cells grew in the present of it, and the entrapment process did not affect NAC stability.

472479 Inexpensive and Rapid Synthesis Unilammelar Liposomal Drugs for Targeted Delivery

Abstract:

Liposomes are rapidly emerging as one of the most advantageous nano-scale drug delivery systems. Their ability to enhance bioavailability through increased solubilities, lengthened clearance rates, and improved targeting capabilities has been well documented. However, current liposome synthesis approaches involve complex and tedious techniques limiting their investigation to specialized facilities. This represents a clear separation between basic liposomal pharmaceutical research, and their translation into therapeutic and theranostic platforms. Here we show a gentle method for reliably synthesizing, purifying, and concentrating unilamellar liposomes using common benchtop equipment. We use the ethanol injection method to encapsulate fluorescent molecules that vary in size (400-20k Da) within liposomes with a mean diameter of less than 150 nm. Using a standard benchtop centrifuge and Amicon Ultra filteration centrifuge tubes, we have characterized a simplified approach to gently and efficiently remove any unencapsulated drug while limiting aggregation or loss of sample, problems commonly encountered in ultracentrifugation and chromatographic approaches. We also demonstrate no loss in bioactivity following the purification process by incubating metastatic breast cancer cells with liposomes containing doxorubicin and monitoring cell viability over several days. It is our hope that with this efficient and accessible approach, the discovery of novel liposomal formulations may be better investigated and more widely implemented in clinical settings.

472629 Fabrication and Characterization of Fluorescently Labeled Polymeric Nanoparticles for Biodistribution Studies of Drug Delivery

Abstract:

The use of nanoparticles in drug delivery can address many of the difficulties faced by traditional methods such as systemic toxicity, transport limitations, and drug solubility. By loading drugs into biocompatible polymeric nanoparticles, drugs that were previously unusable due to their solubility or toxicity can now be used, and by fluorescently labeling the nanoparticles with a small molecule like 6,13-Bis(triisopropylsilylethynyl)pentacene (TIPS pentacene), the particles can be tracked for biodistribution studies in biological systems. TIPS pentacene is a fluorophore that has been primarily used in the optoelectronic device field. However, its high quantum yield and photostability make it a promising candidate as a fluorescent imaging agent for biomedical applications.

Using the Flash Nanoprecipitation process[1], different loadings of this fluorophore as well as a filler homopolymer of poly(L-lactic acid) (PLLA) have been incorporated into amphiphilic diblock copolymer nanoparticles of poly(ethylene oxide)-b-poly(D,L-lactic acid) (PEO-PDLLA). The effects of varying the TIPS pentacene loading as well as the homopolymer to diblock copolymer weight:weight ratio on the characteristics of the resulting nanoparticles such as size, zeta potential, and fluorescent intensity have been studied.

Nanoparticles were fabricated with hydrodynamic diameters in the range of 80-120 nm and polydispersity indices less than 0.3 as measured by dynamic light scattering. The zeta potentials of the particles were typically around -15 mV in phosphate buffered saline diluted 100X. Fluorescence studies show that the nanoparticles containing TIPS pentacene could be excited at wavelengths of 600 nm and below and had a principal emission peak at 650 nm. These nanoparticles were colloidally stable in cell-growth media supplemented with fetal bovine serum (FBS) for periods up to 12 hours. Cell culture studies with macrophage cells show that they were neither immunogenic nor cytotoxic. ELISA tests showed that the nanoparticles induced levels of the inflammatory cytokines interleukin-6 (IL-6) and tumor necrosis factor alpha (TNFa) in both M1 (pro-inflammatory) and M2 (anti-inflammatory) macrophage cells similar to those of mock negative control experiments. Further results will be presented on how these nanoparticles are internalized by macrophages.

1. Gindy, M. E.; Panagiotopoulos, A. Z.; Prud'homme, R. K., Composite block copolymer stabilized nanoparticles: Simultaneous encapsulation of organic actives and inorganic nanostructures. Langmuir 2008, 24 (1), 83-90.

470594 Polyelectrolyte Multilayer Films As Templates for Surface Modification to Design Liposomes Mediated Local and Sustained Therapeutic Delivery

Abstract:

The engineering of drug delivery platforms facilitating spatial and temporal release of a therapeutic is one of the key challenges in biomedical research that can ultimately lead to society-wide improvement in disease management. In recent years, substrate mediated delivery of cargo has shown great promise in applications including drug and gene eluding films/scaffolds, coatings for stents and other implantable devices, and controlling stem cell differentiation. Specifically, the drug delivery kinetics is particularly relevant when it is necessary to achieve effective dose and spatiotemporal release kinetics of the therapeutic agent at the intended site of injury. Delivery via immobilization of the therapeutic cargo by modifying the surfaces demonstrates higher translatable success compared to delivery using the free â€œbolusâ€ form by overcoming unfavorable burst kinetics, toxic offsite effects, and efficacy reduction due to systemic dilution. However, the lack in the development of a substrate mediated delivery platform that can exploit the advantages of nanocarriers to address bolus based therapeutic limitations has greatly hindered application. Here, we report the engineering of novel nanostructured substrate mediated therapeutic delivery system comprising of covalently functionalized liposomes (LNPs) with a surface decoration of hyaluronic Acid (HALNPs) embedded into an engineered Polyelectrolyte Multilayer (PEM) platform (HALNP-PEM) via ionic stabilization (Nature Scientific Reports, 2016). PEM provides an ideal surface modification tool that enables tuning the substrates to facilitate the immobilization of biomolecules including liposomes. The PEM base platform was constructed from poly-L-Lysine (PLL) and poly(sodium styrene sulfonate) (SPS) followed by adsorption of anionic HALNPs. An adsorption affinity assay and saturation curve defined the preferential HALNP deposition density for precise therapeutic dosing. A capping layer on top of the deposited HALNP monolayer was found to promote complete nanoparticle immobilization, support cell adhesion, and provide multi-dimensional nanoparticle confinement for predictable linear release profiles of both the nanocarrier and encapsulated cargo. Doxorubicin (DOX) was chosen as a model hydrophilic cargo and used to verify the structural integrity of the particles following deposition as well as to probe the therapeutic index of the HALNP-PEM platform. Furthermore, a direct comparison between bolus HALNP and the HALNP-PEM exemplified the effect of liposome confinement on intracellular uptake kinetics in patient derived metastatic breast cancer cells and confirmed sustained therapeutic release from the PEM platform. Lastly, we employed micropatterning to pattern both fluorescently conjugated HALNPs and DOX encapsulated in HALNPs into our PEM platform and found cell mediated uptake of HALNPs leading to micro-scale spatial control and sustained therapeutic release for over 60 hrs.

472948 Recognitive Methacrylated Alginate Nanoparticles for Protein Therapeutics

Abstract:

One of the main foci in biomaterial design is the development of novel approaches to specifically recognize biological cues and deliver therapeutics in a targeted mechanism. A variety of recognition molecules have been used to improve protein and drug delivery (e.g. antibodies, carbohydrates, peptides). However, these biomolecules are fragile, temperature and pH sensitive, and highly expensive, thus limiting their application worldwide. An alternative to these, are molecularly imprinted polymers (MIPs), which are biomaterials that recognize a predetermined template (e.g. proteins, small molecules). These systems are synthesized by crosslinking a polymeric matrix around the desired biomarker, therefore creating nanocavities that have complementary physicochemical characteristics to the template.[1] Among these systems are alginate particles, which have been previously used as drug and protein delivery vehicles.[2] Alginates are biocompatible anionic biopolymers, that can be ionically or chemically crosslinked, offering a wide variety of biochemical and physical characteristics.

The work presented herein was developed to provide insights on the protein/carrier interactions, and the importance that surface charge, chemical, and mechanical properties have on the recognition of proteins by molecularly imprinted natural polymers. Sodium alginate was functionalized with different methacrylate molecules (e.g. benzyl methacrylate, 2-hydroxyethyl methacrylate) using a carbodiimide reaction. Successful functionalization of the alginate was confirmed via Fourier transform infrared spectroscopy and NMR spectroscopy. Alginate nanoparticles were synthesized using water/oil microemulsions. Briefly, functionalized-alginate solution (up to 30% (v/v)) was emulsified in toluene with the biomolecular template, 1% (w/w) dioctyl sodium sulfosuccinate as a surfactant, and 1% (w/w)

Irgacure 184 as photoinitiator. Nanoparticle solution was crosslinked for 10 min using a 140 mW/cm^2 UV point-source. Resulting nanoparticles were purified and the template remains were removed by intensive washing steps. Physical characterization of the obtained particles was performed using light and electron microscopy, dynamic light scattering and zeta potential measurements confirming the morphology and charge of the functionalized alginate nanoparticles. Furthermore, the ability of these nanoformulations to recognize their template was evaluated using re-binding studies. Proteins similar to charge and size to the template used, were similarly recognized by functionalized alginate nanoparticles. However, by changing the functionalization molecule, it was shown that the physicochemical characteristics of alginate MIPs affected considerably their binding and recognition. These results exhibit the ability of molecular recognitive natural polymeric systems, such as alginate nanoparticles, to be tailored and enhance their specificity to be used as drug and protein delivery carriers.

[1] Peppas NA, Huang Y. Polymers and gels as molecular recognition agents. Pharmaceutical research. 2002;19:578-87.

[2] Lee KY, Mooney DJ. Alginate: properties and biomedical applications. Prog Polym Sci. 2012;37:106-26.

455389 Gold Nanoconjugates for Spinal Cord Injury Treatment: Recovery and Biodistribution

Abstract:

Due to their versatile surface chemistry, ease of imaging, and biocompatibility, gold nanoparticles (AuNPs) are employed as drug delivery devices in the treatment of spinal cord injury (SCI). In this talk, we will describe the synthesis of a tripartite nanoconjugate comprised of an AuNP drug carrier chemically conjugated to a transporter protein (wheat germ agglutinin coupled to horseradish peroxidase or WGA-HRP) and to theophylline, a SCI drug. The nanoconjugate structure was determined by UV-vis spectroscopy, transmission electron microscopy, dynamic light scattering, thermogravimetric analysis, and mass spectrometry. The therapeutic efficacy and biodistribution were studied in a validated animal model for hemi-diaphragmatic spinal cord paralysis. Our results show that a single administration of the nanoconjugate improved diaphragmatic activity at a much lower dosage than the effective systemic drug dosage and restored the respiratory drive. The effects lasted for 14 days. A biodistribution study using inductively coupled plasma mass spectrometry on the same animal model shows that the nanoconjugate was successful in targeting the respiratory neurons in the medulla. This research demonstrates the translational potential of our nanoconjugate design for SCI patients with respiratory dysfunction.

468582 Development of Drug Delivery Systems Based on a Fructose Polymer and 5-Fluorouracil

Abstract:

DEVELOPMENT OF DRUG DELIVERY SYSTEMS BASED ON A FRUCTOSE POLYMER AND 5-FLUOROURACIL

Alvaro G. Garcinuño, Antonio Tabernero, Miguel A. Galan, Eva M. Martin Del Valle

Department of Chemical Engineering. University of Salamanca.

There are new trends in biomedical engineering related with novel biomaterials which will be more compatible and more biodegradable than those that are being used at this time. For that reason, we use a polymer of fructose (levan) that has been successfully proved for forming nanocapsules by a self-assembled process.

Specifically, it is built a new drug delivery system (DDS) composed by levan and 5-fluouracil (a drug sometimes used in colorectal and hepatic cancer treatment). The work also presents results from a surface-modification of the initial levan (its carboxylate form).

The work shows how some parameters affect levan-nanoparticles synthesis, improving critical values such as size or Zpotential, in order to build a good DDS. The effect of pH and ionic strength are critical for nanoparticles assembly. Basic pHs are needed to obtain the best results (particle size between 300-400 nm with a Zpotential value around -20 mV). Those conditions avoid nanoparticle aggregation because the hydrocyl radicals of the medium take a hydrogen atom from the polymer that gets a negative charge. The ionic strength only affects the aggregation at high pHs (body blood ionic strength offers good results even at high pHs).

Moreover, carboxymethylation of the levan is performed to obtain carboxymethyllevan (CML). That surface-modification maintain good levels of size (between 300-400 nm), and improve Zpotential (-35 mV). That increment in Zpotential value prevents further particle aggregation, which is an interesting advantage for clinical or in vivo essays. Some studies of aggregation at different times were also performed. In both cases, at low ionic strength and basic pH, little aggregation was observed.

Results indicate that 5-fluorouracil is bound to either levan or CML surface by electrostatic interactions, obtaining the best results at basic pHs. According to drug release experiments, between 45-70% of the 5-fluoruracil is released from the levan in 23 hours depending on the pH, whereas only a low percentage of the drug is released at the initial times from the CML because of the stability of the amide bond that is produced on the surface. That fact emphasizes the good strategy of carboxymethylation for creating an effective drug delivery system.

472643 Synthesis, Gas Permeation and Selectivity of Highly Elastic Poly(dimethylsiloxane)/Graphene Oxide Composite Elastomer Membranes

Abstract:

Recently, polymer membranes have been considered ideal materials for large-scale industrial processes and they are currently being used for air separation to produce N_2-enriched air, natural gas treatment to remove acid gas, H_2 separation from hydrocarbons, and many other applications. However, polymer-based membranes still require additional improvements in gas separation performance, which is typically linked to the so-called permeability-selectivity trade-off. In addition, polymers are generally vulnerable to high temperature and harsh chemical environments. Therefore, understanding and enhancing mechanical and chemical stability of polymer-based membranes in industrial environments is crucial.

This study illustrates that amine functional groups on the ends of telechelic poly(dimethylsiloxane) (PDMS) can undergo post-processing reactions with surface epoxy groups on graphene oxide (GO) to form a robust elastomer during simple heating. In these materials, GO acts both as a nanofiller which reinforces the mechanical properties and participates as a multifunctional crosslinker, thereby promoting elastic properties. Experiments indicate that the telechelic PDMS/GO elastomer is highly crosslinked (e.g., more than 75 wt % is a non-dissolving crosslinked gel) but highly flexible such that it can be stretched up to 300% of its original length.

Using the same material, permeability for some common gases was studied as a function of GO concentration. Incorporating only 3.55 vol % GO into the PDMS matrix resulted in a more than 99.9% reduction in gas permeation for various gases, such as H_2, O_2, N_2, CH_4 and CO_2. Moreover, factor of two enhancements in gas selectivities were observed for CO_2/N_2 and CO_2/CH_4 compared to neat PDMS membranes. As a supplement to experimental data from scanning-electron and atomic-force microscopy and x-ray diffraction, theoretical models such as the Nielsen and

Cussler models were applied to comprehend the dispersion and alignment of GO in the polymer. Considering the expected thermal and chemical tolerance of the PDMS/GO composite membrane detailed in this work suggests these membranes could be useful in applications such as post-combustion CO_2 capture, CO_2 removal from natural gas and in other industries that use or process CO_2.

464281 A Molecular Dynamics Study on the Influence of Charge on the Transport of Water and Ions through Carbon Nanotubes (Award submission)

Abstract:

Carbon nanotubes (CNTs) are known to possess smooth hydrophobic walls, can be easily chemically functionalized, and have high surface to volume ratio, thus offering a possibility of unconventionally high flow rates and ion selectivity that makes them potential candidates for nanofluidic and biomimetic applications. However, it is difficult to measure transport processes at nanometer scales and thus much of the experimental data on water and ion transport through nanotubes available so far is limited to larger diameter nanotubes where a predominantly bulk-like behavior is observed. Motivated by the need to understand the superior transport properties of these extremely narrow pores, we perform long time (100 ns) classical MD simulations to investigate in atomic detail the intriguing structural and dynamical features of a NaCl solution under confinement in sub-3 nm diameter carbon nanotubes. Additionally, we assess if Donnan equilibrium theory can be used to ascribe a relation between membrane fixed charges and the concentration of ions within the nanotube.

In this study, the spatial distribution of water and ions and their transport properties in terms of their apparent diffusion coefficients are investigated through uncharged and charged armchair carbon nanotubes ($d = 0.8$ nm – 3.0 nm) embedded between two reservoirs of a 1M NaCl solution. The influence of charges applied to the surface of the CNT on ion-exclusion/screening phenomena is quite apparent and its impact is evident when the pore sizes are comparable to the Debye lengthscale. This can be attributed to the fact that at nanoscopic length scales, because of the large surface to volume ratio as compared with macro and microscopic length scales, surface charge has a pronounced effect on fluid volume in the hydrophobic cavity leading to rejection of ions. We obtain and compare the apparent diffusion coefficients of water and ions in bulk phase to that under confinement and in the vicinity of charges. We find that while the pore sizes that we study are permeable to ions, the axial diffusion coefficient of water and ions decreases under confinement and its effect is most prominent in the smallest diameter charged tube. We also obtain measurements of ion occupancy in the nanotube, coordination numbers of ions, water and ion flux, and ion rejection rates for the various cases considered.

The cases studied give a broad account of ion and water transport properties through uncharged and charged CNTs. The goal is to integrate fast water flow with ion selectivity through modulation of pore sizes, surface charges, and electric fields with potential applications in desalination, as molecular gates for separation of biomolecules, and other nanofluidic devices.

465293 High-Fidelity Single-Column Selective Desorption of Swcnts through the Modulation of Co-Surfactant States Around Carbon Nanotubes

Abstract:

Single-chirality single walled carbon nanotubes (SWCNTs) have unique optoelectronic properties that can be utilized in specific applications, such as photovoltaics and biosensors. However, due to the large variety of SWCNTs and the difficulty of separation, the application of chiral SWCNTs is still limited. The post-synthetic separation of single wall carbon nanotubes (SWCNTs) has been studied with great interest in the past decade. While high-fidelity separations of SWCNTs by their (*n*,*m*) type have been demonstrated, a simple, high-throughput method is still desired. Our previous work has highlighted the importance of surfactant structure along SWCNT sidewalls in the separation mechanism. Here we report the selective desorption of a wide range of single-chirality (*n*,*m*) fractions by a single column process with a single elution profile due to the formation of different thermodynamic co-surfactant states around SWCNTs. High-purity fractions of (*n*,*m*) nanotubes are obtained with high yield once a specific ratio of sodium dodecyl sulfate (SDS)/sodium deoxycholate (DOC) co-surfactant solution is used. By modulating the surfactant concentration of sodium dodecyl sulfate (SDS) and sodium deoxycholate (DOC), these single-chiral nanotubes are also obtained at their specific ratio even with long elution times, different total surfactant background concentrations, and different temperatures. The elution of only one (*n*,*m*) type at a specific co-surfactant ratio while other types are exposed to more surfactant suggests that each (*n*,*m*) type forms a thermodynamically-stable surfactant structure in the co-surfactant solution. These thermodynamic equilibrium states result in entropy-driven desorption at specific SDS/DOC co-surfactant ratios, enabling high-fidelity separations of each (*n*,*m*) type in a single column. These findings provide a promising foundation for the development of large-scale, high-throughput chromatographic separations that can collect each (*n*,*m*) type sequentially.

462904 Novel Energy Sources Based on Excess Thermopower and Carbon Nanotube Fibers (for graduate student award)

Abstract:

As cylindrical forms of graphene sheets, carbon nanotubes are unique in their high surface area for molecular adsorption and exceptional electrical conductivity. The theory describing so called thermopower waves, self-propagating reaction waves that generate electrical current from a thermally driven wavefront, explains the resulting power as arising from two additive components: the thermoelectric effect and the chemical potential gradient across the reaction front or excess thermopower. In this work, we validate this theory and generate unique power sources by creating conduits driven by only excess thermopower. Electrical current (11.9 μA/mg) is reversibly produced across a carbon nanotube fiber by localized chemical doping with acetonitrile. We show that the voltage is driven by a spatial gradient in the Fermi energy across the carbon conduit and can reach an unprecedented 525 mV in magnitude. Chemical modification of the nanotube surface, specifically, oxidation of the conduit is shown to enhance the effect. Closed circuit experiments further reveal an inverse length-scaling of the maximum power as $L^{-1.03}$. The resulting specific powers as large as 30.0 kW/kg for oxidized nanotubes highlight the potential for microscale energy generation and harvesting. These results constitute a strong validation of Excess Thermopower theory and may enable unique power sources capable of continual operation based on repeated solvent exposure.

465692 Crumpled Graphene Nanosheets and Their Assembly into Crosslinked Networks

Abstract:

In this work, aqueous dispersions of pristine graphene nanosheets were prepared using pyrene derivatives as dispersants. Controllable crumpling of these nanosheets was achieved by rapid evaporation of the dispersions in an industrially scalable spray dryer. Morphological transition of 2D nanosheets to 3D crumpled particles was directly observed by sample collection within the spray dryer. The crumpling mechanism depends on parameters such as nanosheet elasticity and surface chemistry. The particle size and morphology of the products can be tuned by altering the spray drying parameters such as temperature and pressure. Also, the unfolding of the crumpled particles upon rewetting depends on the nanosheet type and the solvent choice. Furthermore, 3D networks of highly crumpled graphene oxide were prepared using a low-temperature sol-gel technique. These networks display a remarkable homogenous porous structure with both chemical and physical crosslinks. The electrical conductivity of these samples is comparable to conductivity values reported for the networks prepared from 2D graphene nanosheets.

468792 *Optical Nanosensors for High Spatial and Temporal Studies of Neurotransmitter Imaging in the Brain*

Abstract:

Information processing in the brain occurs over a broad range of spatial and temporal scales, rendering it difficult to produce tools to study brain function. Here, we discuss optical imaging technologies for the detection of neurological biomolecules that have the capacity to function concurrently on the molecular scale and on the macroscopic scale for timescales that match neurotransmitter-based information processing.

We present recent progress in the analysis of optical imaging techniques, specifically the use of infrared nanosensors for neurotransmitter imaging. In this work, we present findings from in-vitro and in-vivo experiments conducted in hydrogel based brain tissue phantoms and acute brain slices, respectively. Furthermore, we present findings from computational modeling of dopamine dynamics in the extracellular space of the brain, and discuss how computational modeling can be used to guide in-vivo neurotransmitter imaging studies. Stochastic simulations reveal that optical nanosensors can capture millisecond-scale changes during phasic and tonic firing of dopaminergic neurons. Our work scales to include a collection of neurons in defined volumes. We couple our findings to experimental results within the context of technological boundary conditions for imaging near-infrared photons. We show that single-walled carbon nanotube-based optical dopamine nanosensors can be used to detect dopamine in biologically complex and optically dense tissue phantoms, and can be rendered non-toxic for in-vivo imaging. These findings suggest that near-infrared (nIR) neurotransmitter nanosensor constructs can relay information about neurotransmission in the tissue-compatible nIR optical window, for timescales that capture single neurotransmitter synaptic release events, and for in-vivo experimentation.

481341 Morphology of Carbon Nanotube Liquid Crystal Solution

Abstract:

Carbon nanotubes (CNTs) have a remarkable combination of electrical, thermal and mechanical properties. In order to realize these properties in applications, CNTs need to be processed into scalable materials such as fibers, coatings, transparent films, and foams. Fluid-phase self-assembly is one of the most promising ways to process CNTs. Chlorosulfonic acid (CSA) has shown to be a true solvent for CNTs. CNTs, similar to other rigid rod molecules, form nematic liquid crystal phase at high concentrations. Low concentration solutions are used for making transparent films, while high concentration solutions can be spun into multifunctional fibers. Successful assembly of CNTs needs a better understanding of the phase behavior and the flow process. Here, we present our recent results on studying the morphology of CNT liquid crystal solutions and how it is affected by the CNT properties, concentration and flow process. We used techniques including polarized optical microscopy and neutron scattering to have a better understanding of the ordering inside the liquid crystal phase. Our polarized optical microscopy results show that for short CNTs at intermediate concentrations, spindle-shaped nematic droplets emerge out of the isotropic phase, while long CNTs show bi-continuous phases. Additionally, high concentrations of long CNTs show higher ordered phases while short CNTs present characteristics of the nematic phase.

481342 Award: Material and Toxicity Evaluations of Nanoclays throughout Their Life Cycle

Abstract:

The incorporation of nanoclays, layered mineral silicates, into polymers leads to the formation of nanocomposites with enhanced mechanical strength, UV dispersion, barrier, and flammability properties. With proper exfoliation of nanoclays into polymers, these nanocomposites can then be applied to different industries, from automotive, to medical, or food packaging. However, with increasing implementation into such industries it is important that we begin to investigate the physical and chemical changes that nanoclays undergo throughout their life cycle as well as evaluate any toxicity during their life cycle to thus ensure both worker and consumer health. Herein, we examined the physical and chemical characteristics of 4 nanomer nanoclays (1 pristine and 3 organically modified) in raw or 'as-received' form to mimic a manufacturing environment, and after their thermal degradation, to mimic end-of life cycle in incineration plants. Material characterization was performed via microscopical and spectroscopical techniques while the toxicity evaluations used *in vitro* testing on human bronchial epithelial cells (BEAS-2B). Dose response curves were obtained via live cell counts to determine the degree of toxicity of each 'as-received' nanoclay and its thermally degraded byproduct. Thermal degradation caused for breakdown of the nanoclay platelet structure and loss of organic modifiers of the organically modified clays. The 'as-received' nanoclays had lower IC50 values relative to their thermally degraded counterparts, with a range in IC50 values of 10-325 µg/ml and 150-400 µg/ml for 'as-received' and thermally degraded nanoclay, respectively. While more studies are needed to examine the mechanism of toxicity of these nanoclays, our results reveal that these nanomaterials have the potential to induce inhalation toxicity in both manufacturing and disposal environments, with the degree of toxicity being dependent on their physical and chemical properties.

456249 Effect of Washing and Drying Process on Particle Growth of Cerium Oxide Nanoparticles

Abstract:

Cerium oxide nanomaterials have been used widely in many applications including catalyst, fuel cell electrolyte, chemical polishing, anti-oxidant due to their property of oxygen vacancy which is occurred by reversible redox reaction between Ce^{3+} and Ce^{4+}. Because this chemical activity of cerium oxide arises from surface of particles, studies regarding surface properties of cerium oxide particles have been carried out up to this day. Particularly, many researchers have studied about synthesis of smaller cerium oxide nanoparticles to extend surface area per unit mass, or particle morphology control using additives and surfactant to expose more catalytic active crystal plane.

However, there are not enough studies how washing and drying processes affects growth of synthesized nanoparticles. This study has focused on washing and drying processes of cerium oxide nanoparticles, which are inevitable procedures for synthesized nanoparticles to be used in industry. Firstly, synthesis of surface modified cerium oxide nanoparticles was carried out using decanoic acid as surfactant and water as solvent under supercritical condition. And then synthesized nanoparticles were treated to wash out residual surfactant and solvent and to dry in the three different following manners; i) Evaporative drying after residual surfactant and solvent washing, ii) Freeze-sublimation drying after residual surfactant and solvent washing, iii) Supercritical CO_2 washing and drying. Samples obtained from different methods were analyzed by Cs-TEM(Spherical aberration corrected Transmission Electron Microscope) to observe particle size and shape. Particle morphology was also characterized by electron diffraction pattern of nanoparticles depending on three post-treatment methods. In result, it was concluded that nanoparticles growth could occur during washing and drying process and kind of post-treatment method also could affect degree of particle growth. Therefore, it is considered that washing and drying processes are also as important to control particle size and morphology as particle synthesis procedure.

457935 Dual-Responsive Plasmonic Behavior of Gold Nanorods@PANI Core/Shell Nanostructures

Abstract:

Reversible tuning of plasmonic properties is becoming important to develop a new series of adaptable devices for plasmonic switches, modulators, and tunable lasers and tunable optical filters. Here, we demonstrate dual-responsive plasmonic systems consisting of gold nanorod (AuNR) cores and responsive polyaniline (PANI) shells. The plasmonic properties of AuNR@PANI core/shell nanostructures can reversibly modulated using two different stimuli: electrical potential and pH change. The PANI shell can change its oxidation and doping state between emeraldine salt (ES), emeraldine base (EB), and leucoemeraldine base (LB) depending on the surrounding chemical environments. The refractive indices of three different PANI states were obtained using spectroscopic ellipsometry, which showed that highest localized plasmonic resonance (LSPR) shift can be obtained by changing local pH surrounding AuNRs. Extremely high LSPR shift around 100 nm was achieved with very thin PANI shells (8 nm) by changing the pH environment. This highly sensitive AuNR@PANI core/shell nanostructures can potentially be used for a variety of applications including photovoltaics, nanophotonics, and optoelectronics.

455447 Adsorption Mechanisms of Palladium (II) Trichloro-Hydroxy Complex on the Tobacco Mosaic Virus Surface

Abstract:

Biotemplating has been successfully employed in nanomaterials synthesis and has been proven to be an efficient means of nanomaterial production. Particularly, templated synthesis, using Tobacco Mosaic Virus (TMV), has produced high quality nanorods and nanowires that found applications in catalysis, memory devices and batteries. Although biotemplating, using TMV and its mutants, has been successfully demonstrated, there is a lack of fundamental understanding of governing molecular mechanisms and processes that happen on the surface of the virus. In this study, the adsorption of the dominant palladium(II) ion (PdCl3H2O)- on TMV Wild type, TMV1Cys and TMV2Cys was investigated using UV-Vis spectroscopy. As a result, adsorption on TMV Wild type and TMV1Cys was found to be successfully modeled by Langmuir isotherm while adsorption on TMV2Cys was modeled by Freundlich isotherm. Using these models, we were able to estimate the maximum adsorption capacities of the viruses and address the thiol group(s) effects on the mechanism of adsorption of the dominant ion. The effect of chlorine ion concentration on adsorption equilibrium on TMV1Cys was also investigated since it was evident from previous studies that ionic strength of the solution had significant effects on chloro-palladate ion adsorption. The adsorption of the chloro-palladate ions on the TMV surface was inversely proportion to total chloride concentration at low palladium ion equilibrium concentrations. Conversely, the loading of chloro-palladate ions on the TMV increased with increasing total chloride concentrations at high palladium ion equilibrium concentrations. This analysis allowed us to get further insight into the molecular mechanism of palladium (II) adsorption on TMV surface and opened new opportunities for future establishment of a more fundamental understanding of the processes behind TMV templated biomineralization.

459716 Discrete Element Model of Nanoparticle Deposition during Electrospraying

Abstract:

Electrospray is one of the devices used in the fast developing sector of the nanostructured materials production. Its advantages: low costs, high efficiency and easy operation as well as the possibility of the deposition of various materials are consequences of the simple principle of its operation: charged nozzle and grounded substrate cause electric field responsible for the flight of highly charged droplets. Liquid evaporates from the droplets during the flight which together with the increasing charge density causes the disintegration of the droplets. But unfortunately the deposition of the particles and also the morphology of the layer strongly depend on the operating conditions and they are therefore hardly predictable.
Our goal is to clarify the influence of various effects taking place during the final part of the dropletâ€™s motion on the morphology of the layer. A convenient way how to do that is the mathematical modelling. We use the Discrete Element Method (DEM) based on the second Newtonâ€™s law and on the summation of the forces affecting the particles. The forces between the particles include non-contact van der Waals and contact elastic adhesive forces, while the drag force acts on a particle surrounded by a fluid. An integral part of the model is the computation of the electric force. The resulting electric force affecting each charged particle depends on many effects: differently charged particles, the permittivity of the particles and the surrounding environment, the electric field between the nozzle and the substrate, etc. Therefore, the force has to be computed numerically. For this purpose, we used the Finite Volume Method (FVM) to discretize the Poissonâ€™s equation and the electric force was evaluated from the resulting field of electric potential.
The predictions of the model were validated by layers created using a custom-built electrospray device. These deposited layers were characterized using a scanning electron microscope and an atomic force microscope.
Simulations in a multi-particle system were initially performed for particles with large electric conductivity, which are discharged immediately after their deposition and we used two configurations. In the first configuration, the so-called frozen deposit, the particles were not allowed to move after their deposition on the substrate or another deposited particle. The second configuration involved interparticle forces acting after the particle deposition. With the assumption of the frozen deposit, the operational parameters caused only minor changes in the final structures. On the other hand, when contact interparticle forces were considered, the resultant layers were more compact and more similar to the layers obtained from the experiments. The influence of each parameter was clearly observed. This clearly shows the importance of the after-deposition processes on the resulting morphology of the layer.
To extend the usability of the model for materials with lower conductivity, the model was equipped with the description of the discharging process. Due to the absence of the established description of the nanoparticles discharging, we had to come up with our own model. It computes the particle discharging mainly from the charge difference between the particle and the objects in the contact with it. The new results prove the importance of this extension and show that the morphology of the layer is significantly altered when particles are discharging slowly.
The obtained results bring more insight into the complicated process of the nanoparticle deposition and allow us to compare the importance of various effects occurring in the system. However, there is still a long and certainly very interesting way to go.

459984 How Protein Corona Affects Cellular Uptake of Nanoparticles

Abstract:

A successful application of nanoparticles(NPs) in targeted drug delivery requires a fine tuning of the interaction between NPs and cells in a physiological environment, where the NPs interact with extracellular proteins, yielding a protein corona due to protein adsorption on the NPs surface. The protein corona thus determines the interactions between NPs and cells and, consequently, the uptake of NPs by the cells. In the present study, we synthesized a carboxylate-modified NPs and amine-modified NPs with similar diameter by emulsion polymerization. The changes in physiochemical properties of NPs before and after protein adsorption were identified and the different protein adsorption of two types of NPs were determined using Isothermal Titration Calorimetry and TEM. Using HepG2 as the model cell lines, we examined the effect of protein corona on the uptake of NPs. The two types of NPs appeared a different protein adsorption thermodynamics and protein corona morphology. However, the two types of NPs saturated with protein adsorption were indistinguishable in terms of size and surface properties. The uptake of NPs was affected by both the protein corona and the nature of NPs. While the protein corona changed the dynamic profile of uptake, the endocytosis pathway and the intracellular transport of NPs, the surface charge of NPs, which impacted on the configuration of the protein corona, affected the uptake percentage significantly.

The present study was sponsored by National Natural Science Foundation of China [NSFC No. 21576148].

463638 Micro-Raman Phonon-Energy Mapping of Cell/Graphene Interface for Non-Invasive Cancer Diagnosis Via Ultrasensitive Cellular Potential and Activity Measurements

Abstract:

Ultra-sensitive detection of signals of cancer and the monitoring of treatment efficacy is critical for patient care. Here we show a graphene-based Raman biosensor capable of sensitively differentiating cancerous (Glioblastoma Multiforme, GBM) from normal cells (Astrocytes) at single cell resolution. Examining the modification of phonon vibrations induced by cellular doping on graphene via Raman Spectroscopy, we found sensitive responses exhibited by GBM cells (highly p doped with a significant 2D peak shift of approximately 6 ± 0.5 cm^{-1}) and Astrocytes (p doped with 2D peak shift of approximately 2.2 ± 0.2 cm^{-1}). The high quantum capacitance of graphene led to a change in the doped carrier concentration and a shift in the phonon energy; which provides a cell surface potential of ~310 mV, twice the potential exerted by astrocytes on graphene. This is attributed to the acidic character and increased surface electronegativity of the GBM cancer cells in comparison to the astrocytes. These results show that the process can be applied to other cancer cell types, particularly metastatic tumors that circulate through the blood stream as a pathological assessment tool. The high sensitivity of the device can be leveraged to characterize different grades of cancer, as in the case of brain tumors, based on their malignancy and aggressive nature.

464288 Correlating Solid-Binding Peptide Structure with Biomimetic Function

Abstract:

While much progress has been made in elucidating the mode of action of biomineralizing proteins and in emulating their function using combinatorially selected solid-binding peptides (SBPs), we remain far from duplicating the sophisticated architectures produced by biological systems. This is in part due to a poor understanding of the relationship between SBP conformation and inorganic adhesion and precipitation. Here, we focus on correlating the structure of Car9, a SiO_2/TiO_2-binding dodecapeptide originally isolated for its ability to recognize the edges of graphitic nanostructures, with silica binding affinity and titania precipitation. Starting with Rosetta and molecular dynamics predictions of the solution and silica-bound structures of Car9, we systematically mutagenize key functional and structural residues, predict the conformation of the mutants, produce them as GFP fusions, quantify the binding of the resulting panel of variants to SiO_2 via SPR and pull-down experiments, and characterize the influence of mutations on TiO_2 mineralization using fluorescence spectroscopy, Ti-depletion, TGA and SEM experiments. We find that the conformational context in which lysine residues are presented to the solvent plays a crucial role in binding and mineralization and that the impact of lysine substitutions can be mitigated through structural rearrangements that reposition basic residues. More broadly, our results suggest that computational predictions of SBP structures can be used to predict subtle changes in binding and mineralization outcomes.

464789 Catalytic CVD Growth of Carbon Nanotube Carpets on Metallic Substrates

Abstract:

To harness the unique properties of carbon nanotube (CNT) carpets for applications in energy storage and thermal management, it is necessary for the CNT arrays to be supported on metallic substrates. Hitherto, growth of high quality, highly dense CNT carpets via chemical vapor deposition (CVD) has been largely limited to catalysts supported on amorphous alumina. To overcome the challenge of conducting CNT carpet growth from catalyst supported on non-alumina substrates, surface modification of substrates using ion beam bombardment is being explored to create surface porosity. In addition, a thin layer of AlxOy is deposited prior to the deposition of the catalyst to create a basic surface. These modifications have resulted in CNT carpet growth on stainless steel. Ion beam bombardment of the substrates has demonstrated the ability to reduce film thickness of alumina required to grow CNT carpets on metallic substrates.

465462 Stacking and Elasticity of Ionic Chromonic Liquid Crystals

Abstract:

Lyotropic chromonic liquid crystals (chromonics) exhibit a unique self-assembly process of theoretical and practical interest, whose characteristic feature is molecular association through stacking at extremely low concentrations. Experimental evidence has suggested that this process is approximately isodesmic across a broad concentration range. Interestingly, these materials also exhibit a strong elastic anisotropy that leads to novel phases in confinement, where imposed deformations exploit an elastically cheap mode. Here, we present a minimal model intended to capture key features of the lyotropic chromonic phase. The model is utilized to interrogate the role of energetic anisotropies, to identify their effects on particle stacking and molecular ordering, and subsequently examine the role of stacking and stack structure in producing the large elastic anisotropies of these materials.

466764 An Asymmetric Dimer in a Periodic Potential: A Minimal Model for Friction of Graphene Flakes

Abstract:

For nanometer sized graphene flakes sliding on graphite, the friction as a function of rotation angle between flake and substrate displays peaks for the perfect commensurate case (flake and substrate have the same orientation) and decreases rapidly for incommensurate cases (superlubricity). In experiments and simulations, superlubric sliding is usually lost due to irreversible rotations to a commensurate contact [1]. However, it has been recently experimentally shown that friction can be increased reversibly by applying load to the incommensurate flake [2]. This reversible transition from superlubric to stick-slip sliding occurs while the flake remains at incommensurate contact and can be attributed to vertical motion of the carbon atoms at the edge of the flake. We propose a minimal model to describe this effects [3]. We discuss the friction and motion of a model of a dimer with asymmetric interactions with a substrate potential. Starting from the consideration that a rigid dimer with spacing equal to half of the period of the potential has exactly zero static friction like the infinite incommensurate Frenkel Kontorova model, we show how stick-slip behaviour and friction arise as a function of asymmetry. We argue that this model can yield a simple yet insightful description of the frictional behaviour of graphene flakes on graphite and of superlubricity. The results can also be of interest for diatomic molecules on surfaces. References: [1] A.E. Filippov, M. Dienwiebel, J.W.M. Frenken, J. Klafter, M. Urbakh, Phys. Rev. Lett. **100**, 046102 (2008) [2] M.M. van Wijk, M. Dienwiebel, J.W.M. Frenken, A. Fasolino, Phys. Rev. B **88**, 235423 (2013) [3] R. Hens, A. Fasolino, *An asymmetric dimer in a periodic potential: a minimal model for friction of graphene flakes*. Manuscript submitted for publication (2016).

465228 High-Fidelity Single Column Separation of Semiconducting Swcnts Using Agarose Gel

Abstract:

The synthesis of single-wall carbon nanotubes (SWCNTs) yields a complex mixture of species with different properties defined by their (*n*,*m*) type. The coexistence of SWCNTs with varied bandgaps hinders their application, especially in high performance electronic devices. Over the past decade, the post-synthetic separation of single wall carbon nanotubes (SWCNTs) has been studied with great interest. While high-fidelity separations of semiconducting (s-) SWCNTs by their (*n*,*m*) type have been demonstrated, a simple, high-throughput method is still desired. Here we report the selective desorption of a wide range of single-chirality (*n*,*m*) fractions in a single column packed with agarose gel. High-purity fractions of (*n*,*m*) nanotubes are obtained with high yield once a specific ratio of sodium dodecyl sulfate (SDS)/sodium deoxycholate (DOC) co-surfactant solution is used. The differences between the co-surfactant ratios needed to separate two (*n*,*m*) species can sometime be very small. The separation can be obtained under a wide variety of co-surfactant background concentrations. However, better separations are achieved at higher background concentrations.

465242 How Ionic Valence and Concentration Affects the Electric Current Generation of in Nanochannel: A Study By Classical Density Function Theory

Abstract:

The nanometer dimension intrinsically accounts for the dramatically different flow behavior and physicochemical prosperities of electrolytes, as compared to that in the conventional channel[1-4]. In this work, we applied classical density function theory (CDFT)[5] and Navier-Stokes Equation to examine the effects of ion valences, concentration and size of channels on current density and efficiency upon fluid flow through nanochannel. A two dimensional nanochannel with 0.6 ~ 4.0 nm in height and surface potential of 0.5 V was set to evaluate the effects of 1- , 2- and 3-valence symmetry ions with concentration of 0.001~5 mol/L. It was shown that energy conversion efficiency was not linearly increased with respect to decrease in height or increase in ion concentration. Instead, an optimal current and energy conversion efficiency exists for given electrolytes. Effects of valence-asymmetry ions and radius-asymmetry ions were also examined respectively. The physical insights of above mentioned results were discussed in terms of charge density profile for co-ions and counter-ions. It was noted here that when we used diameter-asymmetric ion pair, both current and energy conversion efficiency were superior to those of symmetric ion pairs with the same valence. It was indicated that the asymmetric ion pairs, who have more charges opposite to surface charge and small size, can be used as working fluid to convert mechanical energy into electricity. The results presented by this work offer a molecular insight of electric power generation by force flow of fluid flow through nanochannels, which is helpful for the design of nanochannel and electrolyte and the determination of the process parameters for electricity generation.

References:

1. Siria, A., Poncharal, P., Biance, A. L., Fulcrand, R., Blase, X., Purcell, S. T., & Bocquet, L. (2013). Giant

   osmotic energy conversion measured in a single transmembrane boron nitride nanotube. *Nature*, *494*(7438), 455-458.
2. Napoli, M., Eijkel, J. C. T., & Pennathur, S. (2010). Nanofluidic technology for biomolecule applications: a critical review. *Lab on a Chip*, *10*(8), 957-985.
3. Bottenus, D., Oh, Y. J., Han, S. M., & Ivory, C. F. (2009). Experimentally and theoretically observed native pH shifts in a nanochannel array. *Lab on a chip*,*9*(2), 219-231.
4. Van der Heyden, F. H., Stein, D., Besteman, K., Lemay, S. G., & Dekker, C. (2006). Charge inversion at high ionic strength studied by streaming currents. *Physical review letters*, *96*(22), 224502.
5. Li, Z., & Wu, J. (2004). Density-functional theory for the structures and thermodynamic properties of highly asymmetric electrolyte and neutral component mixtures. *Physical Review E*, *70*(3), 031109.

467691 Engineering Supraparticle Assemblies for Catalysis

Abstract:

Nature has developed highly specific and efficient catalysts that can synthesize complex organic compounds, such as the synthesis of sugars in photosynthesis. Due to the efficiency of these reactions, great effort has been put towards developing artificial reaction schemes that can replicate the complexity found in nature. Inorganic nanoparticles are a plausible choice for the development of a biomimetic system due to similarities with proteins found in nature. The advancement of NPs allow for tunable size, shape, charge, and surface functionality, allowing them to self-assemble into terminal superstructures similar to biological components. These superstructures, also known as supraparticles (SPs), can achieve the same catalytic functionality as natural enzymatic systems, and show improved functionality over their NP constituents due to collective and synergistic interactions.

The development of SP assemblies does not only provide a cost-effective and more stable alternative to enzymes but they can improve the activity when compared to current inorganic NP catalysts. The tight structural integration of NPs in SPs present the capacity to enhance the stability of catalysts, and mitigate the aggregation of individual constituents by offering separation while remaining immobilized. SPs are capable of combining the individual properties of constituents to produce enhanced catalysts.

This work will focus on reactions that occur under mild reaction conditions; thereby, the main component of inorganic SPs will be iron sulfide (Fe_xS_x). Iron sulfide is an Earth-abundant, non-toxic, cost effective mineral that is commonly used in solar-based applications. It is a key part of the synthesis/redox of organic compounds occurring via the reduction of CO_2 in the presence of H_2. The ample amount of Fe_xS_x and the ability to tune the bandgap of bulk Fe_xS_x (0.95 eV) to Fe_xS_x NPs (up to values of 2.75 eV), show the potential of iron sulfides in various catalysis applications.

468104 Tumor-Penetrating Aerosol Nanocomposite Microparticles for the Treatment of Lung Cancer

Abstract:

Lung cancer is the leading cause of cancer-related deaths in the United States, representing more than one-quarter of all cancer-related deaths. It is estimated that 433 deaths per day will be due to lung and bronchus cancers in 2016. The five-year relative survival is currently 18%, because most of the cases are diagnosed at a distant stage, for which the 5-year survival is only 4% [1]. Despite the advances in cancer treatment over the past 40 years, the survival rates for lung cancer patients still remains devastatingly low. Therefore, there is an unmet need for the development of more effective forms of treatment for lung cancer.

A chemotherapeutic treatment commonly applied for lung cancer is intravenous (I.V.) paclitaxel (PTX) in the form of Taxol. In general, I.V. therapy results in negative systemic side effects in route to its destination via the bloodstream, which spurs the need for improved delivery techniques. Aerosols have been successfully used for the treatment of pulmonary diseases such as asthma and chronic obstructive pulmonary disease (COPD), and the first use of aerosolized chemotherapy was reported in 1968 [2]. The advantages of using aerosol treatment include better targeting to the lungs and reduced systemic side effects often seen with high systemic drug doses [3].

Cancer cells tend to form solid tumors that are characterized for having hypoxic cells and a necrotic core, which gives them high resistance to the treatments. One reason for the failure of cancer treatment is the low concentration of drug in tumors due to limitations in drug diffusion into the cancerous tissue [4]. Recently, it was identified that tumors expressing Î±v integrins on their surface can be effectively targeted for treatment. The use of the tumor-homing and penetrating peptide iRGD (CRGDKGPDC) as a targeting moeity to integrins helps to overcome the penetration limitations into these tumors via targeting and tumor penetration [5].

In this project we developed a dry powder nanocomposite microparticle (nCmP) aerosol containing PTX-loaded nanoparticles synthesized with a biodegradable polymer, acetalated dextran (Ac-Dex). The byproducts of the degradation of Ac-Dex are harmless to the body and results in controlled release of therapeutic agents under acidic conditions as seen in tumor tissues [6]. In addition to the drug, a tumor-penetrating peptide (iRDG) was conjugated to Ac-Dex to aid in targeting and penetration into the inner layers of solid tumors. Finally, we formulated nCmP in mannitol via spray drying. The physicochemical properties of the nano- and microparticles (size, charge, drug loading) were evaluated. The effectiveness of the complex drug produced was tested in lung cancer cells (A459 lung adenocarcinoma) in two-dimensional (2D) cell culture, which was followed by three-dimensional (3D) cell culture studies that reflect many of the properties of solid tumors and mimic better all the barriers to drug diffusion, transport and distribution in tumors [4, 7]. Overall, the system shows promise in the improved delivery of chemotherapeutic agents to the lungs and allows for a more effective treatment of lung cancer.

References

[1] Siegel, R.L., K.D. Miller, and A. Jemal, *Cancer statistics, 2016.* CA Cancer J Clin, 2016. **66**(1): p. 7-30.

[2] Shevchenko, I.T. and G.E. Resnik, *Inhalation of chemical substances and oxygen in radiotherapy of bronchial cancer.* Neoplasma, 1968. **15**(4): p. 419-26.

[3] Patil, J.S. and S. Sarasija, *Pulmonary drug delivery strategies: A concise, systematic review.* Lung India, 2012. **29**(1): p. 44-9.

[4] Minchinton, A.I. and I.F. Tannock, *Drug penetration in solid tumours.* Nat Rev Cancer, 2006. **6**(8): p. 583-92.

[5] Teesalu, T., K.N. Sugahara, and E. Ruoslahti, *Tumor-penetrating peptides.* Front Oncol, 2013. **3**: p. 216.

[6] Kauffman, K.J., et al., *Synthesis and characterization of acetalated dextran polymer and microparticles with ethanol as a degradation product.* ACS Appl Mater Interfaces, 2012. **4**(8): p. 4149-55.

[7] Sutherland, R.M., *Cell and environment interactions in tumor microregions: the multicell spheroid model.* Science, 1988. **240**(4849): p. 177-84.

468576 Multi-Scale Molecular Modeling of Rosette Nanotubes Derived from a Tetracyclic G∧C Motif

Abstract:

Rosette nanotubes (RNTs) are soft organic nanomaterials self-assembled under aqueous conditions from Watson-Crick inspired guanine-cytosine (G∧C) hybrid building blocks with complementary hydrogen bonding and extensive π-π interaction sites. These materials have substantial design flexibility and a range of applications, which is partly attributed to their diverse surface functionalization and a chemically/physically tunable channel for guest molecule loading. Several studies have established their biocompatibility and applications in nanomedicine such as in coatings for medical devices, materials for tissue engineering and for drug display and delivery.

With novel applications in mind, particularly, drug delivery, a new tetracyclic G∧C motif, the yG∧C, was designed to self-assemble into RNT in water. The G∧C motif was functionalized by two lysine groups, which are known to give RNTs biocompatibility and therapeutic properties. To aid in the characterization of this new RNT, multi-scale molecular modeling techniques were applied to predict its structure in solution. Molecular dynamics (MD), molecular mechanics (MM), and the statistical mechanical theory of solvation, also known as the 3 dimensional reference interaction site model (3D-RISM) theory were applied to predict the self-assembly, conformation, and stability of yG∧C-RNTs. MM was used to determine the possible conformations of the individual motifs. From these, RNT models were built and MD simulations were run to determine the stability and probable structure of the nanotubes. The results suggest that the yG∧C motif can either form a 6-membered or 7-memebered ring stacks. 3D-RISM integral equations were then solved for the system to determine the thermodynamics and to propose a self-assembly pathway for the RNTs. Moreover, a model RNT from yG∧C motifs is proposed.

The yG∧C-RNTs have an inherent advantage of having a large channel, with diameters between 1.8 nm and 2.3 nm that could potentially host guest molecules suitable for various applications. In this study, the model RNT was used to hold guest molecules, phosphorodiamidate morpholino oligomer (PMO), paclitaxel (PTX), and doxorubicin (DOX). These molecules were chosen for their anti-cancer and other pharmaceutical attributes. MD simulations were run to determine the capability of the nanotube to capture and host the guest molecules. 3D-RISM was then used to determine the probable pathway and energies involved in encapsulating the drug molecules. Results indicate that the yG∧C-RNT can stably enclose PMO, PTX, and DOX. This suggests that yG∧C-RNT has high potential in biomedical applications. Experiments are currently being done to determine the structure of the RNTs and test these possibilities.

469210 Characterization and Transport of Polyelectrolyte Membranes Loaded with *in-Situ* Grown Metal-Oxide Nanoparticles

Abstract:

Formation of nanocomposites by *in-situ* growth of metal-oxide nanoparticles (MONP) in polyelectrolyte membranes (PEM) has been quite appealing for several potential applications such as: fuel cells, water purification, and protective barriers against chemical warfare agents (CWA). PEM micro-structure can serve as capping agents for the nanoparticles and depending on PEM conditions (humidity, degree of swelling, solvents present) in-situ grown nanoparticles can adopt various shapes and sizes. Among several potential applications this paper is focused on PEM-MONP composites for self-detoxifying composite PEM barriers against CWA. These PEM have self-assembled interconnected hydrophilic and hydrophobic sub-domains that provide perm-selective properties and good transport of water vapor, while acting as a barrier against dangerous substances. PEM ion-exchange capabilities have been used for the *in-situ* formation of MONP within PEM at various conditions to observe the influence of PEM micro-structure on MONP growth. In addition, transport properties of PEM-MONP composites were analyzed by sorption and permeation experiments with water and chemical warfare surrogates (CWS). Sorption experiments results are compared with mathematical models and show the impact that MONP have on transport properties of PEM. Transmission and scanning transmission electron microscopy (TEM/STEM) were used to observe MONP growth and particle size distribution within PEM. X-ray diffraction (XRD) was used to analyze crystalline structure of MONP. Various spectroscopic techniques (NMR, XPS, EDS) were used to: assess chemical states of MONP within PEM, analyze permeation of CWS, uniformity of MONP distribution within PEM at various conditions.

469489 Cuboidal Colloidal Particles in Nematic Liquid Crystal: Defect Structures and Self-Assembly

Abstract:

The physics of the nematic topological defects around sphere solid particles, and their defect induced self-assembly have been extensively investigated in past years. Here, we examine experimentally and theoretically, the effect of edge curvature of colloidal particles with homeotropic surface anchoring on their defect configuration and self-assembly in a nematic liquid crystal. Our computational studies indicate that the gradual increase of the edges sharpness from sphere to cube modifies the defect structure from a Saturn ring to a distorted ring with sharp bends, which is pinned to the edges of the cube particle. In agreement with our computational results, the polarized light microscopy images show reshaping of the defect ring around the cuboidal particles immersed in 5CB nematic liquid crystal. Moreover, calculation of the pair potential interaction for colloidal cubes indicates that the presence of sharp edges can significantly affect the interaction between particles and their eventual self-assembly in the nemtic liquid crystal. The wide variety of topological defects achievable by changing the curvature could provide new tools to tune colloidal self-assembly.

469606 Material and Toxicity Evaluations of Nanoclays throughout Their Life Cycle

Abstract:

The incorporation of nanoclays, layered mineral silicates, into polymers leads to the formation of nanocomposites with enhanced mechanical strength, UV dispersion, barrier, and flammability properties. With proper exfoliation of nanoclays into polymers, these nanocomposites can then be applied to different industries, from automotive, to medical, or food packaging. However, with increasing implementation into such industries it is important that we begin to investigate the physical and chemical changes that nanoclays undergo throughout their life cycle as well as evaluate any toxicity during their life cycle to thus ensure both worker and consumer health. Herein, we examined the physical and chemical characteristics of 4 nanomer nanoclays (1 pristine and 3 organically modified) in raw or 'as-received' form to mimic a manufacturing environment, and after their thermal degradation, to mimic end-of life cycle in incineration plants. Material characterization was performed via microscopical and spectroscopical techniques while the toxicity evaluations used *in vitro* testing on human bronchial epithelial cells (BEAS-2B). Dose response curves were obtained via live cell counts to determine the degree of toxicity of each 'as-received' nanoclay and its thermally degraded byproduct. Thermal degradation caused for breakdown of the nanoclay platelet structure and loss of organic modifiers of the organically modified clays. The 'as-received' nanoclays had lower IC50 values relative to their thermally degraded counterparts, with a range in IC50 values of 10-325 µg/ml and 150-400 µg/ml for 'as-received' and thermally degraded nanoclay, respectively. While more studies are needed to examine the mechanism of toxicity of these nanoclays, our results reveal that these nanomaterials have the potential to induce inhalation toxicity in both manufacturing and disposal environments, with the degree of toxicity being dependent on their physical and chemical properties.

469638 Ordered Nanoporous Titania Thin Films for Energy Conversion and Storage

Abstract:

The optical and electronic properties of TiO_2 thin films provide tremendous opportunities in several applications including photocatalysis, photovoltaics and photoconductors for energy production. Despite many attractive features of TiO_2, critical challenges include the innate inability of TiO_2 to absorb visible light and the fast recombination of photoexcited charge carriers. In this work, nanostructured TiO_2 thin films are modified by doping using hydrogen, nitrogen and titanium, and sensitization using graphene quantum dot sensitization. For all of these modifiers, well-ordered nanoporous titania films were synthesized by surfactant templated sol-gel process. Two methods: hydrazine and plasma treatments have been developed for nitrogen and hydrogen doping in the nanoporous titania films for band gap reduction, visible light absorption and enhancement of photocatalytic activity. The plasma treated nitrogen doped nanoporous titania showed about 240 times higher photoactivity compared to undoped film in hydrogen production from photoelectrochemical water splitting under visible light illumination. In addition to the intrinsic modification of titania by doping, graphene quantum dot sensitization in nanoporous titania film was also investigated for visible light photocatalysis. Graphene quantum dot sensitization and nitrogen doping of ordered nanoporous titania films showed synergistic effect in water splitting due to high surface area, band gap reduction, enhanced visible light absorption, and efficient charge separation and transport. As a stationary energy storage system, supercapacitors were prepared using hydrogen doped nanoporous titania films. The capacitance of plasma-treated hydrogen doped titania films was 300 and 176 times higher than the pristine film in aqueous and organic electrolytes, respectively, due to increased conductivity, oxygen vacancy formation and increased hydroxyl group formation in titania. This study suggests that plasma based doping and graphene quantum dot sensitization are promising strategies to reduce band gap and enhance visible light absorption of high surface area surfactant templated nanoporous titania films, leading to superior visible-light driven photoelectrochemical hydrogen production and electrochemical capacitance.

469036 Precursor Ion-Ion Aggregation in the Brust-Schiffrin Synthesis of Alkanethiol Nanoparticles

Abstract:

Tetraoctylammonium bromide is used in the Brust-Schiffrin nanoparticle synthesis to phase-transfer chloroaurate ions from the aqueous phase to the organic phase. While it is established that the quaternary ammonium complex self-associates in the organic phase, the actual self-assembled structure is debated. We have confirmed the presence of ion-ion aggregates through quantitative ^{1}H Nuclear Magnetic Resonance spectroscopy (NMR), pulsed field gradient, diffusion-ordered NMR (DOSY-NMR) and density functional theory (DFT) based NMR chemical shift calculations. Tetraoctylammonium complexes (TOA-X, where X = Br, Cl, $AuCl_{4-x}$

investigated to measure the extraction of water into the organic phase. ^{1}H NMR and DFT based NMR shielding calculations indicated that deshielding of water is due to hydration of the anion and not the formation of the aqueous core of a reverse micelle. DOSY-NMR results were consistent with the formation of small aggregates at typical Brust-Schiffrin synthesis concentrations. The extent of aggregation correlated with the size and electronegativity of the anion and was analyzed with a modified, isodesmic, indefinite aggregation model. The substitution of bromoauric acid for chlororoauric acid at conditions emulating the Brust-Schiffrin synthesis increased the aggregation of the quaternary ammonium complex. The increase in aggregation corresponded with an increase in the size of the produced nanoparticles from 4.3 to 4.6 nm. Understanding the self-assembly and supramolecular structure of precursors in the Brust-Schiffrin synthesis will enable further refinement of models that predict the growth of noble metal nanoparticles.

470419 Langmuir-Blodgett Deposition of Anisotropic Nanoparticles

Abstract:

Langmuir-Blodgett coating (LB) is an effective tool for the deposition of nano- and microparticles in mono- or multilayers onto a substrate. During LB deposition, a monolayer of colloidal particles (ex. surfactants, polymers, or nanoparticles) is deposited from a fluid/fluid interface onto a substrate moving through that interface at a specified speed. Although there is a substantial volume of previous work illustrating the potential of this technique to the directed self-assembly of colloidal and nanoparticles, there remain open questions. For instance, it is unknown to what extent the mechanics and rheology of an interface laden with anisotropic particles impacts the structure of deposited material. This poster will summarize our groupâ€™s work on developing a relationship between the interfacial microstructureÂ of anisotropic nanomaterials to the deposited morphology during LB deposition.

470618 Co-Catalysts Incorporated Perovskites Nanofiber for Water Splitting PhotocatalysisÂ

Abstract:

Co-catalysts Incorporated Perovskites Nanofiber for Water Splitting Photocatalysis

Md Moniruddin and Nurxat Nuraje

Department of Chemical Engineering, Texas Tech University

Lubbock, TX, USA

Perovskite materials have tremendous potential for photocatalysis and photovoltaic applications due to many advantageous properties including its low cost, high availability, chemical stability, and excellent opto-electronic properties. Therefore, they have been widely investigated for hydrogen production from photocatalytic water splitting using solar energy. In this study, strontium titanate ($SrTiO_3$) nanofibers were prepared via electro spinning method using the solution of strontium titanium alkoxide and polyvinylpyrrolidone (PVP). Different co-catalysts such as noble metal (Pt), plasmonic metal (Au) and metal oxide (NiO) were deposited on the synthesized $SrTiO_3$ nanofibers by using photo deposition and impregnation method. The nanofibers were then characterized by field emission scanning electron microscopic with energy dispersive spectroscopic observations, X-ray diffraction patterns, BET surface area measurements, and UV-visible spectrophotometry. The photocatalytic performance of the synthesized $SrTiO_3$ nanofibers is being tested for hydrogen production from water splitting using a solar simulator. To further improve the visible light activities, $SrTiO_3$ nanofibers will be doped with nitrogen. High photocatalytic activities of these $SrTiO_3$ nanofibers are expected due to its small crystallite size, high crystallinity, mesoporous nanofiber structure, large surface area, and enhanced visible light absorption. The effects of different types of co-catalysts and N-doping on the photocatalytic performance of the $SrTiO_3$ nanofibers will be discussed.

470933 Assessment of the Exposure of Human Lung Epithelial Cells to Nanoclays

Abstract:

Assessment of the Exposure of Human Lung Epithelial Cells to Nanoclays

A. White[1], A. Wagner[1], R. Gupta[1], S. Agarwal[1], T. A. Stueckle[2,3], Y. Rojanasakul[3], C. Z. Dinu[1]

1. Department of Chemical and Biomedical Engineering, West Virginia University, Morgantown, WV 26506

2. National Institute for Occupational Safety and Health, Morgantown, WV 26505

3. Department of Basic Pharmaceutical Sciences, West Virginia University, Morgantown, WV 26506

Abstract:

Nanoclays are widely considered for applications in plastics industry. Due to their high biocompatibility and controllable surface properties, they are also often incorporated into polymers to form nanoclay-reinforced polymers, or nanocomposites, to be utilized for applications such as food packaging. However, their increased usage and eventual disposal is necessitating investigations into their relative toxicity. The proposed research aims to evaluate the interactions of such nanomaterials with human bronchial epithelial cells (BEAS-2B) with such cells being used to mimic exposure by inhalation. To complete the experiments, we have investigated 4 NanomerÂ® claysâ€™ (pristine and organically modified).Â Chemical and physical properties are assessed using microscopical and spectroscopical assays. Cellular dose responses were extrapolated from inhibitory concentration studies (e.g. live cell counts) and further developed into IC_{50}values. Cellular viability assays were also employed to provide further insights into the mechanism of toxicity of these nanomaterials with respect to cellular fate. Geno- and cytotoxicity pathways are being currently established to further extrapolate the implications that exposure to such nanomaterials could have to workers in both manufacturing and industrial environments.

Disclaimer: The findings and conclusions in this abstract are those of the authors and do not necessarily represent the views of the National Institute for Occupational Safety and Health.

471298 Nanoparticles As Biomolecular Cargo Transporters in Plant Systems

Abstract:

Targeted therapy, molecular imaging, and genetic modification of cells and tissues hinges on controlled transport of biomolecular cargos into their relevant biological domains. Targeted cargo delivery is particularly difficult when targeting sub-cellular organelles, such as plastids, or for traversing dense membranes such as the plant cell wall. Current methods for intracellular transport include but not limited to microinjection, particle bombardment, electroporation and usage of dendrimers, all limited by low transfection efficiency and yielding impaired cell functions or cell death. Nanomaterials have distinct properties that make them useful as transporters of biological cargoes to intercellular locations that have been traditionally difficult to access, and offer significant improvement for both passive and active penetration into biologically important membranes. We present results on the uptake and localization mechanism of nanoparticles within living plants, plant cells, and organelles. In particular, we examine the subcellular uptake of single-walled carbon nanotubes (SWCNT) that are functionalized by genetic materials and small molecule drugs, using the plant as a model system. Our study shows a systematic investigation of nanoparticle transport across the cell wall and to protoplast cells with partial to full enzymatic degradation of the cell wall. Our work improves the current state of art in delivering genetic material to plant cells by proving passive entry and irreversible localization of SWCNT that carries genetic material into mature, living plants and plastids.

471548 Optimization of Aerosol Nanocomposite Microparticles (nCmP) for Deep Lung Delivery of Therapeutics

Abstract:

Pulmonary drug delivery based dry powder aerosols has attracted increasing attention for the treatment of lung diseases, as they are capable of delivering a wide range of therapeutics directly and efficiently to the lungs. Meanwhile, the lungs can also be utilized as a route for systematic drug delivery due to the enormous surface area and plentiful capillary vessels present. In both cases, an intelligent carrier is required to overcome the barriers of pulmonary delivery, which include: (1) dry powder aerosols with aerodynamic diameters smaller than 1 Î¼m will often be exhaled; (2) particles with aerodynamic diameters above 5 Î¼m tend to deposit in the mouth, throat or upper lung mucosa and will then be eliminated due to mucus clearance; (3) particles larger than 260 nm deposited in the deep lung area will undergo macrophage clearance in the alveoli. Our solution to the aforementioned challenges is the use of nanocomposite microparticles (nCmP) for the aerosol delivery of therapeutics. nCmP are microparticles that contain drug-loaded nanoparticles (NP) and an excipient. Upon pulmonary administration, nCmP will deposit on the mucus or surfactant present on the surface of the lungs, decompose into free NP and excipient, and allow the nanoparticles to penetrate the mucus and then release drug to the targeted site at sustained rate.

Many nCmP systems have been developed for various applications, but few comprehensive studies have been done to illustrate how to develop an optimal nCmP system for deep lung delivery of therapeutics. In the present research, we aim to identify the optimal spray drying conditions used to prepare the nCmP with favorable properties including: small aerodynamic diameter, desirable nanoparticle re-dispersity, high drug loading, and low water content. Acetalated dextran was used as the polymer to form nanoparticles due to its biodegradability, biocompatibility and tunable degradation rate. Curcumin was used as model drug and mannitol was applied as excipient of nCmP. The spray drying parameters that were optimized including the nanoparticle-to-total substance ratio, solids feed concentration, and inlet temperature. Boxâ€"Behnken design was applied for experimental design using Design-Expert software. The results indicated that the nanoparticle ratio (NP%) and feed concentration (FC) were significant parameters affecting the aerodynamic diameters of the nCmP. NP% was also the significant parameter affecting nCmP drug loading. FC was the only parameter that influenced nCmP water content significantly. All nCmP can be completely re-dispersed, indicating that the spray drying factors have no influence on the re-dispersity of nanoparticles in the design range. Overall, the optimal spray drying condition that can be used to prepare nCmP with a small aerodynamic diameter, complete re-dispersity of nanoparticles, low water content, and high drug loading is a 80% nanoparticle ratio, 0.5% feed concentration, and inlet temperature lower than 130 Â°C.

471798 Novel Photovoltaic Applications of Photosystem I Multilayer Films

Abstract:

Photosystem I (PSI) is a ~500 kDa photocatalytic membrane protein complex contained within the thylakoid clusters of plant chloroplasts. Boasting a 1.1 V light-induced charge separation, PSI provides high-energy electrons needed for NADPH production within the cells of plants and cyanobacteria. The ubiquity and unique optoelectronic properties of PSI have made it an interesting prospect as a nanoscale photoactive component in nature-inspired â€œbiohybridâ€ solar cell devices.

More recently, multilayer films of PSI have been applied to surfaces for significant improvements in photovoltaic efficiency over monolayer based systems. Purified solutions of PSI proteins are drop cast atop a surface, resulting in a dense, ion-permeable film of stabilized macromolecules. Each protein unit contains a redox-active reaction center internally photo-sensitized by tightly bound core chromophores. The basis for enhancement is rapid photocatalytic cycling of redox-active ions between reduced and oxidized states, culminating in a large and prolonged charge asymmetry within the protein film. When coupled to an electroactive substrate, heterogeneous charge equilibration results in a net exchange of carriers and sustained photocurrent. The enhancement phenomenon is preserved over periods exceeding 180 days (beyond the normal growing season of plants) and increases with the amount of protein deposited.

472541 Characterization of Hydrogel Porous Structure By Using the Differential Scanning Calorimetry Technique

Abstract:

Over the past decades hydrogels have been used widely in various pharmaceutical and biomedical applications such as drug delivery systems, engineering tissue scaffolds, and health care products. As many applications of hydrogels require very accurate measurement and control of the pore network structure i.e. pore size and pore-size distribution, it is important to understand the nature of the hydrogel pore structure and the methods used for the characterization of its network structure. Therefore, numerous techniques have been developed to characterize the internal morphology of pores including swelling, scattering and mercury intrusion porosimetry, among others.

Characterization of the hydrogel pore structure is very important for selecting the gel type for a particular application. In this research we will try to gain a better understanding of hydrogel morphology by using Thermoporometry by "Differential Scanning Calorimetry" (DSC). This technique will be used in this research to characterize both pore size and pore-size distribution for nano-templated hydrogel materials for the applications mentioned above; in particular, we are interested in the separation of bio-pharmaceutical drugs and scaffolds for tissue engineering. Thermoprometry is one of the few techniques available now for determining both pore size and pore-size distribution at the same time. The DSC will take measurements of the temperature shift that occurs between the water confined within the pores of the gel and the bulk water. The temperature difference will be used to generate a heat flow curve. We will present the fundamental aspects of the technique for its adaptation for hydrogel's measurements. We will also present preliminary results for hydrogels and a strategy for its validation with previous results in the literature for regular gels.

472898 Application of Nanofluidics in Understanding Fluid Flow in Tight Rocks

Abstract:

Even with the advent of technologies like hydraulic fracturing and horizontal drilling, the recoveries in shale oil and gas plays tend to be in single digits. There is a dearth of knowledge of dynamics at pore scale. The pores sizes in organic and inorganic parts of rocks are believed to be in nano meters in the shales. The connectivity between these nano pores provide nano flow pathways which contribute towards the flow of oil and gas from matrix to the hydraulic fractures. There is increasing need of understanding the fluid flow in these ultra tight shale reservoirs. Nanofluidic devices with â€˜lab on a chipâ€™ approach provide a fresh perspective to look at the fluid flow in nano channels present in shale rocks.

In the current work, nano channels were fabricated on a silicon wafer and single to multi phase fluid flow was captured using high speed imaging system. The paper talks about the application of traditional fabrication processes like photolithography, Reactive Ion Etching (RIE) and Deep Reactive Ion Etching (DRIE) to make nano fluidic devices for application in understanding flow of oil and gas. An attempt has been made to test and refine Darcyâ€™s law of fluid flow which fails to capture the pore wall interactions in the ultra tight rocks. The paper aims to provide relevant data that can be used to tune traditional flow simulators used in modeling fluid flow in the low permeability rocks.

473684 The Effect of Binder on Volume Variation in Electrodes of Lithium Ion Batteries

Abstract:

The reality of waste gas of automobiles being accounted for a large portion of air pollution and the gradual depletion of fossil fuel has generated a huge demand of the new brand of energy supply for electric vehicle (EV). Although the traditionally designed and manufactured lithium ion battery has served well for devices that requires relatively lower energy and power density, such as laptop and mobile phone, it is far less competitive compared to commonly-used fossil fuel as energy storage and supply solutions in the emerging new technologies of hybrid and electric vehicles. Most of past researchers focused on the thin graphite electrode that contains silicon, while thicker electrode, which means more active material to provide a higher energy capacity, was hardly paid attention on. One of the deep-rooted reasons is that the mechanical failure, including delamination and cracking of electrode from current collector, caused by frequent and large stress-induced volume variation.

In our current research, we report mechanical properties of silicon-contained graphite electrodes with different kinds of binder, novel aqueous material CMC and commonly-used PVDF respectively, in lithium battery via in-situ experimental method by white light interferometry, in order to get a comprehensive understanding of the effect of different binder to maintaining the mechanical integrity of batteries. The two layer cantilever are micro-fabricated by photolithography and laser beam processing and composed of upper copper film acting as current collector and bottom graphite/silicon film being active material. We characterized the relative curvature change and deformation of cantilevers as electrochemical reaction is in progress by using white light interferometry and electrochemical workstation simultaneously. In addition, the scanning electron microscopy, energy-dispersive X-ray spectroscopy and coin cell testing were also used in order to characterize the corresponding surface morphology, element distribution and battery performance of such cantilever structures. The results would be of great significance to quantify the effect of binders to mechanical deformation of electrodes and propose an innovative method to mitigate the deteriorating phenomenon during cycles.

474853 Fabrication of Enzyme-Based Coatings on Intact Multi-Walled Carbon Nanotubes As Highly Effective Electrodes in Biofuel Cells

Abstract:

Carbon nanotubes (CNTs) need to be well dispersed in aqueous solution for their successful use. Most methods to disperse CNTs rely on tedious and time-consuming acid-based oxidation reactions. Here, we report the simple dispersion of intact CNTs by adding them directly into an aqueous solution of glucose oxidase (GOx), resulting in simultaneous CNT dispersion and facile enzyme immobilization through sequential enzyme adsorption, precipitation and crosslinking (EAPC). The EAPC approach achieved high enzyme loading and stability in the form of crosslinked enzyme coatings on intact CNTs, while obviating the chemical pretreatment that can damage the electron conductivity of CNTs. Initial EAPC-driven GOx activity is ca. 4.5- and 11-times higher than those of covalently-attached GOx (CA) on acid-treated CNTs (ox-CNTs) and simply-adsorbed GOx (ADS) on intact CNTs, respectively. EAPC showed no decrease of GOx activity in aqueous buffer at room temperature for 270 days, while under identical conditions the half-lives of CA and ADS were 20 and 19 days, respectively. EAPC was employed to prepare the enzyme anodes for biofuel cells, and the EAPC anode produced 7.5-times higher power output than the CA anode. Even with a higher amount of bound non-conductive enzymes, the EAPC anode showed 1.7-fold higher electron transfer rate than the CA anode, revealing the reduced electron conductivity of CNTs upon acid treatment. The EAPC approach has potential for effective immobilization of various other enzymes on CNTs that can improve both enzyme loading and stability together with key routes of electron transfer important in biosensing and bioelectronics devices.

474854 Conversion of Glycerol to Dihydroxyacetone Using Highly Stabilized Glycerol Dehydrogenase

Abstract:

Biodiesel industry generates glycerol as a by-product, and the conversion of glycerol to useful products is an important issue. An enzyme, called glycerol dehydrogenase (GDH), can catalyze the conversion of glycerol to dihydroxyacetone (DHA), which is used as a tanning agent in cosmetics and a lot more expensive than glycerol. In this study, GDH was adsorbed (ADS) into magnetically-separable mesoporous silica with 38 nm mesocellular pores connected via 18 nm window mesopores (ADS), and further crosslinked via a simple glutaraldehyde treatment to prepare â€˜nanoscale enzyme reactorsâ€™ of GDH (NER). When the stabilities of free and immobilized GDHs were checked under shaking (200 rpm), the residual activities of free GDH and ADS could not be measured due to the inactivation of GDH after 8 days and 22 days, respectively, while NER maintained 64% of initial activity even after 24-day incubation. Magnetically-separable NER maintained 39% of initial activity after seven cycles of repeated uses while the residual activity of ADS dropped to 13% of initial activity only after two recycled uses.

474855 Antibody-Conjugated Nanoscale Enzyme Reactors for Highly Sensitive Immunoassay

Abstract:

A highly sensitive immunoassay method was developed by using antibody conjugated nanoscale enzyme reactors (NERs), termed as NER-linked immunosorbent assay (NER-LISA). Glucose oxidase (GO) was simply adsorbed into amine-functionalized spherical mesoporous silica (ADS-GO), and further treated with glutaraldehyde to crosslink adsorbed GO molecues in spherical mesoporous silica (NER-GO). The activity of NER-GO was 2.5 and 1.5 times higher than ADS-GO and covalently attached GO (CA-GO), respectively. NER-GO was highly stable by maintaining 98% of initial activity under shaking at 200 rpm for 26 days while ADS-GO and CA-GO retained 20% and 4% in the same condition, respectively. For the detection of human IgG (hIgG), highly stable NER-GO with high enzyme loading was conjugated with anti-human IgG. NER-LISA was performed and it was observed that the sensitivity and dynamic range was drastically improved in case of NER-LISA compare to conventional enzyme-linked immunosorbent assay (ELISA). The limit of detection (LOD) in case of NER-LISA was ~0.5 ng/mL hIgG (~3.3 pM), which was 20 times more sensitive than conventional ELISA. The higher sensitivity in case of NER-LISA was due to higher ratio of enzyme/antibody than conventional ELISA with one-to-one ratio of enzyme/antibody.

474869 Chitosan Nanoparticles with Immobilized Glucose Oxidase As Efficient Antimicrobial Agents

Abstract:

We have developed three immobilizations of glucose oxidase (GOx) on chitosan nanoparticles (CSNP), and investigated their antimicrobial activities against *Staphylococcus aureus*. CSNP were synthesized via ionotropic gelation, and GOx was immobilized on CSNP via three different approaches such as covalent attachment (CA), enzyme crosslinking (EC) and enzyme precipitation crosslinking (EPC). The activities of CA, EC and EPC were 8.72, 17.69, and 218.97 Unit/mg CSNP, respectively. According to the turbidity assay, we observed 90% growth inhibition in the presence of EPC when compared to the control with no enzyme added. Furthermore, EPC was the most effective in disintegrating the biofilm network of *Staphylococcus aureus*, resulting in more than 80% reduction in biofilm biomass. EPC on CSNP has shown a great potential in preventing microbial growth and disintegrating biofilms, which can find its applications in preventing microbial infections in medical implants and wound dressing materials.

474872 Enzymatic CO_2 Utilization Using Immobilized and Stabilized Carbonic Anhydrase

Abstract:

Carbonic anhydrase (CA) is a promising biocatalyst that can be used for the CO_2 conversion technologies. CA catalyzes the hydration of carbon dioxide (CO_2) to bicarbonate (HCO_3^-) at high turnover number. Due to its fast kinetics, the introduction of CA in CO_2 capture process can lead to various positive results including the reduction of reactor size as well as lowering the energy input at CO_2 stripper. However, the practical application of CA has been limited by its poor stability. This presentation will introduce nanobiocatalytic approach that can immobilize and stabilize CA using various nanomaterials. By taking advantage of magnetic nanoparticles, nanobiocatalytic CA could be recycled for repetitive enzymatic conversion of CO_2 via facile magnetic separation. The present work has demonstrated a great potential of nanobiocatalytic CA as a stabilized enzyme system that can be employed for the biocatalytic CO_2 conversion and utilization.

474874 Efficient Protein Digestion Using Highly-Stable Enzyme Coatings on Magnetic Nanofibers

Abstract:

Trypsin (TR), commonly used for protein digestion in proteomic analysis, was immobilized and stabilized on polymer nanofibers with superparamagnetic nanoparticles (magnetic nanofibers, NP-NFs). By electrospinning the homogeneous mixture of superparamagnetic nanoparticles (NPs) and polystyrene-poly(styrene-co-maleic anhydride), NPs were effectively entrapped within polymer nanofibers, generating magnetically-separable nanofibers with high surface area as an effective carrier for enzyme immobilization. Trypsin coatings on magnetic nanofibers (EC-TR/NP-NFs, and also called EC-TR), fabricated via simple crosslinking of TR molecules onto NP-NFs, were highly stable and could be easily recycled via facile magnetic separation of NP-NFs. EC-TR showed negligible loss of trypsin activity even after incubation in an aqueous buffer under rigorous shaking for 80 days, while the control samples of covalently-attached TR on NP-NFs (CA-TR/NP-NFs) and free TR lost more than 90 % of their initial activities within 11 and 6 days, respectively. Highly-stable and magnetically-separable EC-TR was employed for the repetitive digestions of enolase under recycled uses for the duration of 50 days and even after treatment with chymotrypsin for 32 hours, and the performance of enolase digestion was successfully maintained. The use of EC-TR for the enolase digestion in the ultra-sonication system resulted in fast (~10 minutes) and efficient digestions with reproducible performance under recycled uses.

477957 Locked Nucleic Acid-Wrapped Single-Walled Carbon Nanotubes Based Optical Sensor for microRNAs Detection

Abstract:

As master regulators of cellular processes, microRNAs have assumed a flagship role in disease diagnosis and treatment, acting both as biomarkers and therapeutic targets for a range of diseases, including early-stage cancer prognosis.[1-2] Single-walled carbon nanotubes (SWCNTs) offer advantageous fluorescence properties for optically detecting microRNA that would enable an in vivo and nearly ideal high-throughput approach to profiling gene expression. Benefiting from indefinite photostability, sensitivities to single-nucleotide polymorphisms, and optical transparency to biological tissue,[3] we developed an optical microRNA sensor based on SWCNT fluorescence. Through non-covalent SWCNT surface functionalization, this sensor interfaces locked nucleic acids (LNAs) with SWCNTs to detect microRNA-21, a key target for the prognosis of several different types of cancers. This sensor exhibited an unprecedented SWCNT fluorescence sensitivity towards RNA, achieving sub-micromolar sensitivities within a 5-minute sensor response time. The SWCNT fluorescence response was shown to vary with microRNA-21 concentration, and it was shown to be both sequence- and oligonucleotide dependent when screened in vitro against alternative microRNA and analogous DNA probes. In the presence of microRNA-21, the LNA-wrapped SWCNTs undergo a chiral-specific fluorescence increase and wavelength shift, with an initial effective binding rate constant of 2.3 x 103 a.u. Î¼M-1 s-1 for the (9,5) SWCNT. Low-energy electron holography images show that LNA assumes a helical conformation on the SWCNT surface similar to the chiral-specific conformation observed for DNA-wrapped SWCNTs, suggesting an analogous hybridization and sensing mechanism. The chirality-dependent molecular fingerprinting of microRNA shown in this study provides a promising platform for multimodal microRNA profiling.

References

(1) T.A. Farazi, J.I. Spitzer, P. Morozov, T. Tuschl, J. Pathol., 2011, 223 (2),102.

(2) E.A. Hunt, D. Broyles, T. Head, S.K. Deo, Annu Rev Anal Chem, 2015, 8, 217.

(3) Boghossian, A. A.; Zhang, J.; Barone, P. W.; Reuel, N. F.; Kim, J.-H.; Heller, D. A.; Ahn, J.-H.; Hilmer, A. J.; Rwei, A.; Arkalgud, J. R.; Zhang, C. T.; Strano, M. S. ChemSusChem 2011, 4 (7), 848.

474877 Enzyme Precipitate Coating of Glucose Oxidase on Electrospun Polymer Nanofibers with Efficient Antibacterial Activity

Abstract:

Polymeric nanofibrous materials in conjugation with antibacterial compounds such as antibiotics, metal nanoparticles have gained attention for the treatment of wounds, burns, post-surgical related problems to reduce microbial infections. In this study, we used electrospun polymer nanofibers consisting of polystyrene (PS) and poly(styrene-co-maleic anhydride (PSMA) for the immobilization of glucose oxidase (GOx) and horseradish peroxide (HRP) enzymes via three approaches, such as covalent attachment (CA), enzyme coating (EC) and enzyme precipitate coating (EPC). The EPC protocol consists of CA, enzyme precipitation and crosslinking while EC represents a control sample without the step of enzyme precipitation. GOx-catalyzed glucose oxidation generates hydrogen peroxide, which can kill both gram-positive and gram-negative bacteria, and the EPC sample showed much more potent antimicrobial activity than CA and EC due to its higher enzyme loading and stability. The antimicrobial potency is further enhanced by the supplement of potassium iodide as well as the addition of HPR-immobilized nanofibers. Further, EPC of GOx showed negligible decrease of antimicrobial activity under recycled uses. EPC with stabilized and highly-loaded enzymes has opened up its great potential as an environmentally-safe agent to kill bacteria.

477958 Interaction of Single-Walled Carbon Nanotubes with Photosynthetic Systems

Abstract:

For years, the distinctive optical and structural properties of single-walled carbon nanotubes (SWCNTs) have inspired the development of promising applications in the field of cell nanobiotechnology(***1***, ***2***). In particular, the synergistic interaction with photosynthetic organisms has attracted an increasing level of attention due to the potential of engineered nanoparticles to enhance the native performance of biological systems, paving the way for novel, renewable, and low cost solutions for light-harvesting, energy conversion and subcellular sensing(***3***, ***4***).

So far, major studies have largely focused on enabling cellular uptake of SWCNTs by engineering the SWCNT surface through non-covalent side-wall functionalization. Non-covalent functionalization with a rich variety of biomolecules and polymers has been shown to potentially increase SWCNT solubility and membrane translocation while endowing these nanostructures with enhanced biocompatibility(***5***).

Most recently, this platform has been successfully applied to intact chloroplasts to augment their photosynthetic capability and improve stability against reactive oxygen species(***6***). Our work builds on these empirical findings to gain an in-depth understanding of the dynamics of the underlying interaction on a molecular level. We performed a systematic investigation of the effect of SWCNT functionalization on membrane penetration properties using a novel biological host. The results of this study offer great promise for the development of a new generation of light-harvesting nanobionic devices.

References:

1. A. A. Boghossian ***et al.***, Near-infrared fluorescent sensors based on single-walled carbon nanotubes for life sciences applications. ***ChemSusChem***. **4**, 848–63 (2011).

2. S. F. Oliveira ***et al.***, Protein functionalized carbon nanomaterials for biomedical applications. ***Carbon N. Y.*** **95**, 767–779 (2015).

3. M. F. Serag, N. Kaji, S. Habuchi, A. Bianco, Y. Baba, Nanobiotechnology meets plant cell biology: carbon nanotubes as organelle targeting nanocarriers. ***RSC Adv.*** **3**, 4856 (2013).

4. M. D. Lambreva ***et al.***, Potential of carbon nanotubes in algal biotechnology. ***Photosynth. Res.*** **125**, 451–471 (2015).

5. S. Vardharajula ***et al.***, Functionalized carbon nanotubes: biomedical applications. ***Int. J. Nanomedicine***. **7**, 5361–74 (2012).

6. J. P. Giraldo ***et al.***, Plant nanobionics approach to augment photosynthesis and biochemical sensing. ***Nat. Mater.*** **13**, 400–8 (2014).

469343 Biomaterials for Human Pluripotent Stem Cell Derived Midbrain Dopaminergic Neuron Generation and Transplantation to Treat Parkinson's Disease

Abstract:

Cell replacement therapies have considerable biomedical potential; however, low survival of cells post-transplantation is a major hurdle hampering their efficacy. For example, cell replacement therapies have strong promise for the treatment of neurodegenerative diseases such as Parkinson's, which is a currently incurable, progressively debilitating disorder affecting ~10M people in the world. However, following striatal transplantation of midbrain dopaminergic (mDA) neurons for the treatment of Parkinson's disease (PD), less than 5% of the neurons typically survive pre-clinically. Such poor neuronal survival requires the resource-intensive generation and implantation of larger numbers of mDA neurons to compensate, and the widespread death of the majority of the graft could reduce the efficacy of the cells remaining. We took a two-pronged approach to resolve this issue by developing methods to first increase the yield of human pluripotent stem cell (hPSC) derived mDA neurons generated in vitro, and second, to increase the post-transplantation survival of neurons in vivo.

Human pluripotent stem cells (hPSCs) were cultured and differentiated into mDA neurons in 3D pNIPAAm-PEG gels. The authenticity of resulting neurons thoroughly characterized by immunocytochemistry, qPCR and voltage-sensitive dye based electrophysiology, and benchmarked against mDA neuronal development on 2D surfaces. Subsequently, a biodegradable, hyaluronic acid (HA) based hydrogel functionalized with ECM derived factors was developed for maturation and transplantation of mDA neurons generated on pNIPAAm-PEG. Post-transplantation survival of hydrogel-encapsulated neurons was then investigated 4.5 months after striatal implantation into rats.

mDA neuron production typically involves undefined components and difficult to scale-up 2D culture formats. Here, we used a fully-defined, 3D thermoresponsive biomaterial platform to derive mDA neurons from PSCs at substantially greater yield, quality, and maturity compared to 2D Matrigel coated surfaces. In particular, after 25 days of differentiation, ~40% of cells differentiated in 3D expressed tyrosine hydroxylase (TH), the rate limiting enzyme for dopamine production and a hallmark of mDA neurons, compared to 20% on 2D. Crucial for functional mDA neurons, ~ 5-fold higher proportion of neurons differentiated in 3D fired mDA-like action potentials. Interestingly, FOXA2 and EN1, markers known to improve survival of mDA neurons, were expressed at significantly higher levels in cells generated in the 3D platform compared to control 2D surfaces.

Low post-transplantation cell survival is often attributed to adverse biochemical, mechanical, and immunological stresses cells experience during and post-transplantation. To address these challenges, we developed a functionalized hyaluronic acid (HA) based hydrogel first to mature pluripotent stem cell (PSC) derived neural progenitors towards functional neurons in vitro in a biomimetic 3D environment, and second to effectively transplant hydrogel-encapsulated neurons into the central nervous system. We demonstrated that incorporating adhesive peptides to promote cell adhesion increased extended neurite density, and adding neurotrophic factors increased dopaminergic differentiation in our 3D HA gels. Furthermore, we observed higher survival post-harvest and injection in vitro for mDA neurons matured within hydrogels compared to those on 2D surfaces. By using an optimized 3D HA biomaterial matrix for neuronal maturation and transplantation, we significantly increased the post-transplantation survival of hPSC derived mDA neurons in the rat striatum by 5 fold compared to the current standard method of bolus injections of unencapsulated neurons.

The combination of a defined, scalable, and resource-efficient cell culture platform and an optimally engineered transplantation biomaterial significantly increased the yield of mDA neurons generated in vitro and improved their post-transplantation viability in vivo, thereby addressing crucial challenges in treatment of neurodegenerative disease.

469101 Engineering an Electroactive Hydrogel for Tissue Engineering Applications

Abstract:

Engineering an Electroactive Hydrogel for Tissue Engineering Applications

Andrew Spencer[1] Hicham Fenniri[1*], Nasim Annabi[1,2,3**]

[1] Department of Chemical Engineering, Northeastern University, Boston, MA, 02115-5000, USA

[2]Biomaterials Innovation Research Center, Brigham and WomenÕs Hospital, Harvard Medical School, Boston, MA, USA.

[3]Harvard-MIT Division of Health Sciences and Technology, Massachusetts Institute of Technology, Cambridge, MA, USA.

*h.fenniri@neu.edu;**n.annabi@neu.edu

INTRODUCTION

Hydrogels are hydrophilic three dimensional (3D) polymeric networks with physical, chemical and biological properties that can be designed to closely mimic the native extracellular matrix (ECM) of mammalian tissues. Because of their biomimetic properties, hydrogels are often used in various tissue engineering and regenerative medicine applications. For tissues whose function is highly dependent on electrochemical signaling between cells and that are responsive to an electric field, such as the heart, it is desired that the hydrogels are electroactive [1]. However, most hydrogels are inherently non-conductive. Numerous approaches have been taken to improve the conductivity of hydrogels using materials that have limited biocompatibility and biodegradability.

There exists a library of conducting organic polymers that offer an attractive alternative to other conductive materials for biomedical applications, due to their ease of processing, versatility, and enhanced biocompatibility both *in vitro* and *in vivo* [2]. Unlike many other conductive materials, such as carbon nanotubes, these conductive polymers can be easily dispersed in aqueous solutions Đ a requirement for biological applications. In addition to their practical advantages, they actuate upon electrical stimulation [3]. This enables simultaneous electrical and mechanical stimulation of cells, which may help to maintain their phenotype and facilitate their growth.

By dispersing a well-established conducting polymer complex, poly(3,4-ethylenedioxythiophene):polystyrene sulfonate (PEDOT:PSS), at various concentrations within a photo-crosslinkable naturally derived polymer, gelatin methacrylate (GelMA), we have synthesized an electroactive and flexible hydrogel that can be used for 3D cell encapsulation. The engineered 3D conductive hydrogel with tunable mechanical and electrical properties can be used for various tissue engineering applications, including muscle tissue regeneration and heart repair.

MATERIALS AND METHODS

Hydrogel formation and characterization: Gelatin methacrylate (GelMA) was synthesized according to a previously reported protocol [4]. PEDOT:PSS was obtained from Heraeus (Germany) in the form of an aqueous dispersion called CleviosTM PH 1000 (PEDOT:PSS ratio 1:2.5). The concentration of the dispersion in GelMA prepolymer after filtering through a 0.22 mm filter was determined to be 0.95 wt% after drying and weighing. Solutions of 7% GelMA at 0%, 0.1% and 0.3% PEDOT:PSS with 0.5% photoinitiator (Irgacure 2959) were prepared for all experiments. The hydrogels (thickness: 2 mm, diameter: 4 mm) were formed by exposing the GelMA/PEDOT:PSS solutions to UV light at 6.9 mW/cm^2 for 360 s. Then, the compressive properties of the engineered hydrogels were tested using an Instron 5944 mechanical tester. Compressive modulus was taken as the slope of the linear portion of the stress/strain curves at 0-5% strain.

3D Cell Encapsulation: C2C12 myoblast cells (ATCC) were cultured in complete media (DMEM + 10% FBS + 1% penicillin/streptomycin) in a humidified 5% CO_2 incubator prior to use. Cell were trypsinized and resuspended in solutions of GelMA/PEDOT:PSS precursor at concentration of 10^6 cells/mL of prepolymer solution. 7 mL of the mixture was then sandwiched between a glass slide and a petri dish with a 150 mm spacer and exposed to UV light for up to 30 s to crosslink to form a cell-laden hydrogel. Live/Dead, Actin/DAPI and PrestoBlue assays were performed on days 1, 3 and 5 to evaluate the cell viability, spreading, and proliferation in 3D hydrogels with varying mechanical and electrical properties.

RESULTS AND DISCUSSION

Mechanical testing demonstrated that the compressive modulus increases with increasing concentration of PEDOT:PSS. This might be due the interaction between the negatively charged PSS and the positively charged arginine or lysine groups in GelMA, adding physical crosslinking to the composite hydrogel in addition to chemical bonds generated during photopolymerization. The *in vitro* data also showed that incorporation of PEDOT:PSS into the hydrogels did not affect the metabolic activity of the cells, as confirmed by PrestoBlue cell viability assay. Live/Dead assay also confirmed that the incorporation of PEDOT:PSS into GelMA did not affect cell viability. In addition, the cells start spreading inside the engineered hydrogels at day 3 of culture. These results suggest that the incorporation of conductive polymer complex PEDOT:PSS created a biocompatible conductive hydrogel that will enable uniform electrical stimulation of electrically responsive tissues.

CONCLUSION

An electroactive hydrogel was developed via the incorporation of a conducting polymer complex into a naturally derived hydrogel. Compressive modulus was improved via the addition of PEDOT:PSS into the hydrogel. Cell viability, metabolic activity, and spreading were not affected by the incorporation of the conductive material into the hydrogel network as shown by PrestoBlue, Live/Dead and Actin/DAPI assays. These results suggest that the engineered composite hydrogel is suitable for various tissue engineering applications where the tissues are electrically responsive.

ACKNOWLEDGMENT

The authors would like to thank Northeastern University for funding this research.

REFERENCE

[1] N. Annabi, A. Tamayol, J. Uquillas, M. Akbari, L. Bertassoni, C. Cha, G. Camci_Unal, M. Dokmeci, N. Peppas, A. Khademhosseini, *Adv Mater* **2014.**

[2] R. Balint, N. Cassidy, S. Cartmell, *Acta Biomaterialia* **2014**

[3] M. R. Abidian, D. _H. Kim, D. C. Martin, *Adv Mater* **2006**

[4] W. Nichol, S. T. Koshy, H. Bae, C. M. Hwang, S. Yamanlar, A. Khademhosseini, *Biomaterials* **2010**

471892 Modulation of Inflammatory Response for Accelerated Tissue Vascularization and Bone Regeneration

Abstract:

There is a profound need for bone regeneration strategies due to trauma and various musculoskeletal diseases. Tissue engineering offer alternative approaches to regenerate tissues. The transfer of tissue engineering from the lab to the clinic is limited by the failure to adequately vascularize scaffolds and control immune response. The goal of this study is to harness the wound healing potential of macrophages and human mesenchymal stem cells (hMSCs) to enhance directed angiogenesis and bone formation. We explored the use of resveratrol, a natural compound commonly used in complementary medicine, to mediate M1 to M2 macrophage plasticity, impart natural release of angiogenic factors by macrophages and enhance osteogenic differentiation of hMSCs. To this end, we mapped the time-dependent response of macrophage gene expression as well as hMSC osteogenic differentiation to varying doses of resveratrol in 2D and 3D biomaterial environments. The results demonstrated that macrophages adopted a pro-healing phenotype in the presence of resveratrol, as confirmed by the expression of IL-6 and TNF-Î±. This effect was coupled with higher expression of anti-inflammatory markers VEGF and IL-10. In addition, hMSCs exposed to resveratrol demonstrated a higher level of calcium deposition and alkaline phosphatase activity. Our results altogether delineate the potential to synergistically accelerate angiogenic factor release and upregulate osteogenic signaling pathways by â€œdialingâ€ the appropriate degree of resveratrol release.

463526 The Impact of Decellularization Agents on Renal Tissue Extracellular Matrix

Abstract:

The impact of decellularization agents on renal tissue extracellular matrix

Nafiseh Poornejad, Lara B. Schaumann, Travis Neuberger, Sarah Chambers, Beverly L. Roeder, Alonzo D. Cook

Brigham Young University, Provo, Utah

The combination of patient-specific cells with scaffolds obtained from natural sources could result in improved regeneration of human tissues. Decellularization of the native tissue is the first step in this technology. Effective decellularization uses agents that lyse cells and remove all cellular materials, leaving intact collagenous extracellular matrices (ECMs). Although complete removal of cellular remnants should be achieved to prevent the induction of an immune response, preserving the underlying structure and basement membrane components are of crucial importance to retain regenerative potential of a decellularized scaffold. Consequently, enough care should be taken in choosing cell-lysing agents and the procedure should be optimized for different tissues.

In this study, the impact of five decellularization agents (0.1 N NaOH, 1% peracetic acid, 3% Triton X-100, 1% sodium dodecyl sulfate (SDS), and 0.05% trypsin/EDTA) on renal tissue was examined to comprehensively compare a broad range of agents including acidic/basic solution, ionic/non-ionic detergents, and protease enzyme. The NaOH solution induced the most efficient cell removal (>90% DNA removal, p-value<0.05), and resulted in the highest amount of human renal cell viability and proliferation after recellularization, although it also produced the most significant damage to collagenous fiber networks, glycosaminoglycans (GAGs), and fibroblast growth factor (FGF). The SDS solution led to less severe damage to the ECM structure, but it resulted in less proliferation of reseeded cells.

Peracetic acid and Triton X-100 preserved ECM components intact. However, these agents alone were not efficient in removing cellular materials; especially peracetic acid, which left more than 80% of cellular remnants within the ECM. Trypsin enzyme didnâ€™t eliminate cellular materials and resulted in significant damage to the collagenous structure of ECM.

In summary, we suggest NaOH as the most efficient agent to remove cellular materials combined with non-ionic Triton X-100 as an optimum procedure for decellularization of renal tissue.

472692 Electrospun Silk with Selenium Nanoparticles for Antibacterial Skin Applications

Abstract:

Electrospun Silk with Selenium Nanoparticles for Antibacterial Skin Applications

Stanley Chung[1], Thomas J. Webster[1-3]
1. Department of Chemical Engineering, Northeastern University, Boston, MA
2. Department of Bioengineering, Northeastern University, Boston, MA
3. Center of Excellence for Advanced Materials Research, King Abdulaziz University, Jeddah, Saudi Arabia

Skin infections have accelerated in the past decade, often caused by *Staphylococcus aureus*, the leading cause of skin and soft tissue infections (SSTI). In addition, strains of *S. aureus* are developing resistance to most conventional antibiotic treatment options. Antibiotic resistance is a growing epidemic and is estimated by the Centers for Disease Control and Prevention to cost $20 billion annually in excess health costs and $35 billion in other societal costs. Thus, it is evident that new strategies are needed to combat this growing public health threat.

Selenium is a novel chemistry in biomedical applications and has shown beneficial properties for reducing bacterial infections. In particular, selenium nanoparticles have shown good antibacterial properties that selectively inhibit bacteria while maintaining mammalian cell growth.[1] In addition, the selenium nanoparticle coating process developed in this lab is facile and versatile, allowing researchers to deposit selenium nanoparticles as an antibacterial additive to a variety of biomaterials.[2,3]

Silk has been shown to possess many beneficial properties for skin regeneration by promoting collagen synthesis, re-epithelialization, wound healing, atopic dermatitis alleviation, and scar reduction. In particular, electrospun silk possesses morphology similar to that of native extracellular matrix in the body and is especially suitable for skin tissue engineering applications. Unfortunately, silk has also been shown to promote bacterial growth. Here, it is proposed that the combination of two biomaterials, selenium nanoparticles and silk, as a nanocomposite scaffold will promote healthy skin cell growth while reducing the bacterial load.

Silk fibroin was extracted from Bombyx mori silk worms and resuspended in formic acid at a concentration of 8% (w/v) for electrospinning. After electrospinning the silk scaffolds, selenium nanoparticles were deposited onto the silk scaffolds. *In vitro* skin cell viability and activity were then assessed by MTS assay (Promega) with passage 3-12 human dermal fibroblasts while *S. aureus* activity was assessed by an ATP assay (Promega). In all experiments, the control group comprised of cells seeded onto polystyrene tissue culture flasks. All experiments were conducted in at least triplicates and N=3.

Recent results on the cellular activity associated with electrospun nanocomposite scaffolds will be presented. Briefly, results show a statistically significant reduction in metabolic activity of *S. aureus*while promoting the metabolic activity of human dermal fibroblasts; for both the MTS and the ATP assay, addition of selenium nanoparticles further enhanced the response of the cells as compared to the control ($p<0.05$ for silk samples without nanoparticles and $p<0.01$ for silk samples with nanoparticles as compared to the control). Further studies involving microscopy and qPCR will elucidate the mechanism of action by which this bactericidal activity is achieved.

Acknowledgement:

We acknowledge the funding provided by Northeastern University.

References:

1. Tran, P. A. & Webster, T. J. Selenium nanoparticles inhibit Staphylococcus aureus growth. *Int. J. Nanomedicine* **6**,1553 (2011).

2. Wang, Q., Larese-Casanova, P. & Webster, T. J. Inhibition of various gram-positive and gram-negative bacteria growth on selenium nanoparticle coated paper towels. *Int. J. Nanomedicine* **10**,2885 (2015).

3. Wang, Q., Mejía Jaramillo, A., Pavon, J. J. & Webster, T. J. Red selenium nanoparticles and gray selenium nanorods as antibacterial coatings for PEEK medical devices. *J. Biomed. Mater. Res. Part B Appl. Biomater.* (2015).

470305 Engineered Cellulose-Based Cell Culture Platforms to Improve Human Health

Abstract:

Two-dimensional (2D) *in vitro* cell cultures do not mimic the microenvironment in the native tissues. Currently, there are several systems that are used to study cell cultures in 3D; however, these systems use complex instrumentation and expensive set-ups, require long and tedious optimization procedures, and do not yield in physiologically relevant structures. Other drawbacks of these constructs include heterogeneous distribution of cells, and potential mass transport limitations. Collectively, it is challenging to fabricate biocompatible constructs in cm-scale. We developed simple cellulose-based cell culture platforms that can be used for a wide range of applications for mammalian cells, bacteria, fungi, and plants cells. These new systems are simple, tunable, and low-cost.

We used cellulose-based materials as 3D scaffolds to support cells and hydrogels. We grew different types of cells including connective tissue cells, osteoblasts, neutrophils, gut cells, tumor cells, stem cells, plant cells, fungi, or bacteria in cellulose-based materials. Subsequently, we characterized the cell behavior by using viability assays, metabolic activity assays, immunocytochemistry, mechanical tests, X-rays scanning, and high-resolution imaging.

In this study, we successfully fabricated cellulose-based scaffolds to (i) investigate migration of primary human cells (tumor fibroblasts and lung cancer cells), (ii) induce template-guided biomineralization, (iii) generate 3D osteoblast cultures that are amenable to high-throughput sample preparation and analysis, (iv) form and control gradients of oxygen and biological molecules, and (v) develop new co-culture systems in multi-layered stacked cultures. In addition to primary cells and mammalian cell lines, we also cultured bacteria, fungi, and plant cells in cellulose-based materials. These cell cultures in cellulose resulted in high cell viability and high metabolic activity, and enabled differentiation of cells. In these experiments, we did not observe mass transport limitations to the cells. Our results have indicated that the cellulose-based materials allow for patterning and easy recovery of cells, can adapt modular configurations, and can provide physiologically relevant tissue models. The scaffolds we fabricated from thin layers of cellulose also yielded in free-standing structures in different scales (micron- to cm-).

To sum up, we developed new cellulose-based cell culture platforms to provide multicellular and compartmentalized tissue-mimetics for clinical applications. To overcome the limitations of the traditional tissue models, we developed a novel layer-by-layer approach to assemble tissue-like structures from cellulose-based low-cost materials. This strategy offers unique opportunities such as understanding fundamental biology, developing disease models for personalized medicine, and assembling different organs together.

471928 Award Submission _ Photo-Electrochemical Characterizations of Photosystem I (PS I) Assembly Under Bio-Mimetic Membrane Confinement

Abstract:

The robust structural and photoactive electrochemical properties of Photosystem I (PSI) make it an ideal candidate for incorporation into bio-hybrid photovoltaic devices. However, the first step towards the rational design of such devices require systematic and oriented assembly of PS I at bio-abiotic interfaces via suitable scaffolding. To this end, our goal is to incorporate and assemble individually oriented PS I complexes into synthetic membrane-bound structures that mimic the natural thylakoid membrane housing of PS I. We investigate the photocurrent generation from such organic-inorganic interfaces in membrane bound PS I systems. Detergent-mediated reconstitution of proteoliposomes is analyzed via absorption and fluorescence spectroscopy measurements. Successful immobilization of membrane bound PS I on gold substrates are mediated by the hydrophillic interfaces of self-assembled monolayers (SAMs). To this end, force-distance spectroscopy during atomic force microscopy (AFM) imaging of lipid bilayer membranes on thiolated Au substrates confirm monolayer formation. Furthermore, topography measurements using liquid-AFM techniques confirm the assembly of membrane-bound PS I. Finally, the current work presents photo-electrochemical (PEC) characterizations demonstrating enhanced photocurrent generation from PS I complexes assembled under membrane confinement as compared to their isolated immobilizations on electrodes. In turn, the aforementioned studies provide valuable insight into future incorporation of bio-mimetic membrane bound protein complexes into novel bio-hybrid devices.

473431 Award Submission: DNA-Programmable Assembly of Enzyme Superlattices

Abstract:

Enzymes are one of Natureâ€™s most diverse and powerful nanoparticles. Assembling them into multicomponent crystals has been of great interest and can be accomplished through a variety of means, typically based on protein surface chemistry. The programmability of DNA makes it an attractive structure-directing ligand for assembling nanoparticles into well-defined materials. Recently, DNA has emerged as an effective ligand by which catalytically active proteins can be assembled into extended crystalline superlattices of CsCl symmetry. In these systems, protein surfaces are covalently functionalized with nucleic acid ligands, which direct interparticle interactions through Watson-Crick base pairing. In this study, it was predicted that under certain conditions the anisotropic shape and surface chemistry of the enzyme could induce directional bonding interactions resulting in the assembly of a different lattice symmetry. To probe this, the enzyme catalase was functionalized with DNA ligands of varying flexibility and co-assembled with gold nanoparticles possessing different numbers of DNA bonding ligands. A phase transformation was observed with decreasing DNA flexibility and increasing gold nanoparticle ligand number from CsCl to Th_3P_4 symmetry, with mixtures of the two lattice symmetries present at intermediate states. Through fitting of the SAXS diffraction patterns, the composition of the mixtures was quantified. Coarse grained molecular dynamics simulations were used to further understand the energetics of the two phases. This is the first instance in these systems that phase transformation has been observed by tuning ligand flexibility, expanding on the means by which well-defined functional materials can be assembled from proteins using DNA.

473175 Award Submission: Bio-Inspired Nanomachines for Biomedical Applications

Abstract:

Robotics deals with automated machines that can locomote themselves and operate tasks in various environments over many orders of magnitudes in scale. One of the most inspiring goals is the construction of smart and powerful nanorobotic systems for operation in the human body. However, viscous forces dominate inertial forces at such small scales, leading to the "low-Reynolds number challenge" for nanoscale propulsion. This presentation will discuss newly created multi-functional nanorobots which can overcome this challenge by utilizing local chemical reactions or external field actuation to achieve efficient movement in biological matrices. With selectively engineered materials, the nanorobots possess numerous attractive properties, including biocompatibility, biodegradability, high loading capacity, and the ability to autonomously release of payloads 'on-the-fly'. The increased capabilities and sophistication of these tiny robots hold considerable promise for a variety of biomedical applications ranging from drug delivery to minimally invasive surgery. Particularly, using zinc-based micromotors as model robots, we reported the first in vivo study of artificial micromotors in a living organism. Such in vivo evaluation examines the distribution, retention, cargo delivery, and acute toxicity profile of synthetic motors in mouse stomachs via oral administration. We demonstrate that the acid-driven propulsion in the stomach effectively enhances the binding and retention of the motors as well as of cargo payloads on the stomach wall. Our work is anticipated to significantly advance the emerging field of biomedical nanorobots and to open the door to in vivo evaluation and clinical applications of these biomedical nanorobots.

471301 Award Submission: Development and Physicochemical Characterization of Tacrolimus-Loaded Nanocomposite Microparticles for the Treatment of Pulmonary Hypertension

Abstract:

.

Pulmonary arterial hypertension (PAH) is a progressive cardiovascular disease that leads to limited exercise capacity, right heart failure, and ultimately death. The primary treatment options currently available for patients with PAH are based on drugs with vasodilatory properties that improve cardiopulmonary function. Unfortunately, these drugs show no effect on prohibiting the progress of obliterative vascular pathology. Because of this, the only option for many patients to improve their health and quality of life is heart-lung transplantation. Therefore, new approaches to utilize compounds that focus on activating cellular mechanisms to reverse vascular remodeling are under investigation. Improving the function of the bone morphogenetic protein receptor-2 (BMPR2) signaling pathway is a potential direction in the development of improved PAH therapies, as the mutations of BMPR2 have been identified as the main cause of inherited PAH, accounting for 60 to 80% of familial cases. Since challenges exist in clinical application of BMPR2 gene therapy, tacrolimus (TAC) can be applied instead, as it increases signaling through the BMPR2 pathway. However, the application of TAC is hindered by its poor solubility, instability, and poor bioavailability. In addition to these limiting pharmacokinetic properties, systemic side effects such as neurotoxicity and nephrotoxicity also complicate the clinical use of TAC.

To overcome this problem, nanocomposite microparticles (nCmP) in the form of dry powder aerosols can be applied to deliver TAC more safely and effectively. This system is comprised of TAC-loaded acetalated dextran (Ac-Dex) nanoparticles (NP) entrapped in microparticle carriers using the excipient mannitol to allow for the delivery of TAC NP to the lungs. Upon pulmonary administration, the nCmP will deposit on the mucus in the respiratory tract, decompose into free NP and mannitol, and allow the nanoparticles to penetrate the mucus barrier and then release drug to the targeted site in a sustained rate. The goal of the described research was the initial development and physicochemical characterization of the nCmP systems with ability of targeted delivery, improved drug solubility, and potential of efficient penetration of mucus barrier and of controlled release via particle engineering.

The NP were approximately 200 nm in diameter with narrow size distribution both before loading into and after redispersion from nCmP. The NP exhibited smooth, spherical morphology and the nCmP were raisin-like spheres. High encapsulation efficacy was achieved both in the encapsulation of tacrolimus in NP and that of NP in nCmP. nCmP exhibited desirable aerosol dispersion properties allowing them to deposit into the deep lung regions for effective drug delivery. Overall, the designed nCmP system is a promising application in pulmonary TAC delivery for the treatment of PAH due to its novel features including targeted pulmonary delivery, improved solubility of tacrolimus, potential of permeation through mucus barrier, and of controlled drug release

469166 Award Submission:Â Photoexcited Quantum Dots Potentiate Antibiotic Activity in Multidrug-Resistant Bacteria

Abstract:

The rise of multidrug-resistant (MDR) bacteria is a growing concern to global health and is exacerbated by the lack of new antibiotics being developed. In order to treat MDR infections, new classes of antibiotics must be developed that are effective against already pervasive MDR infections. While current antibiotics are derived from natural products, investigating true synthetic antibiotics provides opportunity to design therapeutics for antimicrobial targets of interest. In this study we present a novel, nanoparticle based therapeutic that is able to inhibit MDR clinical isolates of *Escherichia coli*, *Salmonella typhimurium*, and *Klebsiella pneumoniae* both as a monotherapy as well as combination therapy with antibiotics that the isolates were previously resistant against. The nanoparticles used in this study generate specific, light activated reactive species (LARS) which alter the drug resistance phenotype and lower the level of resistance of the clinical strains to traditional antibiotics. We show that the redox homeostasis perturbing nanoparticles in combination with protein synthesis inhibiting antibiotics, namely chloramphenicol, streptomycin, and clindamycin, lower the IC_{50} of resistant MDR clinical isolates below the Clinical Laboratory Standards Institute (CLSI) breakpoint. We find that this phenomenon is not only limited to protein synthesis inhibiting antibiotics, but extends to other classes of antibiotics such as fluoroquinolones. Using a multiplicative model, we show that this combination therapeutic effect is additive and the effect is concentration dependent on both the antibiotic as well as the nanoparticle concentration. Interestingly the nanoparticle is able to have this drastic effect on antibiotic resistance at nanomolar concentrations highlighting the strong effect of the LARS produced. The bactericidal effect of the combination therapy is greater than that of each mono-therapy alone highlighting the potential utility of this combination therapy for treating MDR infections.

472307 Award Session: Silk-Gold Nanorod Nanocomposite Films for Rapid Tissue Repair

Abstract:

Laser tissue welding is a tissue repair technique that utilizes a chromophore to convert photons into heat in order to seal apposed tissue edges by localized protein denaturation and interdigitation. Laser tissue welding has emerged as a promising alternative to sutures or staples conventionally used for sealing soft tissues, with the potential to limit surgical complications such as anastomotic leakage and wound dehiscence while also reducing healing times and scarring. However, insufficient closure strength, surgical inconvenience, and extensive thermal damage have hindered the clinical application of this technique, largely due to the unavailability of effective solders (surgical sealants). In this work, we describe the development of novel plasmonic silk nanocomposites for laser tissue welding that overcome these barriers in tissue welding. Leak pressure, burst pressure, and tensile strength measurements indicated that laser welding using plasmonic nanocomposites was able to restore significant mechanical integrity to the welded tissue.

468341 Award Submission:Â Evaluation of the Cancer-Preventive Effect of Resveratrol-Loaded Nanoparticles on the Formation of Tumor Spheroids

Abstract:

Despite the recent advancements in the treatment of cancer, designing advanced therapeutics remains an enormous challenge for scientists. It is estimated in the U.S. that 1,685,210 new cases of cancer will be diagnosed in 2016 and 595,600 people will die from cancer, corresponding to approximately 1,600 deaths per day[1].

Chemotherapy has been extensively used to combat cancer, however its overall success is rather limited for lung cancer. An emerging alternative over the past few decades has been the use of chemopreventive phytochemicals, which are constituents of fruits and vegetables with potential pharmacological and/or nutritional functions. These agents may reverse or inhibit the carcinogenic process resulting in a decreased cancer risk [2]. The FDA has approved several agents for cancer prevention, including Tamoxifen for breast cancer and the Human Papilloma Virus (HPV) vaccine for cervical intraepithelial neoplasia. More than 200 other chemopreventive agents are being evaluated in clinical trials [3].

Resveratrol (3,5,4â€²-trihydroxystilbene, RSV) is a phytoalexin, a substance synthesized by plants to counteract pathogens [4]. It is abundantly found in red grapes and in lower amounts in peanuts and blueberries. In 1997, interest in RSV peaked when Jung *et al.* reported its chemopreventive effects on all stages of neoplastic activity (initiation, promotion and progression) [5]. *In vitro* studies show the promising properties of resveratrol; however, its limited bioavailability in the bloodstream makes it difficult to translate into clinical studies. A study by Dolfini *et al.* [6] showed that RSV induces drastic growth inhibition on a human breast cancer cell line in *in vivo* and *in vitro* assays. Also, RSV-loaded gel nanoparticles enhanced intracellular uptake and cytotoxicity in a NCI-H460 lung cancer cell line [7] and RSV-loaded nanostructure lipid carriers exhibited higher penetration into the skin in comparison to solid lipid nanoparticles [8], demonstrating improvement with RSV in nanoparticle systems.

The aim of this project was to evaluate the chemopreventive effect of RSV and RSV-loaded nanoparticles (RSV-NP) on the formation of tumor spheroids. We synthesized biodegradable nanoparticles comprised of the biodegradable polymer acetalated dextran loaded with RSV and we evaluated the efficacy of pure RSV and RSV-NP using A549 adenocarcinoma lung cancer cell line in 2D and 3D cell culture studies. 3D multicellular spheroids (MCS) reflect many of the properties of solid tumors and better mimic the barriers of drug diffusion, transport, and distribution within tumors [9, 10]. RSV and RSV-NP exerted an inhibitory effect on the growth of MCS, showing a promising alternative as a chemopreventive drug, which could palliate the incidence or mortality from lung cancer.

References

[1] Siegel, R.L., K.D. Miller, and A. Jemal, *Cancer statistics, 2016.* CA Cancer J Clin, 2016. 66(1): p. 7-30.

[2] Sporn, M.B., et al., *Prevention of chemical carcinogenesis by vitamin A and its synthetic analogs (retinoids).* Fed Proc, 1976. 35(6): p. 1332-8.

[3] Davis, J.S. and X. Wu, *Current state and future challenges of chemoprevention.* Discov Med, 2012. 13(72): p. 385-90.

[4] Singh, C.K., M.A. Ndiaye, and N. Ahmad, *Resveratrol and cancer: Challenges for clinical translation.* Biochim Biophys Acta, 2015. 1852(6): p. 1178-85.

[5] Jang, M., et al., *Cancer chemopreventive activity of resveratrol, a natural product derived from grapes.* Science (New York, N.Y.), 1997. 275(5297): p. 218-20.

[6] Dolfini, E., et al., *Resveratrol impairs the formation of MDA-MB-231 multicellular tumor spheroids concomitant with ceramide accumulation.* Cancer Lett, 2007. 249(2): p. 143-7.

[7] Karthikeyan, S., et al., *Anticancer activity of resveratrol-loaded gelatin nanoparticles on NCI-H460 non-small cell lung cancer cells.* Biomedicine & Preventive Nutrition, 2013. 3(1): p. 64-73.

[8] Gokce, E.H., et al., *Resveratrol-loaded solid lipid nanoparticles versus nanostructured lipid carriers: evaluation of antioxidant potential for dermal applications.* Int J Nanomedicine, 2012. 7: p. 1841-50.

[9] Sutherland, R.M., *Cell and environment interactions in tumor microregions: the multicell spheroid model.* Science, 1988. 240(4849): p. 177-84.

[10] Minchinton, A.I. and I.F. Tannock, *Drug penetration in solid tumours.* Nat Rev Cancer, 2006. 6(8): p. 583-92.

455923 Single-Step Synthesis of Aligned High Aspect Ratio Mwcnts Impregnated with Al_2O_3 Particles in an Ultrasonic Atomization Head CVD Reactor

Abstract:

Commercializing the production of functionalized CNTs with distinguish surface properties for nano applications is highly dependent on the development of synthesis and functionalization methods that can be easily implemented and scaled up. Conventional metal oxide CNT impregnation methods involve various steps that are time consuming and difficult to control. In this study, an injection vertical CVD reactor fitted with an ultrasonic atomization head is used in a pilot-plant scale reactor for semi continuous production of MWCNTs *in-situ* impregnated with Al_2O_3 particles. Ferrocene as a source of iron catalyst and aluminum isopropoxide as a source of aluminum oxide nanoparticles are dissolved in p-xylene to make a homogeneous feed solution to the reactor. The Atomization of the feed solution resulted in full and even dispersion of the catalytic solution, which controlled the diameters of the produced MWCNTs. The reactor produced High quality and high aspect ratio bundles, aligned, MWCNTs decorated with Al_2O_3 particles as confirmed by TEM and EDS characterization. TEM images reveled that alumina particles only decorated the outer surface of the tubes and were not present inside the tubes as the hollow structure of the tubes were visible. The use of the support free ferrocene catalyst insured the fully exposer of the metal particles to the hydrocarbon source which resulted in high production yields at optimized conditions. The effect of different reaction parameters on production yields and properties of the produced CNTs were investigated. High yields of pure high surface area MWCNTs were achieved at an optimum temperature of 800 °C as confirmed by SEM images and TGA analysis. The effect of temperature is thought to be due to major alterations in the distribution of atoms at the metal/gas interface. The presented reactor system can be easily controlled to produce MWCNTs impregnated with metal oxides with a wide range of surface properties compared to conventional fluidized bed reactors that are relatively difficult to control.

466931 Growth of Single-Walled Carbon Nanotubes with Rh and Cu Catalysts

Abstract:

Currently the most widely used method for the growth of carbon nanotubes is the catalytic chemical vapor deposition (C-CVD). In this method, a transition metal nanoparticle is used as the catalyst. The nanoparticle is either supported onto a surface or unsupported and floating. Recently, the growth of CNT from the system of Co-Mo supported by MgO was explored using Environmental Scanning Transmission Electron Microscope (ESTEM) [1]. In the computational front, the potential of Ni nanoparticles as the catalyst for C-CVD has been studied using DFT and *ab initio* molecular dynamics (AIMD) simulations [2]. Apart from the traditional Fe, Co and Ni, other transition metals have also been shown to exhibit unique characteristics in terms of decomposing carbon feed gas, dissolving carbon, possessing reasonable strength of metal-carbon bonds, and forming graphitic caps on their surface that could be advantageous for more practical temperature and pressure ranges [3]. In another study [4], Cu has been suggested to be an appropriate catalyst, although for graphene synthesis, due to its low diffusion and nucleation barrier towards carbon.

Subsequently, in this work, using DFT and AIMD simulations we investigate and compare two transition metals Cu and Rh, both of which have low solubility towards carbon. The study is conducted for supported and unsupported nanoparticles. The dissolution of carbon in different sizes of metal nanoparticles is studied. These carburized metal nanoparticles are then brought into contact with graphene fragments and CNT caps with certain chirality to study the interaction and energetics between them. Finally, AIMD simulations of these systems will reveal their dynamics and the changes, if any, in the chirality of these caps. In conclusion, this will shed light on the effect of low carbon-dissolving transition metals on CNT chirality and geometry as possible catalysts for CNT growth.

References

[1] Picher, M., Lin, P.A., Gomez-Ballesteros, J.L., Balbuena, P.B., Sharma, R., 'Nucleation of graphene and its conversion to single-walled carbon nanotubes', Nano letters 14(11): 6104-6108 (2014)

[2] Gomez-Ballesteros, J.L., Balbuena, P.B., 'Structure and dynamics of metallic and carburized catalytic Ni nanoparticles: effects on growth of single-walled carbon nanotubes', Phys. Chem. Chem. Phys. 17: 15056-15064 (2015)

[3] Silvearv, F., Larsson, P., Jones, S., Ahuja, R., Larsson, A., 'Establishing the most favorable metal–carbon bond strength for carbon nanotube catalysts', J. Mater. Chem. C 3: 3422-3427 (2015)

[4] Li, J., Croiset, E., Ricardez-Sandoval, L., 'Effects of metal elements in catalytic growth of carbon nanotubes/graphene: A first principles DFT study', Applied Surface Science 317: 923–928 (2014)

460005 Gaseous Product Mixture from Fischer-Tropsch Synthesis As an Efficient Carbon Source for Low Temperature CVD Growth of Carbon Nanotube Carpets

Abstract:

Low-temperature chemical vapor deposition (CVD) growth of carbon nanotube (CNT) carpets from Fe and Fe-Cu catalysts using a gaseous product mixture from Fischer-Tropsch synthesis (FTS-GP) as a superior carbon precursor is demonstrated. This growth approach addresses a persistent issue of obtaining thick CNT carpets on temperature-sensitive substrates at low temperatures using a non-plasma CVD process. The efficiency of the process is evidenced by the highly dense, vertically aligned CNT structures from both Fe and Fe-Cu catalysts even at temperatures as low as 400Â°C for the conventional alumina substrates and 450Â°C for aluminum foil â€" a record low growth temperature for CNT carpets. The grown CNTs consist of well-aligned graphitic sheets and exhibit a straight morphology with hollow interior and parallel graphitic planes along the tube walls. The low activation energies and the long catalyst lifetimes observed are attributed to the unique composition of FTS-GP. The use of FTS-GP as a carbon precursor has the potential of enhancing CNT quality, catalyst lifetime, and sustainability, as well as lowering the cost of producing scalable amounts of CNTs.

464330 Preparation of Polyacrylonitrile and Polyacrylonitrile/Carbon Nanostructures

Abstract:

Nanocomposite fibers made from polyacrylonitrile (PAN NFs), polyacrylonitrile containing carbon nanotubes (PAN/MWCNTs NFs) and graphene (PAN/G NFs) were fabricated by electrospinning. The polyacrylonitrile were dissolved in dimethylformamide (C3H7NO, DMF) and the carbon nanostructures were successfully dispersed into the PAN matrix. Polymeric solutions were fed through a capillary tip with the diameter of 0.7 mm using a 10 mL syringe. During electrospinning, the applied voltage was 20 kV, the distance between the tip and collector was 13 cm and the flow rate of the spinning solution was 0.5 mL/h. The electrospun fibers were collected on aluminum foil wrapped on a metal drum rotating at approximately 100 rpm and were dried in vacuum at 60 °C before further use. Scanning electron microscopy showed nanocomposite fibers with regular morphologies. Nonwoven mats of uniform fiber with smooth surface were obtained in the absence and presence of carbon nanostructures. As it is obvious, there are no beads in PAN and PAN/MWCNTs nanofibers which indicate a good dispersion of nanostructures in the fiber. It can be seen that the addition of MWCNTs in the polymeric solution caused the increasing of fiber diameter.

458426 Hollow Carbon Nanobubbles: Graphene Related Nanocapsules That Form Stable Dispersions in Water and Can Incorporate a Cargo

Abstract:

Graphene related structures, such as carbon nanotubes and fullerenes attracted tremendous attention in various fields. The central cavity of buckminsterfullerene (C_{60}),[1] the probably most prominent form of hollow carbon, has been used to enclose various atoms and even molecules.[2] The limited space (few atoms per C_{60}) and the difficult fill/release processes, however, do not permit a real container/cargo function. Capsule systems have attracted tremendous attention in catalysis, biomaterials, as electrode material and for the encapsulation of sensitive materials (therapeutics, fluorescent markers and others). For most container-type applications, however, a hollow nanosphere should carry a hydrophobic cargo through an aqueous environment, i.e. the nanocapsule is chemically different at its inner- and outer surface.

We developed multi-walled carbon nanotube related nanocapsules consisting of a few layers graphene-like carbon.[3] Selective covalent chemical derivatization of the outside carbon shells with a negatively charged polymer permits highly stable dispersion in water while the inside's hydrophobicity provides room to accommodate cargo. The synthesis starts from carbon coated cobalt metal nanoparticles (C/Co) with a shell of the above mentioned 3 - 4 layers of sp^2-hybridized carbon.[4] After surface modification the templating metal core can be selectively removed by acidic dissolution at elevated temperature.

The accessibility of the bubble interior was shown with rhodamine B. An unusual intense rhodamine B accumulation in the interior of the nanobubbles (10'000 times more rhodamine B in the bubble interior than in the outside solution) and an unexpected hysteresis in filling and release were found. It was further shown that doxorubicin, one of the most widely used commercial anti-breast-cancer drugs, can be loaded to the hollow carbon nanobubbles.

[1] H. W. Kroto, J. R. Heath, S. C. Obrien, R. F. Curl, R. E. Smalley, *Nature* **1985**, *318*, 162.

[2] a) J. R. Heath, S. C. Obrien, Q. Zhang, Y. Liu, R. F. Curl, H. W. Kroto, F. K. Tittel, R. E. Smalley, *J. Am. Chem. Soc.* **1985**, *107*, 7779; b) S. Iwamatsu, T. Uozaki, K. Kobayashi, S. Y. Re, S. Nagase, S. Murata, *J. Am. Chem. Soc.* **2004**, *126*, 2668; c) K. Komatsu, M. Murata, Y. Murata, *Science* **2005**, *307*, 238; d) K. Kurotobi, Y. Murata, *Science* **2011**, *333*, 613; e) Y. Murata, M. Murata, K. Komatsu, *J. Am. Chem. Soc.* **2003**, *125*, 7152; f) T. A. Murphy, T. Pawlik, A. Weidinger, M. Hohne, R. Alcala, J. M. Spaeth, *Phys. Rev. Lett.* **1996**, *77*, 1075; g) M. Saunders, H. A. Jimenezvazquez, R. J. Cross, R. J. Poreda, *Science* **1993**, *259*, 1428; h) A. Weidinger, B. Pietzak, M. Waiblinger, K. Lips, B. Nuber, A. Hirsch, in *Electronic Properties of Novel Materials - Progress in Molecular Nanostructures: Xii International Winterschool, Vol. 442* (Eds.: H. Kuzmany, J. Fink, M. Mehring, S. Roth), Amer Inst Physics, Melville, **1998**, pp. 363.

[3] Accepted for publication in Angewandte Chemie (10.1002/anie.201602745R1)

[4] R. N. Grass, E. K. Athanassiou, W. J. Stark, *Angew. Chem. Int. Ed.* **2007**, *46*, 4909.

ADDIN EN.REFLIST

465735 Janus Buckypapers: Bilayer Films of Pristine Graphene Stabilized By Pyrene-Functional Copolymers

Abstract:

One of the recurring problems facing graphene-polymer composites is the need for dispersants to allow for stable graphene exfoliation prior to incorporation into composites. These dispersants tend to be counterproductive and are difficult to remove. To meet this need, we have developed a general strategy for modifying target polymer matrices with pyrene-containing side groups; we have previously applied this strategy to siloxanes (*Nanoscale*, 2014). We now show that this technique may be generalized to other polymers. We present a new approach for preparing novel thin films of melt-processable polymers such as polystyrene (PS) and poly(methylmethacrylate) (PMMA), with one graphene-enriched surface. Random copolymers of 1-pyrenemethylmethacrylate with styrene or methyl methacrylate are prepared by simple, bulk, free-radical polymerizations. After dispersing graphene nanosheets in solutions of these polymeric stabilizers, cast films are prepared, and their electrical properties are characterized for the first time. Applying a leaching approach in conjunction with vacuum filtration increases the overall concentration of graphene in the films, while promoting accumulation of graphene at the surface in contact with the filter. The resulting asymmetric films are the first examples of graphene-containing thin films having widely dissimilar electrical conductivities on the top and bottom surfaces. This report describes the preparation, properties, and possible applications of these "Janus buckypapers," which have one electrically conductive surface and one electrically insulating surface.

469360 Mechanism of Direct Growth of Graphene on Si-Based Dielectric Substrates Via Cu Grain Boundaries

Abstract:

Graphene, a one atom thick sheet of carbon atom in a hexagonal network – possesses an extraordinary combination of properties including ultra-high carrier-mobility, ultra-high thermal conductivity, ultra-fast photodetection, ultra-high sensitivity, tunable spintronics, strong/tunable optical absorption, carrier controlled interband/optical-transition, and quantum interference. To incorporate it into the semiconductor industry, most of the recent techniques rely on producing graphene on metal catalyst surfaces, which requires further transfer to desired dielectric substrates for characterizations and applications. Direct growth of wafer scale, high-quality, and monolayer graphene on silicon-based dielectrics (SiO_2/Si and Si_3N_4/Si) is critical for the next step towards realizing graphene electronics. In this talk, we will show a complete mechanism of graphene synthesis at the interface of thin copper (Cu) film and SiO_2/Si or Si_3N_4/Si dielectrics via low pressure chemical vapor deposition (LPCVD). Further, we will discuss the role of kinetics, and process parameters (temperature, CH_4/H_2 flow rate, and total pressure) in the formation of a uniform, continuous, and large-area thin graphene film on the desired dielectric surfaces. Raman spectroscopy, selective area electron diffraction (SAED), and atomic force microscope (AFM) are employed for graphene' growth rate, continuity, and structural characterizations. The Raman D-peak, G-peak, and 2D-peak occur at characteristic ~1350 cm^{-1}, ~1600 cm^{-1}, and ~2700 cm^{-1} respectively. The I_D/I_G ratio (~ 0.25) corresponds to a graphene crystalline sp^2 domain size of 17 nm. The I_{2D}/I_G ratio (~1.4) and SAED intensity profile confirms the possible formation of graphene monolayer with the optimized conditions. The XPS shows the presence of C=C bonding at 284.5 eV, further confirming the formation of graphene. The electrical transport properties of the produced graphene will also be presented. We envision that this work will advance the high-quality growth of high-quality wafer sized graphene on silicon-based insulating substrates without the need of post-synthesis graphene transfer processes.

467926 An Experimental and Computational Study of the Surface Chemistry Effects in the TiO_2 Grafting of Graphene Oxide

Abstract:

Titanium dioxide (TiO_2) and its nanocomposite films are increasingly being used in the making of supercapacitors and photocatalytic decontamination applications. In this work, we have synthesized graphene oxide (GO) platelets with three different surface oxygen densities using the modified Hummerâ€™s method. The GO platelets were further dispersed in a 2:1 water/ethanol mixture using a bath sonicator for 30 min. Then we added the TiO_2 nanopowder to the mixture and maintained it at 120Â°C for 3 h in a Teflon-sealed autoclave. Finally, we recovered the TiO_2-grafted GO powder by filtration and drying and investigated the covalent bond formation between the TiO_2 and GO by the FT-IR method. The unique aspect of our work is to elucidate the effect of GO surface oxygen density on the TiO_2 grafting process. Furthermore, to gain a molecular insight into the kinetics and thermodynamics of the TiO_2 grafting reaction onto the GO sheet, we performed a reactive molecular dynamics simulation of the TiO_2/GO system in the water/ethanol solution at the prescribed reaction condition. We used the ReaxFF force field with a time step of 0.05 fs to first equilibrate the system at room temperature and then increased the temperature to 120°C. We have analyzed the surface assembly of the TiO_2 particles on the GO sheet and quantified the reaction kinetics as a function of GO surface oxygen density.

470004 Effects of Shear on Carbon Nanotube Suspensions That Are Stabilized with Â Surfactants Using DPD Simulations

Abstract:

Effects of shear on carbon nanotube suspensions that are stabilized with surfactants using DPD simulations

Minh D. Vo, Dimitrios V. Papavassiliou

School of Chemical, Biological, and Materials Engineering

The University of Oklahoma,

Norman, OK -73019

Abstract

Due to strong van der Waals attractions, the agglomeration of carbon nanotubes (CNTs) in solvents is one of the first issues that needs to be resolved when using them in many industrial applications [1]. To overcome this problem, non-covalent stabilization of CNTs with surfactants has been shown to work in experiments. In our work, the stability of adsorded surfactant on CNT surface under shear was investigated in the meso-scopic scale with dissipative particle dynamics (DPD) simulation techniques. The CNT was considered as a close ended cylinder with diameter of 10 nm and length of 50 nm. Following our previous work [2-5], all necessary interaction parameters for the system of water, CNT and commercial surfactants (Alfoterra 123-8s and Tergitol 15-s-40) in DPD calculation were determined. A binary surfactant system with different molar ratio was released into the solution in the presence of a CNT at the center of the simulation box. After the equilibrium of surfactant adsorption was reached, all of supernatant surfactant was

removed from the simulation. Then, a single CNT with adsorded surfactants was released into a Couette flow. Observing the status of adsorded surfactants, the threshold shear rate for surfactant desorption was determined with different molar ratios of binary surfactant. Besides, the mechanism of surfactant desorption was explored. The differences with polymer desorption from carbon nanotubes under shear will be highlighted in the presentation. This work can provide insights for the use of surfactants to stabilize carbon nanotube suspensions in practical applications, where the suspension flows and undergoes shear.

ACKNOWLEDGEMENTS

The financial support of the Advanced Energy Consortium (AEC BEG08-022) and the computational support of XSEDE (CTS090017) are acknowledged.

REFERENCES

1. L. A. Girifalco, M. Hodak, R. S. Lee, Phys. Rev. B (2000), 62, 13101-13110.
2. M. D. Vo, B. Shiau, J. H. Harwell, D. V. Papavassiliou, J. Chem. Phys (2016), 144, 19.
3. Vo, M., and D.V. Papavassiliou, Molecules (2016), 21(4), 500.
4. Vo, M., and D.V. Papavassiliou, Carbon (2016), **100**, 291-301, 2016
5. Vo, M., and D.V. Papavassiliou, Molecular Simulation (2016), **42**(9), 734-744, 2016

465039 Controlling Hybrid Structure Assembly, Solid Interactions and Inorganic Mineralization with Solid Binding Proteins

Abstract:

The integration of combinatorially selected solid-binding peptides (SBPs) within functional protein scaffolds is a powerful approach to produce hybrid architectures with unique properties, control biotic-abiotic interactions and study and manipulate inorganic precipitation and morphogenesis. Here, I will illustrate each of these processes with an example from our laboratory. In the first vignette, one of two SBPs incorporated within the green fluorescent protein (GFP) framework is used to mineralize luminescent Mn-doped ZnS nanocrystals, a chloramphenicol-specific DNA aptamer is coupled to the protein-coated quantum dots via click chemistry and the second SBP is used to tether these structures to silica beads. The unique optical properties of the resulting assemblies are exploited for sensitive detection of chloramphenicol in a format that is amenable to field-testing. In a second example, proteins fitted with a silica-binding peptide become stably entrapped within a silica sol-gel that they help condense. The kinetics and extent of protein release is finely controlled by chemically disrupting the interaction between silica-binding extension and surface and by changing the size and shape of the particles. In the last example, I will describe how mutagenesis of an SBP and its insertion at multiple locations of the GFP scaffold influences titania precipitation and affects non-equilibrium formation of titania patterns.

461745 Programmable, Chemically Mediated Control of Hydrogel Patterning and Chemomechanical Response

Abstract:

Hydrogels are networks of chemically crosslinked polymers containing water as the dispersion medium. The ability to spatially pattern hydrogels with chemical groups or biomolecules and to control their response to stimuli enables their application in areas including chemical separation, controlled release, tissue engineering and the development of therapeutics.

I will describe our group's efforts to develop methods to use biomolecular stimuli such has nucleic acid sequences to direct the formation of heterogeneous patterns of chemical functionality and mechanical stiffness. In each of these methods, designed chemical reaction networks interpret environmental stimuli and induce a response that can be either the instigation of a reaction diffusion process that can produce a stable final pattern of concentrations with a designed, potentially complex shape, shape change or a change in a material's mechanical properties. These methods could allow hydrogels to respond in complex ways to a combinatorial variety of different signals or to respond to biomolecular signals autonomously, making them of potential interest for in vivo applications.

461881 Lipid Nanoemulsions Impact Transport Across Mucosal Surfaces

Abstract:

Mucus is a natural hydrogel that lines the bodyâ€™s wet epithelial surfaces, providing a finely tuned barrier to particulate and pathogen transport. Intestinal mucus, with a nominal pore size of approximately 100 nm, can act as a major barrier to nanocarrier drug delivery systems, yet enables efficient absorption of nutrients and water.

Lipids are endogenously present in mucus at a mass ratio up to 2% and contribute significantly to mucus rheological properties of mucus. Our lab has demonstrated that ingested lipids present in the intestinal lumen, originating from food or lipid-based nanoemulsion delivery systems, for example, significantly strengthen the intestinal mucus barrier to nanoparticle and microbe transport. These observations motivated exploration of the impact of exposure to lipid nanoemulsions on transport of molecular species across the intestinal mucosal surface. We utilized electron paramagnetic resonance (EPR) to study the diffusion and microenvironment characteristics of low molecular weight paramagnetic probe molecules serving as model drugs. Interestingly, EPR spectra indicate that molecular transport in mucus gels is considerably impacted by exposure to lipids, with neutral molecules remaining associated with lipid phases within mucus, positive species being immobilized at the interface of hydrophilic and lipid domains, and negative species being expelled into the aqueous phase within the mucus gel. These results support the conclusion that intestinal mucus can pose a significant barrier to penetration on the molecular scale (< 1 nm), and that significance of mucus barriers to molecular species, as with nanoparticles, is strongly dependent upon charge and dramatically altered by food-associated lipids. To further explore the potential significance of intestinal mucus impact on drug absorption, we utilized ion-selective electrodes to investigate the behavior of supersaturated drug solutions, as are frequently created by certain drug delivery technologies, in intestinal mucus. Intestinal mucus significantly delays the onset of drug precipitation from a supersaturated solution, enabling a prolonged high driving force for drug absorption. Taken together, our results indicate that the intestinal mucus hydrogel can significantly impact the access of nano-scale drug carriers as well as molecular species to the underlying epithelium, and that the barrier posed by this hydrogel is considerably altered in the presence of lipids characteristic of ingested food or nano-emulsion drug delivery systems. The findings strongly support investigation of the impact of the intestinal mucus barrier in design of novel oral delivery technologies.

464036 Observation of Extreme Phase Transition Temperatures of Water Confined inside Isolated Carbon Nanotube Nanopores

Abstract:

Fluid phase transitions inside single, isolated carbon nanotubes (CNT) are predicted to deviate substantially from classical thermodynamics and also allow the study of ice nanotube (ice-NT) properties. Herein, we measure, using two different techniques, the diameter dependent phase boundaries of ice-NTs within 6 isolated CNTs 1.05, 1.06, 1.15, 1.24, 1.44 and 1.52 nm in diameter using Raman spectroscopy. The results reveal both an exquisite sensitivity to diameter and substantially larger temperature elevations of the melting transition than theoretically predicted by as much as 100 °C. Dynamic water filling and reversible freezing transitions were marked by 2 to 5 cm^{-1} shifts in the radial breathing mode (RBM) frequency, revealing reversible melting at 138 and 102 °C for 1.05 and 1.06 nm single and double-walled CNTs, respectively. A near-ambient phase change at 31 and 15 °C was observed for 1.44 and 1.52 nm CNTs, respectively, whereas freezing inside 1.15 nm tube was suppressed at -35 °C. We find that the interior aqueous phase also decreases the axial thermal conductivity of the CNT reversibly by as much as 500%, allowing digital control of the heat flux. These extreme phase transitions enable the study of ice-NT at high temperatures and their potential utilization as novel phase-change materials.

464333 A Molecular Dynamics Study of the Influence of Charge on the Transport of Water and Ions through Carbon Nanotubes

Abstract:

Carbon nanotubes (CNTs) are known to possess smooth hydrophobic walls, can be easily chemically functionalized, and have high surface to volume ratio, thus offering a possibility of unconventionally high flow rates and ion selectivity that makes them potential candidates for nanofluidic and biomimetic applications. However, it is difficult to measure transport processes at nanometer scales and thus much of the experimental data on water and ion transport through nanotubes available so far is limited to larger diameter nanotubes where a predominantly bulk-like behavior is observed. Motivated by the need to understand the superior transport properties of these extremely narrow pores, we perform long time (100 ns) classical MD simulations to investigate in atomic detail the intriguing structural and dynamical features of a NaCl solution under confinement in sub-3 nm diameter carbon nanotubes. Additionally, we assess if Donnan equilibrium theory can be used to ascribe a relation between membrane fixed charges and the concentration of ions within the nanotube.

In this study, the spatial distribution of water and ions and their transport properties in terms of their apparent diffusion coefficients are investigated through uncharged and charged armchair carbon nanotubes ($d = 0.8$ nm – 3.0 nm) embedded between two reservoirs of a 1M NaCl solution. The influence of charges applied to the surface of the CNT on ion-exclusion/screening phenomena is quite apparent and its impact is evident when the pore sizes are comparable to the Debye lengthscale. This can be attributed to the fact that at nanoscopic length scales, because of the large surface to volume ratio as compared with macro and microscopic length scales, surface charge has a pronounced effect on fluid volume in the hydrophobic cavity leading to rejection of ions. We obtain and compare the apparent diffusion coefficients of water and ions in bulk phase to that under confinement and in the vicinity of charges. We find that while the pore sizes that we study are permeable to ions, the axial diffusion coefficient of water and ions decreases under confinement and its effect is most prominent in the smallest diameter charged tube. We also obtain measurements of ion occupancy in the nanotube, coordination numbers of ions, water and ion flux, and ion rejection rates for the various cases considered.

The cases studied give a broad account of ion and water transport properties through uncharged and charged CNTs. The goal is to integrate fast water flow with ion selectivity through modulation of pore sizes, surface charges, and electric fields with potential applications in desalination, as molecular gates for separation of biomolecules, and other nanofluidic devices.

463511 Electrical Transport and Network Percolation in Graphene and Boron Nitride Mixed-Platelet Structures

Abstract:

Percolating network of mixed 2D nanomaterials (2DNs) can potentially leverage the unique electronic structures of different 2DNs, their interfacial doping, manipulable conduction pathways and local traps. Here, we present the percolation mechanism and electro-capacitive transport pathways of mixed-platelet network of hexagonal boron nitride (hBN) and reduced graphene oxide (rGO) – two isostructural and isoelectronic 2DNs. The transport mechanism is explained in terms of electron hopping through isolated hBN defect traps between rGO (possibly via electron 'funneling'). With optical bandgaps of 4.57 eV and 4.08 eV for the hBN-domains and 2.18 eV for the rGO domains, the network of hBN with rGO exhibits Poole-Frenkel emission based transport with mean hopping gap of 1.12 nm (~hBN trilayer) and an activation barrier of ~15 ± 0.7 meV. Further, hBN (1.7 pF) has 6 folds lower capacitance than 1:1 hBN:rGO, which has two orders of magnitude higher resistance than rGO (1.46 MΩ). These carrier transport results can be applied to other multi-2DN networks for development of next-generation functional 2D-devices.

465130 Modulation of Thermodynamically-Stable Surfactant Structures for Selective Desorption of Single Wall Carbon Nanotubes

Abstract:

Over the past decade, the post-synthetic separation of single wall carbon nanotubes (SWCNTs) has been studied with great interest. The most promising techniques include the selective adsorption of SWCNTs onto hydrogel stationary phases and aqueous two-phase extraction. While several researchers have successfully achieved separation of the semiconducting fractions by (*n*,*m*) type by manipulating the selective adsorption of the nanotubes, these processes often require multiple columns or experimental conditions that are difficult to control. However, the high-fidelity desorption of SWCNTs would provide a simpler approach that minimizes dilution and provides higher yields. Here we report the selective desorption of a wide range of single-chirality (*n*,*m*) fractions in a single column packed with agarose gel. High-purity fractions of (*n*,*m*) nanotubes are obtained with high yield once a specific ratio of sodium dodecyl sulfate (SDS)/sodium deoxycholate (DOC) co-surfactant solution is used. In the selective desorption from hydrogels, the co-surfactant structure surrounding the SWCNTs has been shown to play an important role in the separation. The elution of only one (*n*,*m*) type at a specific co-surfactant ratio while other types are exposed to more surfactant suggests that each (*n*,*m*) type forms a thermodynamically-stable surfactant structure in the co-surfactant solution. These thermodynamic equilibrium states result in entropy-driven desorption at specific co-surfactant ratios, enabling high-fidelity separations of a single (*n*,*m*) type in a single column. Moreover, the differences between the co-surfactant ratios needed to separate two (*n*,*m*) species can sometime be very small. The separation can be obtained under a wide variety of co-surfactant background concentrations. An improved separation sensitivity can be achieved at higher background concentrations.

468942 Scaling the Separation of Single-Wall Carbon Nanotubes through Countercurrent Chromatography

Abstract:

Aqueous two-phase extraction (ATPE) is a technique recently demonstrated to successfully separate both surfactant-coated and DNA-wrapped single-wall carbon nanotubes (SWCNTs) by electronic structure, diameter and helical handedness. The differential partitioning of the carbon nanotubes between the two aqueous polymer phases is governed by structure-dependent solvation free energy and can readily be tuned by a number of parameters including polymer composition, surfactant structure, and the presence of salts. Additionally, ATPE is robust, cost-effective and does not require complicated instrumentation. Recent progress will be presented on the improvement, scalability, and automation of the ATPE method for SWCNT separation through the use of countercurrent chromatography.

472038 Ultra-Breathable Carbon Nanotube Pores

Abstract:

Previous reports for pressure-driven transport through CNT membranes demonstrated CNT permeability values that were orders of magnitude larger than those predicted by Knudsen diffusion theory for gases (~10^2 fold enhancement) and by Hagen-Poiseuille equation for liquids (10^3-10^5 fold). While these results spurred great interest in CNTs for efficient membrane separations, it remains an open question if driving forces other than pressure could result in similar transport rate enhancements. A positive answer would greatly extend the promises and application space of CNTs as fluidic channels.

In this work, we provide the first experimental evidence of enhanced gas transport in CNTs driven by a concentration rather than a pressure gradient. We fabricated cm^2, free-standing, flexible, 1-5 nm SWNT/parylene membranes with well-aligned nanotubes as only transporting pores, and we measured the water vapor diffusion rate through the membrane when each surface is exposed to a different relative humidity. Our measurements demonstrate that these membranes exhibit rates of water vapor transport (~8000 gr/m^2day) that surpass those of commercial breathable fabrics, even though the CNT pores are only a few nm wide and the overall porosity is less than 5.5%. Measured permeability of our CNT channels is 24 times larger than Knudsen diffusion prediction, and this flow enhancement is close to that measured for pressure-driven transport of nitrogen.[1] Membranes made from 1-3 nm SWNT forests with higher number densities (> 10^{12}/cm^2) display even larger gas-transport enhancements.

This ultrafast rate of water vapor transport in CNTs suggests that CNT membranes hold great potential for pervaporation, membrane distillation, and as building block of breathable and protective fabrics. For the last application, a membrane shall be able to block dangerous components while permitting perspiration. By demonstrating complete rejection of 3-nm charged dyes, 5-nm uncharged gold (Au) nanoparticles, and ~40-60-nm Dengue virus from aqueous solutions during filtration tests, we provide evidence that, in addition to outstanding breathability, our CNT membranes provide a high degree of protection from bio-threats by size exclusion.[1]

This work was supported by the Defense Threat Reduction Agency (DTRA) D[MS]^2 project under Contract No. BA12PHM123 and was performed under the auspices of the U.S. Department of Energy by Lawrence Livermore National Laboratory under Contract DE-AC52-07NA27344.

[1] N. Bui, E. R. Meshot, S. Kim, J. Pena, P. W. Gibson, K. J. Wu, F. Fornasiero, *Adv. Mater.* **2016**, *accepted*.

472343 Nucleation and Growth of Spontaneously Aligned Regions in Carbon Nanotube Thin Films: A Morphological Analysis

Abstract:

In the ongoing search for scalable organization of carbon nanotubes, deposition of dispersed nanotubes on a filter membrane via pressure-driven flow has long been a promising candidate. Thin films of uniform thickness, from sparse sub-percolation networks to thick freestanding buckypapers, can be fabricated by this method. Minimal apparatus is required and the films can be readily transferred to various substrates.

The films constructed by this method do not generally exhibit any organization or ordering (aside from uniform thickness). However, recent work [1-4] has shown that under specific process conditions, short and long-range ordering of the nanotubes does occur.

As fluid moves across the membrane, the dispersed nanotubes are unable to pass. This results in elevated concentration of tubes near the membrane surface. Furthermore, the velocity of the fluid medium presumably confines the tubes to orient themselves parallel to the plane of the membrane. Tubes near the membrane eventually form domains of uniform local alignment. While the fluid velocity and concentration of the nanotube dispersion are important factors for the emergence of aligned domains, it seems that formation time (time required for the entire volume of fluid to pass through the membrane, leaving dry nanotube membrane behind) is the critical parameter, with organization in the films increasing for longer formation times.

We prepared films with a uniform nanotube density of 0.46 µg/mm2 – approximately enough to form a monolayer of nanotubes on the membrane. Formation times were varied by altering fluid velocity, and by altering the sample concentration. In the first case, the solution volume was held constant at 100 mL and vacuum pressure was varied. In the second, vacuum pressure was held constant (near the maximum level

used in constant volume series) and the sample was diluted to volumes of 100 to 800 mL.

Domains – regions of uniform nanotube orientation – were identified from SEM images by a two-step algorithm. SEM images of the sample were converted to orientation maps, then k-means segmentation was applied to the maps, and continuous regions of the clusters which met threshold criteria for shape and coherency were identified as domains.

Visual inspection of the images and trends in domains identified by the algorithm show that in all cases, the film morphology transitions from many small domains to fewer, and larger domains. However, the effect is more pronounced in the constant fluid velocity case.
Nearest neighbor statistics of the domains show that, for the constant fluid velocity samples, increasing formation time causes the film morphology to transition from one with small domains which are relatively different from their neighbors in terms of orientation angle and size, to films of larger domains, more similar to their neighbors. The same trends in nearest neighbor statistics were not present in the samples of varying fluid velocity, though average domain area did still increase.

Thus it appears that the formation of films with large domains is in part due to the reconciliation of orientation vectors among the small domains which occur when the process is first initiated. This effect was present in the constant fluid velocity samples, but absent in the long formation-time samples with very low fluid velocities.

Further investigation with these methods into other factors such as surfactant concentration will help pinpoint the driving forces behind organization of the nanotubes and guide development of large-scale aligned nanotube films.

References

1. Dan, B., Ma, A. W. K., Hároz, E. H., Kono, J., & Pasquali, M. (2012). Nematic-Like Alignment in SWNT Thin Films from Aqueous Colloidal Suspensions. Industrial & Engineering Chemistry Research, 51(30), 10232–10237. doi:10.1021/ie3001925

2. King, B., & Panchapakesan, B. (2014). Vacuum filtration based formation of liquid crystal films of semiconducting carbon nanotubes and high performance transistor devices. Nanotechnology, 25(17), 175201. doi:10.1088/0957-4484/25/17/175201

3. Oh, J. Y., Yang, S. J., Park, J. Y., Kim, T., Lee, K., Kim, Y. S., ... Park, C. R. (2015). Easy preparation of self-assembled high-density buckypaper with enhanced mechanical properties. Nano Letters, 15(1), 190–197. doi:10.1021/nl5033588

4. He, X., Gao, W., Xie, L., Zhang, Q., Lei, S., Li, B., ... Adams, W. W. (2016). Wafer-scale monodomain films of spontaneously-aligned single-wall carbon nanotubes, (April). doi:10.1038/nnano.2016.44

458657 Understanding the Intrinsic Water Wettability of Graphitic Carbon

Abstract:

Understanding the intrinsic water wettability of graphitic carbon

Andrew Kozbial and Lei Li

Department of Chemical and Petroleum Engineering, University of Pittsburgh, 3700 Oâ€™Hara Street, Pittsburgh, PA 15261

The intrinsic water wettability of graphitic carbon, e.g., graphite, graphene and carbon nanotube, has attracted a lot of research interests in the past three years due to the recent finding that the long-established hydrophobicity of graphitic surfaces actually results from airborne hydrocarbon contamination. Due to the existence of defects, determining intrinsic water wettability of pristine graphitic carbon remains challenging. In the current study, water wettability of freshly exfoliated graphite of varying quality was studied using static, advancing, and receding water contact angle (WCA) measurements. The defect densities were characterized by Raman spectroscopy, XPS and AFM. A qualitative model is proposed to explain the effect of defects on measuring the advancing, static, and receding contact angles. Based on the model and the experimental results, the pristine graphitic carbon is mildly hydrophilic.

471634 Wrapped up in Nanotubes: Probing Photoluminescence Dynamics of Wrapped Single-Walled Carbon Nanotubes (SWCNTs) for Sensing Applications

Abstract:

The last decade has realized an onset of novel optical, SWCNT-based sensors that have been crucial in shaping a distinct generation of unconventional analytical tools. These optical nanosensors exploit the advantageous fluorescence properties of semiconducting SWCNTs, which benefit from stable and sensitive near-infrared emissions capable of deep-tissue sensing and imaging applications. SWCNTs are non-covalently wrapped in a variety of chemical moeities, including DNA, polymers, and proteins, that impart the SWCNT with selectivity towards specific analytes of interest. Recent research endeavors have largely focused on designing wrappings that control the sensor's selectivity by modulating the surface coverage of the nanotube in the presence of an analyte. In addition to selectivity, these wrappings also modulate fluorescence properties such as emission intensity; properties that play an integral role in determining the sensitivity limits of these sensors. We have developed a model that explores the perturbations of SWCNT fluorescence emissions in the presence of a wrapping by de-coupling the structural and chemical contributions of the wrapping. The results of this model have been corroborated by experimental measurements and extended to explore the development of unprecented biohybrid sensors. Taken together, the compoutational and experimental contributions of this work enable a comparative approach towards achieving design rules for the rational engineering of sensor wrappings.

470426 Unlocking Intracellular Therapeutic Targets through Novel Nanostructured Biomaterials

Abstract:

Nucleic acid cargoes offer unmatched diversity in gene regulatory potential and therapeutics, and understanding of nucleic acid functionality continues to expand rapidly and dramatically through seminal discoveries including RNA interference approaches and gene editing technologies. In nature, the basis for gene regulation is ultimately encoded by the exquisite specificity with which cells are able to control both the location and accessibility of nucleic acid constructs to govern their activation states. My research program seeks to mimic these activities through the design of stimuli-responsive synthetic materials whose interactions with nucleic acids and cells can be controlled dynamically by specific intracellular or external triggers. We exploit our ability to regulate binding/release and cellular processing to gain new mechanistic insights over nucleic acid delivery mechanisms.

This presentation will outline recent work in my laboratory to develop peptide and polymer constructs that respond to specific intracellular signals (e.g. enzymes) or exogenous stimuli (e.g. light) to control the release of plasmid DNAs and siRNAs, respectively, and thereby enhance nucleic acid activity while enabling spatiotemporal control over gene suppression/activation. Specifically, in one case, we capitalized on newly recognized and highly pivotal roles for histone tails in native gene regulatory control to develop a gene transfer method that utilizes native, histone-based processing pathways *via* incorporation of post-translationally modified (PTM) histone tails within controllably-assembled DNA vehicles (polyplexes). Our efforts proved that polyplexes displaying PTM-modified histone tails promote nuclear accumulation, DNA release, transcription, and enhanced transfection. Moreover, our group has combined detailed nanostructure engineering with sophisticated cellular imaging to identify novel aspects in the cell biology framework regulating polyplex transport to the nucleus. In another example, we have developed versatile and highly tunable light-based strategies to release siRNAs with control at cellular length scales. Light-triggered delivery materials exhibit innate advantages in spatiotemporal control as compared with chemically responsive delivery materials or strategies to control nucleic acid exposure based upon "stamping" or liquid-liquid patterning, yet existing light-triggered nucleic acid delivery materials have minimal dynamic range in gene modulation. We demonstrated a new class of light-sensitive polymers that provide biocompatibility and rapid application with proven on/off initiation of gene silencing and the ability to locally "dial-in" the level of siRNA deployment and gene regulation. The ability to control the spatial placement and induction of biomolecular signaling programs through this on/off switch will be essential in developing and modeling complex tissues such as cardiovascular and neural networks.

464239 Improving Selective Targeting to Macrophage Subpopulations through Modifying Liposomes with Arginine Based Materials

Abstract:

Introduction: Two pathways for activating macrophages (MΦ) exist. One of these routes is termed the classically activated M1 pathway is achieved through exposure to lipopolysaccharide (LPS). M1 MΦs are part of the type 1 T helper (Th1) response and are known as pro-inflammatory cells. The other pathway is reached through interleukin-4 (IL-4) and is known as the alternatively activated M2 pathway. M2 MΦ produce pro-angiogenic factors. Tumor-associated MΦ (TAMs) are of the M2 pathway and promote tumor growth through the release of angiogenic molecules. Our goal is to use liposomal drug delivery systems to selectively deliver drugs to TAMs. These polymers will be eventually used to delivery anti-cancer therapeutics to the tumor.

Materials and Methods: We fabricated liposomes and modified their surfaces with 14 molecules that are chemically similar to arginine in order to investigate the effects of surface modifications on internalization by macrophages. Also, we characterized the size and surface charge of both modified and unmodified liposomes. Finally, the liposomes were loaded with doxorubicin, an anti-cancer drug, and were examined for their loading efficiency and release kinetics at pH 7.4. Changes in the $IC_{50}$of doxorubicin entrapped in the unmodified and modified liposomes was compared to that of free doxorubicin in M(LPS), M(IL-4), and M(0) macrophages.

Results and Discussion: Liposomes were loaded with doxorubicin and incubated with M(LPS), M(IL-4), and M(0) cells. Dose response curves were fit to a Sigmoidal curve and the half-maximum of inhibitor concentration (IC_{50}) was obtained. The lower IC_{50}values for M(IL-4) cells compared to M(LPS) cells suggests that modifications H and I could be used to improve targeted delivery to TAMs.

PCA was applied to the dataset to determine the relationships between physicochemical properties of the modifiers and the IC_{50} values. The multidimensional dataset was reduced to a two-dimensional plot to better facilitate analysis of latent relationships. The physicochemical properties were chosen based on previous reports describing attributes of drug molecules such as Lipinski's rule of five, polar surface area, flexibility, enthalpy, lipophilicity, and charge. Macrophages were observed to have different correlations with these physicochemical properties based on their phenotype. The IC_{50} values for both M(LPS) and M(0) cells were well aligned on the projection map and were situated between the number of hydrogen bond donors and the number of freely rotating bonds of the modifiers. The M(IL-4) cells were dependent on the zeta potential of the liposomes and the logP (the partition coefficient of the modifier in octanol and water) of the modifiers. Taken together, these insights may elucidate design principles in drug delivery targeted to specific macrophage phenotypes. Further work on larger library is necessary to determine if these relationships between the identified materials properties and IC_{50}values hold for different macrophage polarizations.

Conclusions: Cellular uptake of the liposomes was found to be dependent upon macrophage phenotype and surface modifications. There were also differences in trends between internalization of liposomal FC and the IC_{50} of liposomal doxorubicin, which were attributed to changes in the ability of doxorubicin to escape the endosome. Two modifications were able to increase the toxicity of encapsulated doxorubicin for M(IL-4) cells over M(LPS) cells, improving targeted deliver to specific macrophage subpopulations.

458472 Harnessing Biomaterials to Study and Engineer Lymph Node Function

Abstract:

Vaccines and immunotherapies have generated some of the largest impacts on human health in history, but a fundamental challenge now facing the field is how to direct the specific properties of immune responses that are elicited. This idea of tuning response is termed â€œimmunomodulationâ€, and is critical in designing more efficacious and specific vaccines and immunotherapies. In this seminar I will discuss two strategies we are developing to study and exploit the interactions between biomaterials and immune cells and tissues. One approach involves direct delivery of synthetic vaccine carriers to lymph nodes, key tissues that coordinate immune response. We have combined direct lymph node injection with biomaterials to establish a platform to study the link between local lymph node function and systemic immunity by probing the roles of signal density and material properties. In addition to these ideas, we are exploiting directed delivery for therapeutic vaccination in the areas autoimmunity and cancer. The second focus area is the design of new modular materials we have created using polyionic immune signals to form stable vaccine capsules. These immune polyelectrolyte multilayers (iPEMs) are self-assembled entirely from antigens and adjuvants to allow selective activation of pro-inflammatory signaling pathways without other carrier components such as polymers or lipids. In mice, iPEMs injected along traditional vaccination routes enhance the function of dendritic cells in draining lymph nodes, potently expand antigen-specific T cells against antigens used to build iPEMs, and provide protection during tumor challenge. Ultimately, these strategies could contribute to better understanding of the interactions between biomaterials and the immune system, and improve the rational design of materials that serve not only as carriers, but also as agents that actively direct immune response.

471175 Injectable Hydrogels for Tandem Cell/Gene Transplantation

Abstract:

Cell transplantation by direct injection is a minimally invasive approach for the treatment of many ischemic diseases including stroke, peripheral arterial disease, and myocardial infarct. For example, adipose-derived stromal cells (ASCs) have been shown to promote neovascularization in preclinical models of ischemic tissues. Unfortunately, the direct injection of ASCs into ischemic tissue is limited by low cell viability and poor cell function. The transplanted cells are exposed to strong mechanical forces during injection (which results in cell damage) and low oxygen levels after injection (which results in further cell death and compromises cell function). To overcome these challenges, we have developed a family of injectable hydrogels that protects cells from mechanical forces during injection and provides mechanical scaffolding and biochemical cues to promote long-term cell survival and potency within ischemic tissues. At room temperature, all gel formulations are weak hydrogels (Gâ€™ <50 Pa) that form due to peptide-peptide heteroassembly to form a supramolecular network. These weak hydrogels undergo rapid shear-thinning to be easily hand injectable and rapid self-healing to localize transplanted cargo at the desired target site. Upon heating to 37Â°C, the hydrogels undergo a secondary crosslinking mechanism due to lower critical solution temperature (LCST) behavior. Simply tuning the weight fraction of the LCST polymer from 0-5 wt% results in hydrogels with tunable stiffness, Gâ€™ ~ 100 - 1000 Pa. Current research efforts focus on tuning the biochemical and mechanical cues provided by this delivery platform to optimize the hydrogel for tandem delivery of ASCs and therapeutic gene vectors to significantly enhance the angiogenic potential of transplanted cells.

469094 Tailoring the Mechanical Properties of Multi-Functional Polyampholyte Hydrogels for Tissue Engineering Applications

Abstract:

The development of novel polymeric scaffolds for tissue engineering depends on the successful demonstration of three critical features: resistance to nonspecific protein adsorption, the capability to deliver bioactive signaling molecules over both short and long term timelines, and the ability to easily tune the mechanical properties to match the tissue of interest. While there are numerous polymeric platforms that can address these three features individually, there are few, if any platforms that can adequately meet all three. Our work is focused on demonstrating the capability of polyampholyte polymers composed of equimolar concentrations of positively and negatively charged subunits to meet all three criteria. Polyampholyte hydrogels composed of varying ratios of [2-(acryloyloxy)ethyl] trimethylammonium chloride (TMA, positively charged), 2-carboxyethyl acrylate (CAA, negatively charged), and 3-sulfopropyl methacrylate (SA, negatively charged) monomers were synthesized. Regardless of the underlying composition, the hydrogels were demonstrated to be resistant to nonspecific protein adsorption from both negatively charged fibrinogen and positively charged lysozyme. All of the different hydrogel compositions were also demonstrated to be capable of covalently attached bioactive molecules using EDC/NHS conjugation chemistry. Finally, the mechanical properties of the resulting hydrogels were demonstrated to be dependent upon the composition, allowing for tunability in this property. When these results are coupled to the literature regarding polyampholyte polymers, it can be concluded that these systems represent a promising approach for tissue engineering applications.

465836 Cell Delivery Systems Via Complex Emulsion Templated Hydrogels

Abstract:

Multiphase emulsions with co-continuous and tunable internal microstructure are formed via colloidal silica self-assembly during spinodal decomposition in partially miscible binary solutions. Silica particles tuned to mutually wet the respective fluid phases strongly absorb at the interface at the onset of temperature-driven phase separation. Demixing comes to a kinetic halt once the interfacial area is sufficiently reduced to only accommodate particles, and the mixture undergoes a remarkable gelation transition. This now kinetically arrested, bicontinuous interfacially jammed emulsion gel (bijel) is comprised of co-continuous, fully percolating channels separated by a silica monolayer. Bijels provide a tunable template for synthesizing uniquely structured soft materials containing penetrating microchannels with consistent curvature. By selecting biocompatible polymers, the inherent properties of these templated materials can be utilized for probing applications in regenerative cell delivery and tissue engineering. We have synthesized a novel cell delivery system that contains fully penetrating and uniform void channels with tunable diameter for cell loading and unloading; exhibits robust mechanical properties brought about by the repetitive structure of the percolating polymer phase; and can be filled with common cell-friendly, soft matrix environments. These particular properties are exclusively afforded by this new class of soft materials because of their unique microstructure. In this talk, we will demonstrate how these bijel-templated hydrogels are utilized as cell delivery systems, with particular emphasis on cell seeding and unloading via chemoattractant gradients. Further, we will discuss the broader applications of these bijel-derived scaffolds for tissue engineering, particularly their application as biocompatible, densely vascularized membranes for therapeutic tissue implants.

463543 Non-Invasive Structural Investigation of Renal Scaffold By Magnetic Resonance Imaging (MRI)

Abstract:

Non-invasive structural investigation of renal scaffold by Magnetic Resonance Imaging (MRI)

Nafiseh Poornejad, Jonathan J. Wisco, Beverly L. Roeder, Alonzo D. Cook

Brigham Young University, Provo, Utah

Naturally obtained scaffolds combined with patient-specific cells could result in improved regeneration of human tissues. Scaffold fabrication is the first step in this technology, which is decellularization of the native tissue and sterilization of resulting scaffold. Decellularization, as the most effective procedure, could be done by exposing the native tissue to cell lysing agents that remove all cellular materials leading to the intact collagenous structure of extracellular matrices (ECMs). The majority of cellular remnants should be withdrawn to avoid immune response induction while the ECM should remain intact. In complex organs such as the kidney, the ultrastructure is of significant importance since it is essential for the filtration function of the kidney. Examining the fabricated ECM for structural integrity and intactness of the vascular network was pursued in this study via a novel method.

In this study, we decellularized whole porcine kidneys via our previously developed fabrication method (1-3) and investigated the ultrastructure of renal tissue by magnetic resonance imaging (MRI). Initially, we intended to perfuse a solution of nano iron particles, as contrast medium, through the vascular network to obtain high-resolution imaging with MRI. However, the nanoparticles would diffuse out of vessels in a decellularized ECM because of the lack of endothelial cells lining the vessels. So instead, human renal cells were loaded with nano iron particles and perfused through the renal artery (to image the vascular network) and renal ureter (to examine tubule structures) of native and decellularized porcine kidneys. The MRI scanning was performed on a 3 Tesla MR scanner (Trio, Siemens, Erlangen, Germany) with a 32-channel head matrix coil. The obtained results confirmed structural intactness of decellularized ECM compared with native tissue.

In summary, we developed a feasible method to investigate the vascular network and ultrastructural integrity of any decellularized organ including kidney, heart, liver, and lung by perfusing the organs with cells loaded with nano iron particles, and imaging the organs using MRI.

1. Poornejad, N., Frost, T.S., Scott, D.R., Elton, B.B., Reynolds, P.R., Roeder, B.L., andCook, A.D. Freezing/thawing without cryoprotectant damages native but not decellularized porcine renal tissue. Organogenesis **11**, 30, 2015.

2. Poornejad, N., Nielsen, J.J., Morris, R.J., Gassman, J.R., Reynolds, P.R., Roeder, B.L., andCook, A.D. Comparison of four decontamination treatments on porcine renal decellularized extracellular matrix structure, composition, and support of human renal cortical tubular epithelium cells. Journal of biomaterials applications, 0885328215615760, 2015.

3. Poornejad, N.M., N.; Salehi, A.S.M.; Scott, D.R.; Fronk, C.A.; Roeder, B.L.; Reynolds, P.R.; Bundy, B.C.; Cook, A.D. Efficient decellularization of whole porcine kidneys improves reseeded cell behavior. Biomedical Materials **11**, 025003, 2016.

467849 Bi-Modal Porous Poly(Îµ-caprolactone) Scaffolds Fabricated Via Two-Step Depressurization Supercritical CO_2 Foaming

Abstract:

Poly(Îµ-caprolactone) (PCL) is widely used as scaffold in bone tissue engineering due to its lasting degradation rate that matches the rate of natural bone growth. To mimic the extracellular matrix environment of natural bone tissue, emphasis has been laid on the pore morphology and structure of PCL scaffolds, and it demonstrates that a macroporosity of a few hundred microns is beneficial for cell adhesion, proliferation and differentiation; while a microporosity of tens of microns is necessary for nutrients and waste transportation. Therefore, formation of a bi-modal porosity is of particular interest in the fabrication of novel PCL scaffolds. Supercritical carbon dioxide foaming (sc-CO_2 foaming) is an alternative as solvent-free green process for scaffold fabrication. In this paper, a novel two-step depressurization sc-CO_2 foaming was designed, in which two different depressurization rates were utilized in one single foaming process to induce pores formation at different scales and attain bimodal porous scaffolds. Effects of important processing parameters, such as foaming temperature, saturation pressure, intermediate pressure, saturation time, holding time and depressurization rate etc., on pore morphology and structure of PCL scaffolds were investigated in details. Results indicated that intermediate pressure had dominant impact on the final pore morphology. Decreasing the intermediate pressure from 12 MPa to 5 MPa, the macropore size would increase from 130 Î¼m to 404 Î¼m. Interestingly, when intermediate pressure reached 9 MPa, only monomodal porosity was observed. Studies implied that nucleation size and macropore density was mainly determined in the first depressurization step by intermediate pressure and pore growth was more affected by the second depressurization. In summary, two-step depressurization sc-CO_2 foaming is verified available to produce bimodal porous PCL scaffolds with suitable pore structure and fine porosity. The plasticizing effect of sc-CO_2 during foaming causes the reduction of PCL polymer glass transition temperature (T_g) and melting point (T_m), which avails processing polymers at a relatively low operating temperature,

making it potential in protein growth factor incorporation for cells growth.

470724 Controlled Released Antibacterial Ag/Poly (L-lactic acid)/Poly(vinyl alcohol) (Ag/PLLA/PVA) Coreâ€"Shell Nanofibers Prepared By Cold Atmospheric Plasma (CAP) Treatment and Electrospinning

Abstract:

Introduction: Electrospinning is a well-established process for the fabrication of nanofiber mats of high surface areas and high porosities. This process has been widely used for low-cost wound dressing production [1]. Silver is another widely-used and recognized effective anti-bacterial agent. Generally, the simplest and most commonly used method for incorporating silver nanoparticles into electrospun nanofibers is by the suspension of silver nanoparticles into polymer solutions. However, nanofibers produced by this method have reduced antibacterial properties due to silver nanoparticle aggregation. In order to avoid this problem, the objective of the present study was to utilize cold atmospheric plasma (CAP) [2], which is an ionized gas with ions, electrons, and excited atoms, as a reducing agent for photosensitive silver salts.

Materials and methods: Highly porous core-shell nanofibers were generated by rapid phase separation during the electrospinning process. A 10% PLLA solution was prepared using poly (L-lactic acid) (Sigma, U.S) with 1% silver nitrate (w/w) (Sigma, U.S). Then, the core-shell nanofiber was electrospun with a two-fluid coaxial spinneret (Inovenso Ltd. Co, Turkey), inner tube diameter of 1.2 mm, and outer diameter 1.4 mm, at a voltage ranging from 12-15 kV applied at the tip of the spinneret. The spun fibers were collected on a rotating collector at a speed of 300 rpm. Samples were then treated by cold atmospheric plasma at different times (30s, 60s, 90s, and 180s). Treated and untreated samples were observed by scanning electron microscopy (Hitachi, S-3700N, US). Anti-bacterial experiments and cell cytotoxicity studies will be conducted and reported later. All experiments were run in triplicate and repeated three times for each group with statistics analyzed by ANOVA followed by student t-tests.

Results and discussion: The topography of the fibers plays an important role in regulating initial cell behavior (such as cell adhesion and proliferation) due to changes in surface energy. SEM images of untreated and CAP treated nanofibers showed that the diameter of electrospun nanofibers ranged from 400 nm to 900 nm. Nanopores could be observed on the nanofiber surface, which indicated higher surface areas compared to nanofibers without nanopores. The diameter of nanopores on the untreated surface ranged from 5 nm to 20 nm. However, the nanopores became larger as CAP treated time increased. For 30s of CAP treatment, nanopores on the surface enlarged to about 30 nm in diameter, and 40 nm with 60s of CAP treatment. In addition, it is clear that the color of the electrospun polymer matrix gradually turned darker with increased durations of CAP exposure, indicating that silver nanoparticles were generated during CAP treatment.

Conclusions: In this study, a nanoporous core-shell nanofiber matrix was fabricated by an electrospinning method. Anti-bacterial silver nanoparticle reduction was successfully prepared by CAP treatment of the electrospun scaffold. In addition, results here showed for the fist time that the CAP treatment process is an effective tool to increase nanopore size on the surface of the nanofibers, resulting in increased surface areas, which will affect surface energy, initial protein adsorption events and cell adhesion as well as proliferation.

Acknowledgement: This study was supported by Northeastern University.

References: [1] D. Reneker, A. Yarin. Polymer, 2008, 49 (10), 2387â€"425.

[2] M. Wang, X. Cheng, W. Zhu, et al. Tissue engineering: part A, 2014, 5(20), 1060-1071.

461879 Peptide-DNA Hybrid Nanomaterials for Biology and Regenerative Medicine

Abstract:

Peptide-DNA Hybrid Nanomaterials for Biology and Regenerative Medicine

Ronit Freeman[1], Nicholas Stephanopoulos[2], Samuel I. Stupp[1,3,4,5]*

(s-stupp@northwestern.edu)*

1. Simpson Querrey Institute for BioNanotechnology, Feinberg School of Medicine, Northwestern University, Chicago, IL 60611, United States

2. Current address: School of Molecular Sciences, Center for Molecular Design and Biomimetics, The Biodesign Institute Arizona State University, Tempe, AZ 85287, United States

3. Department of Materials Science and Engineering, Northwestern University, Evanston, IL 60208, United States

4. Department of Chemistry, Northwestern University, Evanston, IL 60208, United States

5. Department of Medicine, Northwestern University, Chicago, IL 60611, United States

Peptides and DNA represent two of the most attractive categories of molecules for the construction of nanomaterials for biology and medicine. Peptides provide a rich palette of biological functionality and self-assembly behavior, and DNA can be used to construct complex nanostructures with programmable, dynamic properties. We sought to merge the advantages of these two molecular platforms through the use of peptide-DNA (P-DNA) hybrid materials. Here we will outline recent progress in three distinct areas, each harnessing the unique potential of DNA as a novel biomaterial.

The first system used a DNA nanotube scaffold to present cell-adhesive peptides in a multivalent fashion (Figure 1A). Neural stem cells adhered to these DNA-peptide structures and differentiated selectively into neurons. Both the nanostructure and the biological signal could be independently controlled.

The second system involves the synthesis of peptide-DNA hybrid molecules that can be linked to non-bioactive substrates through Watson-Crick pairing to an immobilized complementary strand. The use of these molecular modifications on cell substrates has enabled us to manipulate different matrix functions, (Figure 1B): (1) induce reversible biological cell adhesion over "multiple cycles" using three orthogonal mechanisms; (2) optimize at the nanometer scale the distance between two signals in the matrix for synergistic signaling of the cell; and (3) probe the individual roles of two localized signals in the matrix. This system is the first to combine all three of these matrix functions using a single molecular platform. In particular, the platform described demonstrates the possibility of dynamic features on cell matrices not possible with previous approaches, which are either not reversible or use potentially harmful stimuli to the cell such as ultraviolet light.

This system provides a fundamentally new paradigm for DNA as a functional linker to mimic diverse ECM properties, and the orthogonality and programmability of DNA makes this platform highly attractive for a wide range of biological applications. In addition, this system can be readily interfaced with other DNA nanotechnology-based constructs, which have as of yet found only limited application in regenerative medicine and related fields.

A third project appended DNA or PNA(Peptide Nucleic Acids) to peptide amphiphiles (Figure 1C). These hybrid materials merge the dynamic properties of DNA with the bioactive and self-assembly properties of peptides. The resulting constructs were tunable and reversible in both their physical and biological properties due to the functionality of the DNA, allowing for dynamic signal presentation, reversible gelation and programmable hierarchical self-assembly of peptide-based filament structures.

The formation of 3D peptide amphiphile gels using the nucleic acids as cross-linkers involved the formation of long and dense bundled fibers that are enriched with nucleic acids while the non-bundled fibers are DNA-poor. The dimensions of the fiber bundles, and as a result, the mechanical properties of the gel network can be dynamically tuned by the number of DNA crosslinks. This dynamic 3D microenvironment provides new opportunities for dynamically altering different features and testing their effect on cell behavior.

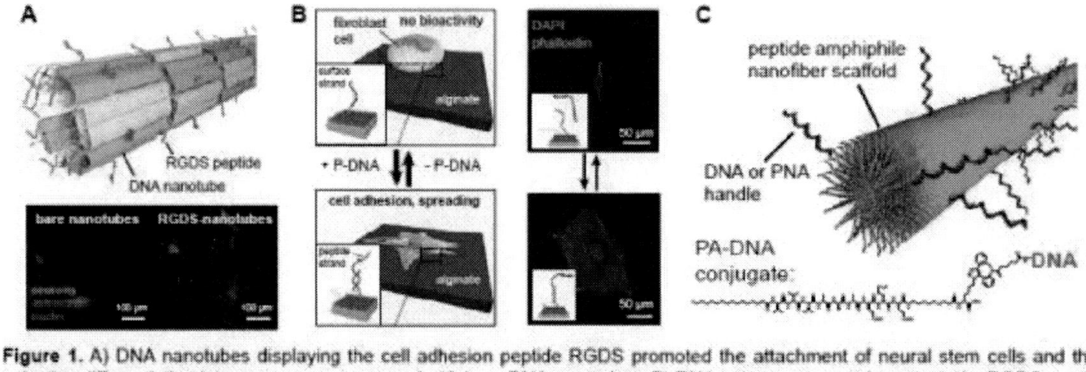

Figure 1. A) DNA nanotubes displaying the cell adhesion peptide RGDS promoted the attachment of neural stem cells and their selective differentiation into neurons, as compared with bare DNA nanotubes. B) DNA tethers were used to attach the RGDS peptide to non-bioactive alginate surfaces. The DNA allowed for reversible, dynamic presentation of the peptide by using toehold-mediated strand displacement. C) Peptide amphiphile (PA)-DNA molecules were co-assembled with unfunctionalized PAs to create hybrid nanofibers with handles for cross-linking into hydrogels or reversible presentation of biological signals.

463446 Vapor-Phase Eta-6 Functionalization of Graphene with Retained Charge Carrier Mobility

Abstract:

To widen the spectrum of its applications, it is important to functionalize graphene, while preserving its superior properties, and retaining its planar lattice (for high mobility) and its carbonsâ€™ sp^2 hybridized state (for high carrier density). Such a functionalization mechanism, when conducted in compliance to the needs of semiconductor manufacturing processes will enable grapheneâ€™s incorporation into diverse applications. Here, we develop a unique eta-6 organometallic approach to functionalize graphene in a vapor-phase process, which retains the structural and electrical properties, while offering chemical sites for interaction and interfacing with other chemical or biochemical systems. In contrast to other functionalization processes, the eta6-functionalized graphene maintained its high charge carrier mobility (1000 $cm^2V^{-1}s^{-1}$ at 300 K). We will discuss the mechanism of charge transfer in eta-6 functionalization of chromium carbonyl on graphene. This process will unveil grapheneâ€™s previously unknown potential to hierarchically interface with physical and biological components to produce novel systems and applications. Results will also facilitate gate-fabrication for FETs *via* atomic-layer-deposition (currently a major challenge).

463474 Electrical Properties of Controlled, Longitudinal Wrinkles on Graphene Produced Via Bacterial-Scaffold Shrinkage

Abstract:

Graphene interfaced with biological cells is an important system with applications in cell actuated sensors, cell-driven field-effect-transistors (FETs), cell-excretion based FETs, and cell electrochemical transponders. However, little is known about cell induced mechanical actuation (such as wrinkling) of graphene. For example, local pi-orbital stretching, dipolar doping, and/or carrier puddling caused by wrinkles in graphene directly influence its electronic and phononic properties. Here, we show that bacteriumâ€™s high surface energy, transportable volatile content and shrinkable microstructure can induce controlled and confined wrinkles on interfaced graphene sheets. The relaxation of pre-stretched bacterial cell in vacuum results in graphenic wrinkles to orient in the longitudinal direction to the *strepto-bacillus cereus* cells with a texture aspect-ratio of 0.125. Coarse-grained molecular dynamics (CGMD) simulations suggest that tension in graphene prompts wrinkle formation with wavelength of ~34 nm, consistent with the observed wavelength of 32.4 â€“ 34.3 nm. This talk will (a) demonstrate directed electrophoresis of bacterial cells between electrodes for position-controlled 2D-wrinkle placement, and (b) discuss the electron density distribution and transport properties through wrinkled graphene. These graphenic bio-interfaced wrinkles can lead to novel nano/bio microelectromechanical systems with applications in electrical cell actuation, dehydration, sensing, restraining, retention, electro-microfluidics, and controlled delivery.

468891 Size-Controlled, Surface Functionalized Graphene Sheet Deposit Films for Enhanced Performance of Supercapacitors

Abstract:

Abstract for 2016 AICHE

Size-controlled, Surface Functionalized Graphene Sheet Deposit Films for Enhanced Performance of Supercapacitors

Gyoung G. Jang and Michael Z. Hu

Energy and Transportation Science Division, Oak Ridge National Laboratory, Oak Ridge, TN 37831

ABSTRACT

Sub-micron-sized graphene oxide flake particles were grafted with an aromatic amine and hydroxyl group molecule (2,5-diamino-1,4-dihydroxylbenzene dihydrochloride (DDDC)) and was then reduced to form DDDC ligand-grafted reduced nano-GOs (nano-DDDC-rGO). DDDC serves as an interlayer spacer prohibiting stacking/aggregation and provide hydrophilicity, resulting in well dispersed slurry that is suitable for deposit uniform thin films on a circular disc substrate of carbon fiber paper (as current collector). The surface-modified graphene sheets enhanced surface usability and diffusion and accessibility of electrolyte ions by shortening transport paths (compared with horizontally stacked tenth micron size graphene sheets). By controlling the graphene size and surface property, nano-DDDC-rGO film electrodes showed enhancement in electrochemical performance (i.e., higher specific capacitance, faster electron and ion transport kinetics, better contact between current collector and active material), compared to the baseline films with larger sized (tens of micrometers) graphene flake particles. With aqueous electrolytes (e.g., 1M H_2SO_4), up to ~2.4 times higher specific capacitance and ~67 % reduced equivalent series resistance compared to the micron-sized DDDC-rGO were achieved. With a large molecule sized organic electrolyte (e.g., 1M tetraethylammonium tetrafluoroborate in acetonitrile), the size control enhancement was more effectively achieved by ~4.2 times higher capacitance, 4.0 times lower IR drops and order of magnitude enhanced response time. Therefore, controlling the particle size in the graphene film deposits could be an effective technique to tailor the property of graphene film-based electrodes for high performance supercapacitor.

Keywords: Nano sized graphene, Ligand grafting, Reduced graphene oxide, Supercapacitor, Graphene functionalization

459466 Electronic, Mechanical, and Thermal Transport Properties of Hydrogenated Irradiated Graphene

Abstract:

Defect engineering through irradiation processes and chemical functionalization of graphene are promising routes for fabrication of carbon nanostructures and two-dimensional metamaterials with unique properties and function. In previous computational studies, we reproduced experimentally observed structures of irradiated graphene sheets through introduction of random distributions of vacancies in the honeycomb lattice of graphene and proper structural relaxation. We found that a vacancy-induced crystalline-to-amorphous transition in graphene occurs for an inserted vacancy concentration between 5% and 10%. This amorphization transition is accompanied by a brittle-to-ductile transition in the mechanical response of the irradiated graphene sheets as well as introduction of localized electronic states near the Fermi level.

In this presentation, we use hydrogenation of irradiated, including irradiation-induced amorphous, graphene as a means of studying the effects of chemical functionalization on its electronic structure and thermomechanical response.. We use molecular-dynamics simulations according to a reliable bond-order potential to prepare the hydrogenated configurations and to carry out mechanical tests of dynamic deformation of the prepared configurations at constant strain rate and temperature. We find that hydrogenation does not affect the ultimate tensile strength (UTS) of the irradiated graphene sheets if the hydrogenated C atoms remain sp^2-hybridized. However, upon inducing sp^3 hybridization of these C atoms, UTS decreases by about 10 GPa; in spite of this decrease, these hydrogenated defective structures remain remarkably strong. We also find that the fracture strain of the irradiated structures decreases by up to 30% upon hydrogenation independent of the hybridization type. We characterize the fracture mechanisms and analyze the embrittlement effects of hydrogen. In addition, we report results from thermal transport analysis of the defective graphene sheets with emphasis on the effects of defect density and the extent of hydrogenation and the resulting hybridization of the hydrogenated C atoms on the thermal conductivity of the graphene sheets, computed based on an equilibrium Green-Kubo approach. Finally, based on first-principles density functional theory calculations of the electronic structure of unpasssivated as well as hydrogenated irradiated graphene structures, we assess the potential for tuning the electronic properties of these defective, functionalized graphenes.

458101 Characterization of Effects of Systematically Increased Graphene Dip Coating Durations on the Surface Morphology

Abstract:

The unique electrical and thermal properties of graphene has triggered numerous applications in optics, electronics, sensors and solar energy. These properties have resulted in devices that cater to water filtration, aerospace components, flexible electronics and photovoltaic cells. However the application of graphene for heat transfer enhancement is still relatively unexplored. In the recent years, graphene based coating materials as a result various graphene manufacturing techniques has gained a lot of attention. In this paper, we present a) detailed characterization of electrochemically generated solution consisting of graphene oxide (GO) and graphene and, b) dip coating of GO and graphene solution on copper substrate at various durations, and correlating the dip coat times with resultant surface morphology c) heat transfer performance of dip coated samples. The graphene oxide (GO) and graphene solution was characterized using dynamic light scattering and Fourier Transform Infrared (FTIR), and thermogravimetric analysis (TGA). The graphene oxide (GO) and graphene solution dip-coated copper surfaces were using X-ray diffraction (XRD), Fourier Transform Infrared (FTIR) spectroscopy, scanning electron microscopy (SEM) and X-ray spectroscopy (EDS).

465750 Microwave Induced Welding of Carbon Nanotube-Thermoplastic Interfaces for Enhanced Mechanical Strength of 3D Printed Parts

Abstract:

Additive manufacturing or 3D printing is a burgeoning method for manufacturing mechanical components; these parts however, are typically not intended for load-bearing use due to incomplete bonding between successively deposited layers. We report a novel method for improving the bond strength of 3D printed thermoplastic interfaces using intense localized heating of carbon nanotubes (CNTs) upon microwave irradiation. The microwave heating mechanisms of CNT-polymer composites were thoroughly investigated by analyzing the microwave dielectric properties of a range of CNT contents both above and below the percolation threshold. These results were coupled with in-situ thermal imaging of the composites during microwave irradiation and validated with computational simulations.

470522 Morphology of Carbon Nanotube Liquid Crystal Solutions

Abstract:

Carbon nanotubes (CNTs) have a remarkable combination of electrical, thermal and mechanical properties. In order to realize these properties in applications, CNTs need to be processed into scalable materials such as fibers, coatings, transparent films, and foams. Fluid-phase self-assembly is one of the most promising ways to process CNTs. Chlorosulfonic acid (CSA) has shown to be a true solvent for CNTs. CNTs, similar to other rigid rod molecules, form nematic liquid crystal phase at high concentrations. Low concentration solutions are used for making transparent films, while high concentration solutions can be spun into multifunctional fibers. Successful assembly of CNTs needs a better understanding of the phase behavior and the flow process. Here, we present our recent results on studying the morphology of CNT liquid crystal solutions and how it is affected by the CNT properties, concentration and flow process. We used techniques including polarized optical microscopy and neutron scattering to have a better understanding of the ordering inside the liquid crystal phase. Our polarized optical microscopy results show that for short CNTs at intermediate concentrations, spindle-shaped nematic droplets emerge out of the isotropic phase, while long CNTs show bi-continuous phases. Additionally, high concentrations of long CNTs show higher ordered phases while short CNTs present characteristics of the nematic phase.

472104 Selective Extraction of Carbon Nanotube Enantiomers By Specific DNA Sequences

Abstract:

Aqueous two-phase extraction (ATPE) is a robust and cost-effective technique recently demonstrated to separate a broad range of semiconducting and metallic single-wall carbon nanotube (SWCNT) species. We find that the partitioning of DNA-wrapped carbon nanotubes is highly sequence dependent and can further be controlled by the addition of salts and polymers. A comprehensive search of over 300 individual sequences was conducted, leading to the isolation of 20 individual SWCNT species. We apply circular dichroism to additionally show that each purified species is enantiomerically enriched. Enantiomeric differentiation by DNA is highlighted by a specific sequence which is able to isolate both enantiomers of the (6,5) species in record high purity.

465467 Controlling the Spatial Distribution and Orientation of Carbon Nanotubes and Graphene Nanoribbons in Polymer Nanofibers for Future Application in Li-Ion Battery Anode

Abstract:

Spatial distribution and orientation of nanoparticles (NPs) in polymer matrices has a vital effect on intrinsic properties of advanced polymer nanocomposites. Here, we demonstrate that without surface modification, we can control the spatial orientation and dispersion of 1-D and spherical NPs in polymer matrices by applying a circumferentially uniform air flow through the sheath layer of a coaxial nozzle during gas-assisted electrospinning (GAES). The ability to tune the dispersion and orientation of various NPs was demonstrated by varying the air flow rate in GAES of 10 wt.% aqueous solution of Poly(vinyl alcohol). First, carbon nanotubes (CNTs) and their unzipped counterpart graphene nanoribbons (GNRs), and later spherical SiO_2 and Si were investigated in the current study.

To quantify the dispersion, TEM images of microtomed longitudinal sections of NFs were examined using the dispersion area analysis (DAA), calculating the area occupied by NPs relative to the total NF area. Then, the fast Fourier transform (FFT) was performed to obtain the highest probability separation between individual NPs and/or agglomerates. Finally, the alignment analysis (AA) was carried out to study the orientation of 1-D NPs in the flow direction as a response to higher deformation.

Our analysis indicated a 90% and 350% improvement in dispersion area and a 50% and 75% decrease in separation between NPs and/or agglomerates, with the application of additional extensional air flow, for CNTs/GNRs and spherical NPs respectively. This behavior is well explained through a rigorous time scale analysis that related deformation and diffusion rates at the initial stage of fiber formation and revealed the existence of two dispersion regimes. During the first, the diffusion-enhanced dispersion mechanism is leading and competing with the deformation-induced dispersion mechanism ($0.2 < Pe < 6.77$). However, it was empirically observed that at the intermediate assisting air flow rate ($Pe = 6.77$), there is an apparent increase in dispersion area and a sharp decrease in the most probable separation, which is the evidence of the second dispersion regime, where deformation-induced dispersion is a dominant mechanism. At the same intermediate air flow rate there was a 60% improvement in the alignment of 1-D NPs with the flow direction. This behavior can be well comprehended by studying the orientational component of NP motion in polymeric fluid, the time scale of which will significantly decrease when the additional deformation is applied. CNTs have higher rigidity and would faster and more uniformly respond to deformation and align with the flow. On the other hand, more flexible GNRs tend to inhomogeneously respond to additional deformation, aligning along the flow direction at lower air flow rates, however forming coils at higher.

Our experimental observations are further supported by the results from the Li-ion battery performance study. We tested the life cycle of half-cells of anodes based on Si NP/CNT/nanofibers at 0.2 A/g for 100 cycles, fabricated via ES and GAES. It is evident that GAES fibers with the same composition have more than 600 mAh/g improvement in capacity over the life of the battery that is a direct result of enhanced NP distribution. Finally, we performed a CGMD simulation for different planar elongational rates and concentrations of 1-D and spherical NPs. Similar to the experimental observations there is a significant improvement in NP dispersion and a good organization towards the extensional direction with the increase of elongation rates.

469278 Filled Single-Wall Carbon Nanotubes: Endohedral Volume Control for Improved Nanotube Properties

Abstract:

Liquid-phase processing of single-wall carbon nanotubes (SWCNTs) generally results in the exposure of their core to the environment (opening) due to the energy input necessary for purification and solubilization, and this often affects the SWCNT properties. With aqueous processing, SWCNTs are routinely fill with water, leading to significant redshifts and detrimental broadening of their electronic transitions and fluorescence quantum efficiencies. Selecting empty (end-capped) SWCNTs has historically been the only strategy to avoid these effects, but is a natively low yield process. In this work, adverse filling is prevented by intentionally prefilling the SWCNTs with various linear and cyclic alkanes to serve as passive and highly homogeneous spacers, or with molecules of varying composition to specifically modulate optical properties or the particle density. In contrast to water ingestion, filling the SWCNT core with low dielectric alkanes only weakly affects the electronic structure of the SWCNTs, effectively simulates empty core conditions but at near 100% yield. This yields highly resolved optical spectra with strongly shifted peak positions compared to water filled SWCNTs, and fluorescence efficiencies approaching those of empty SWCNTs. The effects of endohedral control and its applicability to different SWCNT materials and downstream separations will be presented.

ChemComm

COMMUNICATION

View Article Online
View Journal

Cite this: DOI: 10.1039/c5cc07692g

Click and release: fluoride cleavable linker for mild bioorthogonal separation†

Elia M. Schneider, Martin Zeltner, Vladimir Zlateski, Robert N. Grass and Wendelin J. Stark*

Received 14th September 2015, Accepted 16th November 2015

DOI: 10.1039/c5cc07692g

www.rsc.org/chemcomm

Herein, we present a water dispersable, magnetic nanoparticle supported "click and release" system. The cleavable linker has been synthesized by using a strain-promoted copper-free "click" reagent to establish the specific link and a fluoride cleavable silane moiety for mild cleavage. Small organic molecules, azide-bearing dyes and functionalized enzymes have been bound to the magnetic particle and released in a bioorthogonal way.

Bioorthogonal chemistry[1] opened the way for a detailed description and understanding of many biological processes, namely by the possibility to label and track specific biomolecules such as enzymes, lipids and sugars *in vivo*.[2] However, these kind of reactions, such as the Staudinger ligation or the 1,3-dipolar Huisgen cycloaddition of azides and alkynes ("click"),[3] are not reversible and cannot be cleaved in a specific and bio-compatible way (*i.e.* low temperatures, at neutral pH and in water). This would be beneficial for a large set of applications, for example native protein analysis,[4] protein glycosylation analysis[5] and analysis of protein complexes[6] *via* mass spectroscopy. There exist only a few reports of such linker systems;[7] however, the cleaving conditions are often quite harsh or not fully bioorthogonal. Our idea was to implement a cleavable silane moiety, one of the most used protecting groups for alcohols in organic chemistry.[8] The main advantage of this system is the cleavage *via* fluoride ions, which has already been used in solid peptide synthesis.[9] However, instead of using anhydrous HF we opted for buffered oxide etch (mild buffered fluoride solution) which has been used to uncage DNA and RNA from silica without damaging the biomolecules.[10] Our aim was to combine this mature technique to the modern copper-free "click" chemistry. Since the Huisgen-cycloaddition usually needs a high amount of copper catalyst, which hinders proper cell function,[11] a variety of strain-promoted copper-free click reagents have been developed.[12] These reagents promote the relatively fast dipolar cycloaddition without the need of any copper catalyst, thus leading to biocompatible chemistry. By combining the mature and biorthogonal silane-chemistry with the state-of-the-art copper-free "click"-chemistry, creation of a cleavable "click and release" system is anticipated. Additionally, a suitable way to get in and out of a biological system has to be considered. A promising approach is provided by magnetic nanoparticles, since they can be separated quickly from any solution by using a permanent magnet. Recently, a plethora of magnetic nano-sized reagents has been developed,[13] based on magnetic carbon-coated metal nanoparticles. Advantages of these nanoparticles are a higher saturation magnetization and thus faster separation compared to silica coated iron oxide nanoparticles.[13a] Furthermore, the graphene layer protects the particle from degradation by

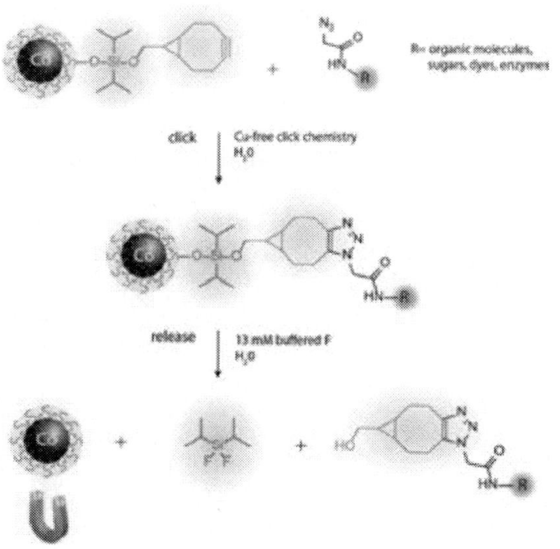

Scheme 1 Copper-free dipolar cycloaddition (click) on the magnetic nanoparticle followed by the fluorine-mediated release.

Institute for Chemical and Bioengineering, ETH Zurich, Vladimir-Prelog-Weg 1, 8093 Zurich, Switzerland. E-mail: Wendelin.stark@chem.ethz.ch

† Electronic supplementary information (ESI) available: Experimental procedures, protein adsorption experiments, XRD spectra and TEM micrographs. See DOI: 10.1039/c5cc07692g

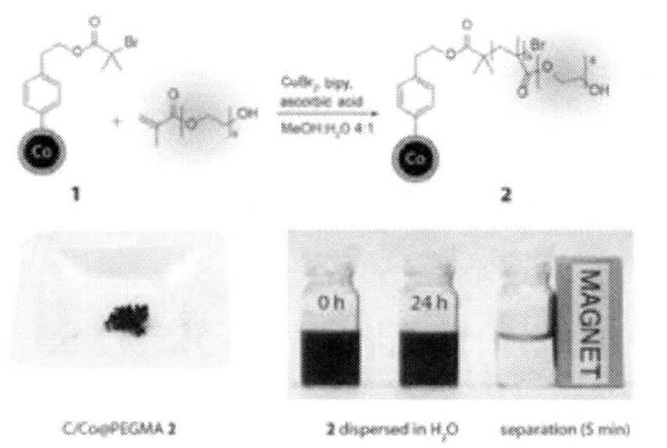

Fig. 1 PEG-methacrylate polymerisation *via* SI-ATRP. Sufficient dispersion stability in aqueous media is obtained while magnetic separation is still possible. Amount of solid (left) 150 mg, concentration (right) 10 mg mL^{-1}.

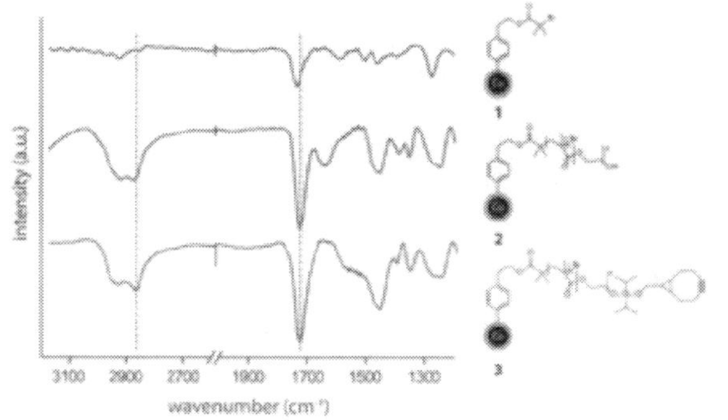

Fig. 2 FT-IR spectra of the magnetic compounds **1–3**. Red lines highlight the following vibrations: 2876 cm^{-1} C–H stretch, 1725 cm^{-1} carbonyl stretch.

buffered oxide etch and can be functionalized in a covalent and stable manner.[13d] However, one of the main drawbacks of this material is the unspecific protein adsorption ("fouling") on the highly hydrophobic, graphene-like surface.

To provide a solution for this problem, defined coatings have been developed, often PEG-chains or analogues.[14] A practical route to coat nanoparticles in an uniform and defined way is surface initiated atom transfer radical polymerization (SI-ATRP), a "grafting from" approach.[15] Herein, we present the synthesis of a bioorthogonal "click" and release system (Scheme 1) on a magnetic, PEG-coated support.

The preparation of the PEG-coated magnetic nanoparticles was conducted *via* standard surface initiated atom transfer radical polymerization using $CuBr_2$ and bipyridine (bipy) as catalyst and ligand, respectively (Fig. 1). First, the magnetic carbon-coated cobalt nanoparticles (C/Co) were functionalized according to literature,[14b] yielding covalently bound starter particles **1**. Subsequently, **1** was polymerized using ascorbic acid to reduce Cu(II) to the active Cu(I) form (known as activator regenerated by electron transfer (ARGET) ATRP),[16] resulting in the novel PEG-coated magnetic nanoparticles C/Co@PEGMA **2** (Fig. 2). Using this polymerization technique, control of the polymer length and thus the desired dispersion stability could be obtained.[14a] Indeed, the so obtained polymer-coated particles were sufficiently dispersible in an aqueous solution and the formation of a polymer layer around the magnetic particles was confirmed by transmission electron microscopy (TEM, see Fig S1 in ESI†). Comparing the unfunctionalized C/Co particles to the polymerized C/Co@PEGMA **2**, the first stack while the latter do not due to steric repulsion of the transparent polymer layer. Compound **2** was characterized *via* elemental microanalysis (ΔC = 24.4%, see Table S1 in the ESI†) and IR (Fig. 2). Since an extensive polymer layer can significantly lower the saturation magnetization, a vibrating sample magnetometer measurement (VSM) was executed. 151 emu g^{-1} was obtained for non-functionalized carbon coated cobalt nanoparticles C/Co, while for **2** 61.1 emu g^{-1} was measured. Despite the lower magnetization and high dispersion stability, fast separation by using a permanent magnet was possible (Fig. 1). To introduce the cleavable silyl linker, dichloro-diisopropylsilane was reacted with a hydroxyl-bearing moiety for 6 h, and subsequently added to well-dispersed C/Co@PEGMA particles **2** thus affording the fluoride-cleavable linker system (Scheme 2). In order to verify that the silane linking system is indeed cleavable with dilute F^- solution (buffered oxide etch, BOE) and to determine the ideal cleaving conditions, test reactions were run using fluorescent 1-naphthol as the leaving group (see Fig. S2 in the ESI†). The linker system was stable in water, 5 mM HCl solution and pH 4 sodium phosphate buffer (SPB). Notably, 13 mM of BOE pH 4 proved to be an optimal amount of fluoride to cleave the silane bond in quantitative yields. The formation of 1-naphthol and difluoro-diisopropylsilane upon treatment with BOE was confirmed by LC-MS and GC-MS, respectively. Subsequently, the more sophisticated copper free click reagent ((1*R*,8*S*,9*S*)-bicyclo-[6.1.0]non-4-yn-9-ylmethanol), which contains an accessible hydroxyl group, was reacted with dichloro-diisopropylsilane and *in situ* added to C/Co@PEGMA **2** to afford a magnetic, cleavable copper-free click reagent C/Co@PEGMA-Si-octyne **3** (Scheme 2). Characterization *via* elemental microanalysis and IR (esp. C–H stretch at 2876 cm^{-1}) confirmed the establishment of the desired linker system. The loading was determined by clicking the commercially available azide functionalized dye carboxy-tetramethylrhodamine (TAMRA-azide) to **3** (Fig. 3). Therefore the dye and **3** were shaken for 6 h, resulting in a 1,3-Huisgen dipolar cycloaddition. Subsequently, the absorbance of the supernatant

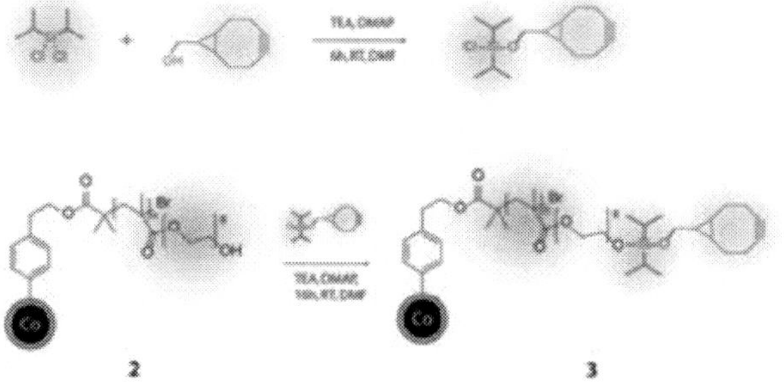

Scheme 2 Synthesis of C/Co@PEGMA-Si-octyne.

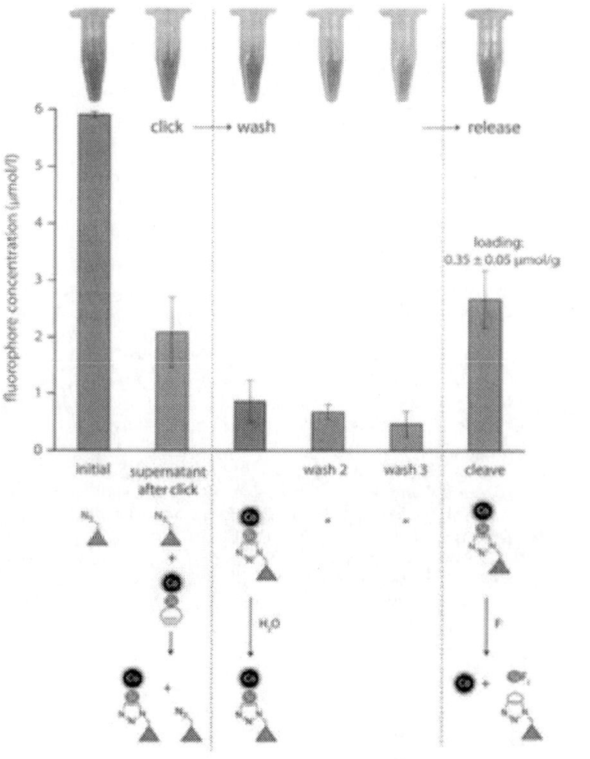

Fig. 3 Click and release of an azide-functionalized dye (TAMRA-azide). Conditions: in water, click for 3 h at RT, wash with water (3×, 1 h), cleave for 16 h using 13 mM BOE.

was measured at 550 nm, containing the remaining dye species. The particles were washed with water three times for one hour, resulting in no absorbance at 550 nm, *i.e.* no color. However, if dispersed in 13 mM BOE, a clear peak at 550 nm appeared, also visible as a purple coloring. It should be noted that 13 mM buffered oxide etch contains a relatively low amount of fluoride (247 ppm), comparable to common toothpaste (1000–1500 ppm).[17] The amount of cleaved dye was used to calculate the loading, determined as 0.35 ± 0.05 µmol g^{-1}. The loading was intentionally set lower (by controlling the amount of mono-chlorinated species added to C/Co@PEGMA) than in the case of 1-naphthol (49 ± 0.4 µmol g^{-1}) because many biomolecules are large in diameter (1–6 nm)[18] and therefore hinder a high loading onto nanoparticles due to size-constraints. It should be mentioned that a higher loading could possibly be achieved, but since copper-free click reagents are expensive in large quantities (*e.g.* grams) and the resulting benefit (*i.e.* higher biomolecule loading) is low, this was not further pursued. Furthermore, unspecific protein binding onto the nanoparticles ("fouling") was investigated, since the aim of this work is to catch and release specific (*i.e.* azide-labelled) compounds. It is crucial that non-labelled proteins do not adhere to the particle surface and that bound proteins can be released effectively. This means that there is no unspecific interaction between the particle and a bound protein. Therefore, rhodamine labelled BSA was dissolved in water (0.5 nmol mL^{-1}) and the absorption at 555 nm was measured at $t = 0$ h, then after 2, 18, 32 h using naked particles, **2** or **3** (20 mg each, see Fig. S3 in the ESI†). While the naked C/Co particles irreversibly adsorbed over 90% of the protein over time, the PEG-functionalized particles **2** and **3** showed much lower unspecific protein adsorption (*i.e.* <5% after 32 h). Since PEG-coating has been known to prevent unspecific protein adsorption, these results are in line with recent literature.[19] To demonstrate applicability in the biological field, the protease enzyme α-chymotrypsin (from bovine pancreas) has been functionalized with an azide group by using a commercially available azide–PEG_4–NHS reagent. The well-studied reaction of NHS-ester with unprotonated, free amine groups of proteins (*e.g.* lysine) proceeded smoothly in basic buffer solution. The activity of the azide-functionalized enzyme was then compared to the immobilized enzyme (*i.e.* bound by click-reaction to the magnetic particle) using a standard enzymatic assay (Fig. 4).

As anticipated, the immobilized enzyme exhibited a lower specific activity (*i.e.* <5% of the free enzyme). Compared to other reports of covalent[20] or biospecific[21] immobilization of enzymes on magnetic nanoparticles (15–35%, resp. 106% retained activity) the herein reported retained activity is low. A possible explanation for this drop can be the random orientation of the immobilized enzyme. Also, the covalent linking strategy might negatively influence the enzyme activity.

However, if the clicked enzyme was released by the use of 13 mM BOE, the activity of the released enzyme was substantially increased (*i.e.* to 67%, compared to the free enzyme). It should be noted that the destructive effect of buffered oxide etch on chymotrpysin is marginal (93 ± 3.1% retained activity after 16 h in 13 mM BOE at RT). As mentioned before, the here used dilute F^- solution proved to be harmless to DNA and even RNA.[10] Therefore buffered oxide etch can be assumed to

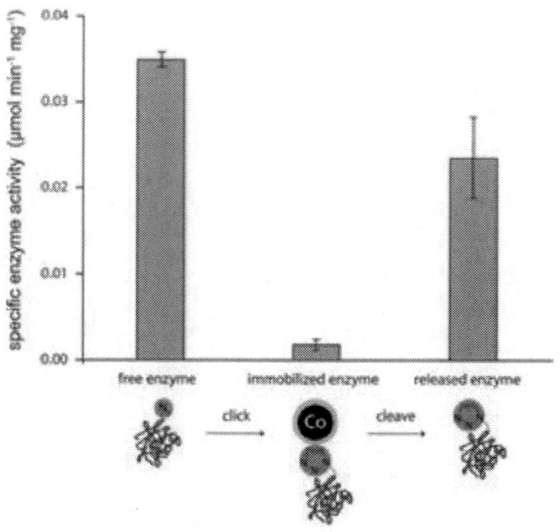

Fig. 4 Specific activity of azide-functionalized α-chymotrypsin. Enzymatic assay was performed measuring the formation of nitroaniline with UV spectroscopy at 390 nm, using *N*-benzoyl-L-tyrosine *p*-nitroanilide as a substrate. Amount of bound enzyme was determined by elemental analysis.

be bio-orthogonal and mild, making it a suitable reagent for a wide use in biochemistry. However, there are also limitations to this cleavage system: pH 4 for 16 h can be too harsh for certain proteins, resulting in denaturing. The long time-span could be reduced by optimizing the conditions (*e.g.* higher fluoride concentration), if necessary. Furthermore, fluoride can interact with certain metal ions, (*e.g.* calcium, forming CaF_2) forming stable compound and thus deactivating an enzyme by binding the cofactor.

In summary, a well-dispersed, magnetic and cleavable copper-free click-nano-reagent has been synthesized and characterized. An azide-functionalized dye has been clicked to the magnetic particle and released using 13 mM dilute F^- solution. Furthermore it was shown that also enzymes could be bound and released without a substantial loss of activity.

The authors declare the following competing financial interest(s): two authors (W. J. S. and R. N. G.) declare financial interests, as they are shareholders of Turbobeads LLC, a company producing magnetic nanoparticles.

Financial support was provided by ETH Zurich and the EU/ITN network Mag(net)icfun (PITN-GA-2012-290248).

Notes and references

1 E. M. Sletten and C. R. Bertozzi, *Acc. Chem. Res.*, 2011, **44**, 666.
2 (*a*) A. B. Neef and C. Schultz, *Angew. Chem., Int. Ed.*, 2009, **48**, 1498; (*b*) T. Plass, S. Milles, C. Koehler, C. Schultz and E. A. Lemke, *Angew. Chem., Int. Ed.*, 2011, **50**, 3878; (*c*) E. M. Sletten and C. R. Bertozzi, *Angew. Chem., Int. Ed.*, 2009, **48**, 6974.
3 (*a*) H. C. Kolb, M. G. Finn and K. B. Sharpless, *Angew. Chem., Int. Ed.*, 2001, **40**, 2004; (*b*) E. Saxon and C. R. Bertozzi, *Science*, 2000, **287**, 2007.
4 K. Lorenzen and E. V. Duijn, *in Current Protocols in Protein Science*, John Wiley & Sons, Inc., 2001.
5 X. Zhang and Y. Zhang, *Molecules*, 2013, **18**, 7145.
6 H. Zhang, X. Tang, G. R. Munske, N. Tolic, G. A. Anderson and J. E. Bruce, *Mol. Cell. Proteomics*, 2009, **8**, 409.
7 (*a*) C. Sibbersen, L. Lykke, N. Gregersen, K. A. Jorgensen and M. Johannsen, *Chem. Commun.*, 2014, **50**, 12098; (*b*) S. Wang, W. Xie, X. Zhang, X. Zou and Y. Zhang, *Chem. Commun.*, 2012, **48**, 5907.
8 (*a*) E. J. Corey and B. B. Snider, *J. Am. Chem. Soc.*, 1972, **94**, 2549; (*b*) S. J. Danishefsky, K. F. McClure, J. T. Randolph and R. B. Ruggeri, *Science*, 1993, **260**, 1307.
9 S. Sakakiba, Y. Shimonis, Y. Kishida, M. Okada and H. Sugihara, *Bull. Chem. Soc. Jpn.*, 1967, **40**, 2164.
10 (*a*) R. N. Grass, R. Heckel, M. Puddu, D. Paunescu and W. J. Stark, *Angew. Chem., Int. Ed.*, 2015, **54**, 2552; (*b*) D. Paunescu, R. Fuhrer and R. N. Grass, *Angew. Chem., Int. Ed.*, 2013, **52**, 4041; (*c*) M. Puddu, D. Paunescu, W. J. Stark and R. N. Grass, *ACS Nano*, 2014, **8**, 2677; (*d*) M. Puddu, W. J. Stark and R. N. Grass, *Adv. Healthcare Mater.*, 2015, **4**, 1332; (*e*) D. Paunescu, M. Puddu, J. O. B. Soellner, P. R. Stoessel and R. N. Grass, *Nat. Protoc.*, 2013, **8**, 2440.
11 A. J. Link, M. K. S. Vink and D. A. Tirrell, *J. Am. Chem. Soc.*, 2004, **126**, 10598.
12 (*a*) N. J. Agard, J. A. Prescher and C. R. Bertozzi, *J. Am. Chem. Soc.*, 2004, **126**, 15046; (*b*) J. M. Baskin, J. A. Prescher, S. T. Laughlin, N. J. Agard, P. V. Chang, I. A. Miller, A. Lo, J. A. Codelli and C. R. Bertozzi, *Proc. Natl. Acad. Sci. U. S. A.*, 2007, **104**, 16793; (*c*) J. Dommerholt, S. Schmidt, R. Temming, L. J. A. Hendriks, F. P. J. T. Rutjes, J. C. M. van Hest, D. J. Lefeber, P. Friedl and F. L. van Delft, *Angew. Chem., Int. Ed.*, 2010, **49**, 9422; (*d*) C. S. McKay, J. Moran and J. P. Pezacki, *Chem. Commun.*, 2010, **46**, 931.
13 (*a*) E. M. Schneider, R. A. Raso, C. J. Hofer, M. Zeltner, R. D. Stealer, S. C. Hess, R. N. Grass and W. J. Stark, *J. Org. Chem.*, 2014, **79**, 10908; (*b*) W. J. Stark, *Angew. Chem., Int. Ed.*, 2011, **50**, 1242; (*c*) V. Zlateski, R. Fuhrer, F. M. Koehler, S. Wharry, M. Zeltner, W. J. Stark, T. S. Moody and R. N. Grass, *Bioconjugate Chem.*, 2014, **25**, 677; (*d*) A. Schaetz, M. Zeltner, T. D. Michl, M. Rossier, R. Fuhrer and W. J. Stark, *Chem. – Eur. J.*, 2011, **17**, 10565.
14 (*a*) C. J. Hofer, V. Zlateski, P. R. Stoessel, D. Paunescu, E. M. Schneider, R. N. Grass, M. Zeltner and W. J. Stark, *Chem. Commun.*, 2015, **51**, 1826; (*b*) M. Zeltner, R. N. Grass, A. Schaetz, S. B. Bubenhofer, N. A. Luechinger and W. J. Stark, *J. Mater. Chem.*, 2012, **22**, 12064.
15 (*a*) S. Edmondson, V. L. Osborne and W. T. S. Huck, *Chem. Soc. Rev.*, 2004, **33**, 14; (*b*) M. Kato, M. Kamigaito, M. Sawamoto and T. Higashimura, *Macromolecules*, 1995, **28**, 1721; (*c*) J.-S. Wang and K. Matyjaszewski, *J. Am. Chem. Soc.*, 1995, **117**, 5614.
16 K. Matyjaszewski, H. Dong, W. Jakubowski, J. Pietrasik and A. Kusumo, *Langmuir*, 2007, **23**, 4528.
17 S. Twetman, S. Axelsson, H. Dahlgren, A. K. Holm, C. Kallestal, F. Lagerlof, P. Lingstrom, I. Mejare, G. Nordenram, A. Norlund, L. G. Petersson and B. Soder, *Acta Odontol. Scand.*, 2003, **61**, 347.
18 H. P. Erickson, *Biol. Proced. Online*, 2009, **11**, 32.
19 I. Banerjee, R. C. Pangule and R. S. Kane, *Adv. Mater.*, 2011, **23**, 690.
20 (*a*) J. Lee, Y. Lee, J. K. Youn, H. B. Na, T. Yu, H. Kim, S.-M. Lee, Y.-M. Koo, J. H. Kwak, H. G. Park, H. N. Chang, M. Hwang, J.-G. Park, J. Kim and T. Hyeon, *Small*, 2008, **4**, 143; (*b*) H. J. Park, J. T. McConnell, S. Boddohi, M. J. Kipper and P. A. Johnson, *Colloids Surf., B*, 2011, **83**, 198.
21 J. Garcia, Y. Zhang, H. Taylor, O. Cespedes, M. E. Webb and D. Zhou, *Nanoscale*, 2011, **3**, 3721.

457380 Synthesis and Hyperthermia Application of Drug Loaded Magnetic Nanostructured Lipid Carriers (MNLC)

Abstract:

Magnetic nanoparticles (MNPs) have been used in many biomedical applications, such as magnetic hyperthermia, magnetic resonance imaging and drug delivery systems. [1] Magnetic hyperthermia (MHP) is based on induction heating of MNPs under the application of an alternating magnetic field (AMF) and the temperature increase in body tissues leading to cellular structure change. In high temperature, such as 40-43°C cancerous cells would be damaged but healthy cells survive. Basically, MHP is focused on the destruction of tumors by heat, which is obtained from magnetic nanoparticles under AMF [2].

In literature, it is seen that MNPs are often used in conjuction with different drug delivery systems, such as micelles [1], liposomes [2], hydrogels [3], solid lipid nanoparticles (SLN) [4] and magnetic/gold nano-hybrids [5]. SLN have advantages, such as good biocompatibility and low toxicity, when compared with the emulsions and polymeric nanoparticles in drug delivery systems. However, low drug loading and leaking cause some problems in terms of drug delivery in SLN [4]. To overcome these drawbacks, second generation lipid based nanoparticles, which called as Nanostructured Lipid Carriers (NLC) are developed. NLC contain solid and liquid lipid blending, which consist less ordered crystal structure with imperfections and this structure provide high drug loading and prevent drug leaking. Also, NLC have some other advantages such as high biocompatibility, controlled drug release profiles and scale up production possibility [6].

Magnetic nanoparticles containing NLC systems are not used in the drug delivery application to our knowledge. Using NLC and magnetic nanoparticles together for hyperthermia application is a new type of drug delivery system and has a big potential in the treatment of cancer. In accordance with this purpose, MNPs embedded NLC were developed, synthesized and hyperthermia potential of these particles were investigated.

In this work, iron oxide nanoparticles (MNPs) were synthesized via a well known co-precipitation method. Magnetic nanoparticles were stabilized by using oleic acid. These MNPs show superparamagnetic behavior with high magnetization saturation value. To obtain MNPs embedded NLC, magnetic nanoparticles are added to lipid particles during the NLC synthesis. In NLC synthesis, stearic acid and oleic acid were used as solid and liquid lipids; as organic phase acetone-ethanol mixture and as an emulsifier Pluoronic F-127 were used. To determine optimum amount of embedded MNPs, different ratios were studied. Synthesized particles were characterized by Dynamic and Static Light Scattering (LS), Atomic Force Microscopy (AFM), Differential Scanning Calorimetry (DSC), Transmission Electron Microscopy (TEM), Vibrating Sample Magnetometer (VSM) and Thermo-gravimetric Analysis (TGA). The nanostructures in NLC were determined via TEM as multiple oil in fat in water (o/f/w) type with oil nano-compartments. Two different size distributions were observed from TEM images; the size of NLC without MNPs were 57±16 nm and 124±17 nm with spherical shape and the size of NLC with MNPs were 65±15 nm and 125±21 nm. To investigate the hyperthermia potential of these particles, different parameters were studied such as lipid concentration, magnitude of magnetic field and application time period. AMF were applied to the particles and temperature increase were observed. Furthermore, Specific Absorption Rate (SAR) or Specific Loss Power (SLP) were calculated, which is an another important factor for hyperthermia application. SAR is defined as, the amount of heat released by a unit weight of material per unit time during AMF [5]. Compared to other particles in literature, SAR values are relatively higher. Additionally, temperature increase of particles was studied under AMF and maximum temperature increase (ΔT_{max}) was obtained as 16°C after 30 minutes. Also, simulating drug encapsulated MNPs embedded NLC were used as drug delivery agent. The drug release performance of these NLC were studied with applied AMF for 5, 15 and 30 minutes at 37°C to stimulate the body temperature. Besides, drug release properties of the samples at which the AMF has not been applied were investigated as a control group. From the study it is observed that, drug release performance of NLC was increased with increasing application time period of magnetic field. This work demonstrated that, magnetic nanoparticle embedded NLC can be used in hyperthermia application successfully and also they can be thought as drug delivery systems.

References

[1] H.Y.Kim, E.Kim,S.W.Jeong, T.L.Ha,S.I.Park,S.G.Lee, S.J.Lee and S.W.Lee,Nanoscale,2015,7,16470.

[2] D.Qiu,X.An,Z.Chen and X.Ma,Chem.Phys. Lipids,2012,165,563.

[3] G.Giani, S.Fedi and R.Barbucci, Polymers,2012,4,1157-1169.

[4] S.Zhao, Y.Zhang, Y.Han, J.Wang and J.Yang, Pharm Res,2015,32,482-491.

[5] M.M.Ramos-Tejada, J.L.Viota, K.Rudzka and A.V.Delgado, Colloids and Surf. B: Biointerfaces, 2015,128,1-7

[6] R.H.Muller,M. Radtke, S.A.Wissing, Adv. Drug Deliver. Rev. 54 (Suppl.1) (2002) S131–S155.

467840 Development of Magnetic Alginate Microcapsules As Dual Treatment (chemotherapy and hyperthermia) for Human Lung Cancer

Abstract:

Human lung cancer is one of the most common damaging tumors and it has the highest death rate. Magnetic hyperthermia has been successfully used for many years in cancer treatments. It is a non-invasive approach for tumor ablation and is based on heat generation by magnetic materials when subjected to an alternating magnetic field. Many studies have shown a significant reduction in tumor size when hyperthermia is combined with other treatments.

In this study, a novel magnetic polymeric microparticle with a core of magnetite and polymeric shell of alginate is reported. Alginate microparticles with magnetic core allow the combined treatment (chemotherapy and hyperthermia) in the same carrier and at the same time. These microparticles have been prepared using a fan-jet nozzle. The resulting alginate-magnetite microparticles are linked to an antitumor drug, cisplatin, and an antibody, epidermal growth factor (EGF).

Hyperthermia treatment must reach around 42ºC in cells area. The control of these temperatures in Petri plates have been done with a thermography camera. This system has been used to check the temperature profile.

The cytotoxicity of the microparticles have been tested with H460 lung cancer cell line. In vitro validation tests were developed in Petri plates and the cells viability was controlled by MTT assay. These experiments show that magnetic polymeric microparticles present a better effect of cancer cell mortality compared to the same drug delivery system without hyperthermia treatment previously reported in other studies.

As a result, this work reports the development and in vitro validation of magnetic microparticles, which in the future may be useful for various biomedical applications.

470770 Simulation of Dynamic Magnetic Drug Carrier Particle Capture and Accumulation Around a Ferromagnetic Wire

Abstract:

A new approach for modeling implant assisted-magnetic drug targeting (IA-MDT) systems during the time-dependent capture and accumulation of magnetic drug carrier particles (MDCPs) by a ferromagnetic wire was developed. This new approach assumes the fluid (slurry) viscosity, comprised of water and MDCPs, is a function of the MDCP concentration in the fluid, with imposed maxima on both the particle concentration and fluid viscosity to avoid unrealistic limits. In 2-D, the unsteady-state Navier-Stokes equations for compressible fluid flow and the unsteady-state continuity equations applied separately to the water and magnetic particle phases in the slurry were solved simultaneously, along with the Laplace equations for the magnetic potential applied separately to the slurry and wire, to evaluate the velocities and concentrations around the wire in a narrow channel using COMSOL Multiphysics. The results from this model revealed very realistic magnetically attractive and repulsive zones forming in time around the wire. These collection zones formed their own impermeable viscous phase during accumulation that was also magnetic with its area and magnetism impacting locally both the fluid flow and magnetic fields around the wire. These collection zones increased with an increase in the applied magnetic field. For a given set of conditions, the capture ability peaked and then decreased to zero at infinite time during magnetic particle accumulation in the collection zones. Predictions of the collection efficiency from a steady-state, clean collector, trajectory model could not show this behavior; it also agreed only qualitatively with the dynamic model and then only at the early stages of collection and more so at a higher applied magnetic field. Also, the collection zones decreased in size when the accumulation regions included magnetic particle magnetization (realistic) compared to when they excluded it (unrealistic). Overall, this might be the first time a mathematical model was shown to be capable of realistically predicting the dynamic nature of MDCP capture and accumulation around a wire in IA-MDT systems.

470775 Preparation of Thermally Responsive Magnetic Nanocomposites for the Removal of Environmental Pollutants

Abstract:

Magnetic nanocomposite based sorbents have gained great attention for environmental remediation (especially water treatment) due to their magnetic separation properties and high adsorption capacities. In addition, smart magnetic sorbents that are responsive to surrounding stimulus have also been developed through the combination of magnetic nanoparticles and thermo-responsive polymers. N-isopropylacrylamide (NIPAAm) is one of the most widely studied thermo-responsive materials, which undergoes reversible phase transition at its lower critical solution temperature (LCST) around 32Â°C. NIPAAm-based thermally responsive materials can be synthesized with various functionalities, which can provide specific interactions with environment pollutants (e.g., polychlorinated biphenyl, PCBs). Our groupâ€™s recent efforts have focused on applying naturally derived polyphenols, such as curcumin and quercetin, to develop materials with binding affinities to PCBs. The presence of aromatic rings in polyphenols is expected to mimic the binding domains as observed in PCB antibodies (e.g., S2B1).

The overall goal of this work was to develop thermally responsive materials for on/off PCB binding purpose, aiming at provide an easy-recovery and cost effective strategy for the removal of PCBs from the environment. Specifically, two types of thermally responsive magnetic nanocomposites were developed, including magnetic nanocomposite microparticles (MNMs) and core-shell magnetic nanoparticles (MNPs). The physicochemical properties and PCB capture/release performance of the composite particles were characterized. Their size and morphology were characterized by scanning electron microscopy (SEM), the magnetic nanoparticle loading was determined by thermogravimetric analysis (TGA), the thermo-responsive was characterized by dynamic light scattering (DLS), and the magnetic separation property was characterize by a static magnetic fields.

470910 Insights into Magnetic Particle Imaging Obtained from Modeling

Abstract:

Magnetic particle imaging (MPI) is an emerging biomedical imaging technology enabled by the non-linear response of magnetic nanoparticle tracers to a scanned magnetic gradient field.[1] This tracer-based technology provides a safer alternative to other radioactive imaging modalities like single photon emission computed tomography (SPECT) and positron emission tomography (PET). MPI has been demonstrated in applications such as real-time cardiovascular imaging[2], cell labeling and tracking[3,4], and has potential in magnetic fluid hyperthermia.[5] The original MPI theory assumes that tracers respond instantaneously to the magnetic fields encountered in MPI. In practice, this is not the case, and finite magnetic relaxation or response time has been implicated in secondary blurring of the image, even limiting the attainable resolution in MPI.[6,7] To understand the effect of response time, or relaxation, on the MPI signal and resolution, we applied rotational Brownian dynamics (RBD) simulations[8] and we observed a drop in resolution and signal. To investigate the effect of field-dependent relaxation we used ferrohydrodynamic modeling to obtain predictions of the point spread function (PSF) in an x-space relaxometer[7] and the nanoparticle harmonic spectra in a magnetic particle spectrometer (MPS)[9,10]. These instruments are employed to assess the performance of tracers for use in MPI. Through this study we observed considerable agreement between the model predictions and experiments without the need of empirical parameter fits. Finally, we explored the potential of MPI field gradients in magnetic fluid hyperthermia to achieve selective heating of nanoparticles. Here, we used the ferrohydrodynamic equations to calculate theoretical predictions of specific absorption rate (SAR) and obtain millimeter scale spatial distribution of SAR by tuning the MPI field gradients. Through our modeling efforts we were able to understand the effect of relaxation in MPI and show the potential of MPI field gradients to achieve spatially selective magnetic fluid hyperthermia.

[1] B. Gleich and J. Weizenecker, Nature **435**,1214 (2005).

[2] J. Weizenecker, B. Gleich, J. Rahmer, H. Dahnke, and J. Borgert, Physics in Medicine and Biology **54**,L1 (2009).

[3] B. Zheng, M. P. von See, E. Yu, B. Gunel, K. Lu, T. Vazin, D. V. Schaffer, P. W. Goodwill, and S. M. Conolly, Theranostics **6**,291 (2016).

[4] B. Zheng, T. Vazin, P. W. Goodwill, A. Conway, A. Verma, E. Ulku Saritas, D. Schaffer, and S. M. Conolly, Sci Rep **5**,14055 (2015).

[5] K. Murase, H. Takata, Y. Takeuchi, and S. Saito, Physica Medica-European Journal of Medical Physics **29**,624 (2013).

[6] L. R. Croft, P. W. Goodwill, and S. M. Conolly, IEEE Transactions on Medical Imaging **31**,2335 (2012).

[7] P. W. Goodwill, A. Tamrazian, L. R. Croft, C. D. Lu, E. M. Johnson, R. Pidaparthi, R. M. Ferguson, A. P. Khandhar, K. M. Krishnan, and S. M. Conolly, Applied Physics Letters **98**,

262502 (2011).

[8] R. Dhavalikar and C. Rinaldi, Journal of Applied Physics **115**,074308 (2014).

[9] R. Dhavalikar, L. Maldonado-Camargo, N. Garraud, and C. Rinaldi, Journal of Applied Physics **118**,173906 (2015).

[10] S. Biederer, T. Knopp, T. F. Sattel, K. Ludtke-Buzug, B. Gleich, J. Weizenecker, J. Borgert, and T. M. Buzug, Journal of Physics D-Applied Physics **42**, 205007 (2009).

470970 Utilizing Magnetic and Diffusive Properties to Improve Size Homogeneity of Superparamagnetic Nanoparticles

Abstract:

Superparamagnetic iron oxide nanoparticles (SPIONs) have enormous potential in biomedical applications, including drug delivery, hyperthermia, and as a contrast agent for magnetic resonance imaging (MRI). SPIONs with diameters greater than about 20 nm are generally synthesized with multi-core structures held together by an external matrix. This typically yields particles with a very broad size distribution, having a dynamic light scattering polydispersity index (PdI) of about 0.23. This broad size distribution hinders clinical translation of these particles due to safety and performance variability. While several size fractionation techniques have been employed, including magnetic fractionation, centrifugation, gel chromatography, and vacuum filtration to improve nanoparticle size homogeneity, PDI values remain much higher than gold, silica and other nanoparticles.

A new size fractionation method was based on the difference in particle mobility between different sized particles from a controlled periodic magnetic field. Separated SPIONs (SePs) had an average size distribution at least three times narrower than all four conventional methods. The method had an overall PdI 0.072 ± 0.0056, independent from five different scale-up conditions. A strong relationship was observed between a theoretically based mathematic model and experimental results, which confirmed the methodology toward a true homogenous size separation. Starch coated SPIONs with a PdI of 0.23 were separated into eight fractions with average sizes of 70.7, 72.3, 75.5, 78, 86, 100.6, 109.2, and 116.9 with PdIs of 0.061, 0.08, 0.07, 0.085, 0.074, 0.088, 0.073, and 0.094 respectively. Results indicated that the PdI of the SPION suspension could have a dramatic effect on MRI imaging and tissue penetration of these particles. SePs significantly increased the particle penetrating through 1 μm membrane and had distinct differences in MRI contrast enhancement and relaxivities from their polydispersed original sample and different conventional separation methods.

472772 Multispectral Optoacoustic Tomography (MSOT) for Imaging of Pharmacokinetics and Biodistribution of Pegylated Superparamagnetic Iron Oxide NanoparticlesÂ

Abstract:

Multispectral optoacoustic tomography (MSOT) has emerged as a leading technology that uses photoacoustic imaging for high-resolution imaging of cancer. MSOT combines the high contrast of optical imaging with the high resolution and great penetration depth of ultrasound imaging. It can detect sound waves that are produced following illumination of target tissue by multiple wavelengths to characterize the distribution of multiple endogenous and exogenous molecules simultaneously. MSOT is noninvasive and fast; therefore, it can image dynamic processes in real-time with high temporal resolution.

We used MSOT imaging to compare the *in vivo* pharmacokinetics, i.e. disappearance rate from circulation, and biodistribution profiles of Cy 7 fluorophore conjugated superparamagnetic iron oxide nanoparticles (Cy7-SPIONs) coated with three different sizes of polyethylene glycol (PEG). We used image reconstruction and spectral unmixing algorithms to visualize and quantify the concentration of Cy7-SPION in circulation and in organs of elimination. To validate the results, we collected blood and tissue samples, and quantified Cy7-SPIONs concentrations using electron spin resonance (ESR). The size of PEG, which alters the surface properties of SPIONs, significantly affected the pharmacokinetics and biodistribution profiles of Cy7-SPIONs.

In another study, we developed RGD-functionalized Cy7-SPIONs (RGD-Cy7-SPIONs) for *in vivo* MSOT imaging of Î±vÎ²3 integrin positive MCF-7 breast tumor-bearing mice. A significant increase in MSOT signal was observed at the tumor site with RGD-Cy7-SPIONs, compared to nontargeted Cy7-SPIONs. MSOT demonstrated superior sensitivity in tumor detection compared to two-dimensional fluorescent imaging.

Abstract:

Fe_3O_4 promoted metal organic framework MIL-100(Fe): Can be a promising candidate as a drug carrier

Abhik Bhattacharjee and Sasidhar Gumma

Department of Chemical Engineering, Indian Institute of Technology Guwahati

Email: s.gumma@iitg.ernet.in

In this study, porous metal organic framework (MOF) MIL-100(Fe) and magnetic nanoparticle MOF composites i.e. Fe_3O_4@MIL-100 were investigated as drug delivery vehicles for anticancer drug doxorubicin hydrochloride. Magnetite is an inorganic metal oxide was chosen for its nontoxicity and high magnetic saturation. MIL-100 (Fe) was synthesized by solvothermal route at about 150 °C from its organic linker like trimesic acid (BTC). Magnetite were synthesized according to standard procedures in literature to yield magnetic nanoparticles (MNPs) of about 10-30 nm. A series of MNP and MOF composites i.e. Fe_3O_4@MIL-100(Fe) were prepared by adding magnetite in different molar ratio in the raw precursor of MIL-100 at temperature of 150 °C. All porous adsorbent samples were characterized by X-ray diffraction, Nitrogen adsorption-desorption isotherm and Transmission electron microscopy (TEM) methods. Doxorubicin hydrochloride (DOX), a well-known anticancer drug for the treatment of various cancer syndromes like leukaemia, lymphoma, and carcinoma was chosen as a model drug for current study. The drug was loaded on to the MOF composite by equilibrating the aqueous DOX solution with the porous MNP incorporated MOFs. For release study, "DOX" loaded carriers, were dispersed in simulated phosphate-buffered saline (PBS) solution at a pH of 7.4 and incubated in a shaker at 37 °C in the dark. The sample was centrifuged and the supernatants were collected for analysis at periodic time intervals. A fresh PBS which is equal in volume to that of the sample withdrawn for analysis is added to the contents of the incubation flask. The results indicate that the drug loading capacities of the Fe_3O_4@MIL-100 for the composites studied in this work are higher than both pure Fe_3O_4 and pure MIL-100. The release kinetics also indicate sustained and stable release of "DOX" for more than ten days by incorporating MNPs into MOF structure.

Table 1: Molar compositions and drug loading capacities of different carriers

Drug carriers	Molar ratio Fe:BTC:Fe_3O_4	"DOX" Loading (wt%)
Fe_3O_4	0:0:1	11.37
MIL-100	50:33:0	11.09
Fe_3O_4(50)@MIL-100	50:33:1	18.31
Fe_3O_4(100)@MIL-100	50:33:2	18.45
Fe_3O_4(125)@MIL-100	50:33:2.5	19.69
Fe_3O_4(150)@MIL-100	50:33:3	19.00
Fe_3O_4(200)@MIL-100	50:33:4	15.14
Fe_3O_4(400)@MIL-100	50:33:9	13.81

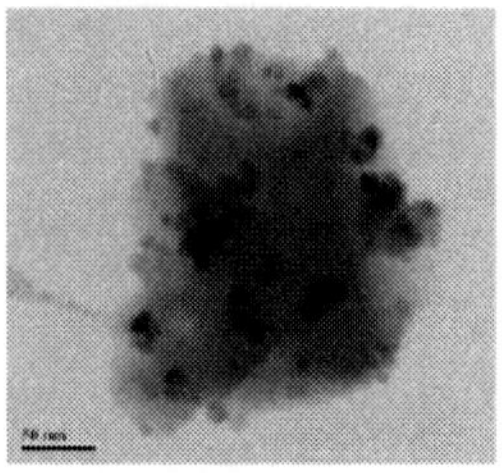

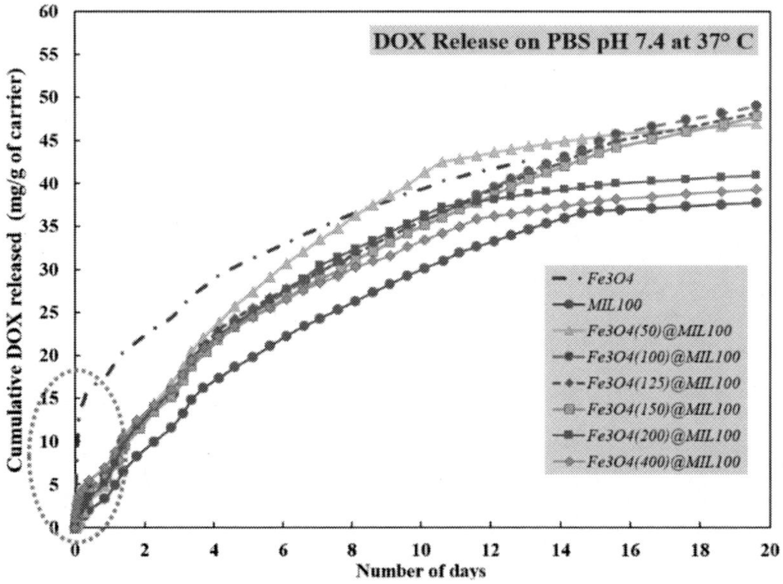

Fig. 1: TEM micrograph of Fe_3O_4@MIL-100 and "DOX" release study on different carriers

References:

1. R. Massart, IEEE Trans.Magn., (1981), 17,1247.

2. P. Horcajada et.al. *Nature*, (2010), 9, 172.

470009 High Throughput Block Copolymer Micelle Assembly Methods and Morphologies

Abstract:

Block copolymer micelles hold tremendous potential in the biomedical and materials science fields, with uses ranging from drug delivery to new methods of lithography. Micelles are an attractive technology for these applications because of their ease of assembly and tunable properties. Typical micelles are nanoscale structures with hydrophobic cores and hydrophilic coronae. Smaller nanomaterials can be loaded in the core, thus allowing aqueous delivery of otherwise hydrophobic payloads. To date, most micellar research has focused on bench-scale development, rather than translation of this technology to industrial processes. Flexible, high throughput manufacturing strategies for micellar production are necessary for commercialization.

Here, we describe and compare two promising semi-continuous methods for synthesis of inorganic-polymer composite micelle nanoparticles. The first combines a scalable electrospray technique with an interfacial instability process to form micellar structures. The second is an adaptation of a flash nanoprecipitation technique originally developed by Prudâ€™homme and colleagues as a way to more rapidly generate polymer aggregates.

In the electrospray method, separate and immiscible aqueous and polymer-in-organic solutions flow through a thin coaxial needle assembly, and voltage is applied until a cone-jet spray forms. Aerosol particles are ejected from the spray, travel through the air, and are collected in water as emulsion droplets that undergo transient interfacial surface tension changes, resulting in micelle formation. Here, we explored the limits of the coaxial electrospray process by varying operational parameters to achieve higher polymer processing and throughput, thus accessing greater nanoparticle yields. We independently evaluated the influence of electrospray volumetric flow rates, organic block copolymer loading, and system processing temperature on micelle formation and processing rates for the polymer polystyrene-block-poly(ethylene oxide) (PS-b-PEO). Using this approach, we increased production rates by a factor of 40 over previous reports and accessed different assembly morphologies.

We compared this process to the flash nanoprecipitation method. In this method, water is rapidly mixed with block copolymers in a water miscible solvent (i.e., tetrahydrofuran). We studied polymer nanoparticle formation for both PS-b-PEO and poly(ethylene glycol)-block-polycaprolactone (PEG-PCL), and demonstrated micelle production at different mixing speeds, allowing flexibility of production at rates higher than observed using micellar electrospray.

Both micelle production techniques hold promise for the formation of empty and loaded micelles of different morphologies. Micellar electrospray is preferred for use with longer block copolymers because of the high mixing rates required in flash nanoprecipitation. However, flash nanoprecipitation is a faster method for appropriate molecules. Each synthesis system has potential application in drug delivery, imaging, diagnostics, and separations.

467179 Shrinking Anodized Aluminum Oxide Pores Beyond 10 Nanometers

Abstract:

Anodization of aluminum has piqued research interest because of the self-ordering porous nanostructure created during anodization. This porous structure can be tuned to be a specific diameter and packing density based on macroscopic properties, such as voltage, acid type, and concentration. Recent work has achieved smaller pore sizes with the addition of an organic solvent that enables a reduction of temperature. Another study demonstrated rapid fabrication of 10 nm pores by utilizing selenic acid as the electrolyte. However, no study has successfully fabricated sub-10 nanometer pores through anodization. Shrinking the pore diameter below 10 nanometers is problematic because the driving force for oxidation is insufficient at lower anodization potentials. In this work, we report a novel, self-ordering method that produces sub-10 nanometer pores during anodization in selenic acid/ethylene glycol. The pores fabricated in this method have a mean pore size of five nanometers, half the size of those reported from previous approaches. These sub-10 nm pores have the potential to be used in a variety of applications, including photonics, membranes for gas separations, and as templates for fabricating nanowires.

468004 Development of a New Molecular Dynamics Force Field to Model Intermolecular Interactions at MoS_2 Interfaces: Application to Liquid-Phase Exfoliation

Abstract:

The controlled synthesis of molybdenum disulfide (MoS_2) using liquid-phase exfoliation, as well as several of its proposed applications, involve liquids coming into intimate contact with MoS_2 surfaces. Molecular dynamics (MD) simulations offer a robust methodology to investigate nanomaterial/liquid interactions involving weak van der Waals forces. However, MD force fields for MoS_2 currently available in the literature incorrectly describe not only the cohesive interactions between layers of MoS_2, but also the adhesive interactions of MoS_2 with liquids such as water. Here, we develop a set of force-field parameters that reproduce the properties of bulk 2-H molybdenite, with special attention to the distinction between the covalent, intra-layer terms and the non-covalent, inter-layer Coulombic and van der Waals interactions. The resulting model is compatible with MD force fields for organic compounds, and can correctly describe the interactions of MoS_2 with liquids, yielding excellent agreement with experimental contact angles for water and diiodomethane. Potential of mean force (PMF) calculations using various solvents, including isopropanol (IPA), water (H_2O), dimethyl sulfoxide (DMSO), dimethylformamide (DMF), and N-methyl-2-pyrrolidone (NMP), demonstrate that the use of our force field can explain the current selection of solvents used in experimental studies of the liquid-phase exfoliation of MoS_2 flakes, as well as the observed colloidal stability of the resulting dispersion. Our new force field enables an accurate description of MoS_2 interfaces, and will pave the way for the simulation-aided design of solvent media for the exfoliation of MoS_2, as well as its applications in applications including membranes, microfluidic devices, and sensors.

470736 From Batch- to Continuous Flow Synthesis of Noble Metal Patches on Supports from Colloidal Particles up to Porous Bulk Materials

Abstract:

The preparation of nanoparticles of gold, silver or platinum have subject of intense study for several decades. A wide range of potential applications can be found e.g. in the fields of biomedicine, plasmonics, sensors or catalysis. Achieving controlled and scalable fabrication of asymmetrical particle morphologies by simple methods is a major hurdled to exploiting such materials, which are expected to have enhanced functionality. In the special case of patchy particles, the surface shows an inhomogeneous distribution of chemical and physical properties. Such situations might be achieved by depositing a material like gold, silver or platinum asymmetrically onto an initially homogenous surface. A well-known example of these relatively new types of particles are Janus particles which show a 50% surface contribution of different material and/-or functionality. Most of the fabrication methods for combining noble metals and non-metallic co-particles are based on the immobilization of spherical core particles onto a substrate to shadow a desired fraction of the particles surface against further chemical or physical treatments. Such approaches include several steps which lead to a high complexity and a limited scalability. In contrast, in our group we have developed a facile route to produce patchy particles which is based on the cooperative interaction of metal precursors and other reagents with a core particleâ€™s surface. We have shown that this leads to a strong preference for heterogeneous nucleation and surface diffusion, leading to the surface conformal growth of noble metal patches. This general method can be applied to several patch- and co-particle combinations like Au@PS (polystyrene particles), Pt@PS and Ag@SiO_2. In this presentation it will be shown how this method has been introduced on the small scale in batch processes and further developed into a T-mixer based continuous flow process at different scales. Our results show that the mixing of the educts plays an essential role in the final particles properties like the distribution of metal loading at the co-particles surface â€“ measured by analytical ultra-centrifugation (AUC), morphology and related plasmonic properties in the UV-VIS-NIR spectrum. By an increase of the core particle complexity by an increase of size and porosity as in case of catalyst support materials like porous glass particles, the limits of continuous flow synthesis at lab scale are given by the colloidal instability of the corresponding suspensions. We will show how these limits can be expanded by the use of a CSTR based process which allows the dispersion of the stock- and reactive suspensions by stirring. In addition, by the use of a mixing chamber where substrates can be immobilized, our general method allows the synthesis of novel metal patches on macroscopic porous materials such as silver patches on bio glass scaffolds in the millimeter scale.

468296 Surface and Structural-Modified Pyroxene Nanoparticles for Adsorption and Catalytic Thermal Decomposition of Visbroken Residue Asphaltenes

Abstract:

The energy demand is growing worldwide due to the increase in population and quality of life. Oil will continue to be a major source of current and future non-renewable energy. However, the current upgrading and recovery technologies are energy and water intensive. Hence new cost-effective technologies for enhancing upgrading and recovery of conventional and unconventional oils are needed and of paramount importance.

Asphaltenes are the largest, densest, most polar and surface-active compounds present in crude oil. Asphaltene association and destabilizing problems affect the utilization and recovery of oil from reservoirs, resulting in major complications in the pipelines and wellbores. Consequently, operational costs will be vastly increased and the production rates will be adversely influenced. Various techniques are currently used for removing asphaltenes after deposition. These methods involve production scheme alteration techniques, chemical treatment, mechanical methods, thermal and biological methods. These treatment techniques are expensive and may cause many environmental problems.

To this end, nanotechnology, in the form of nanoparticles, has emerged as an attractive area for the oil industry. The unique physical and chemical properties of the nanoparticles promote their adsorptive and catalytic behaviors towards heavy hydrocarbons. Thus, nanoparticles could be employed as adsorbents and catalysts "nanosorbcats" for in-situ and ex-situ application processes.

This study presents a new environmentally sound and low-cost, yet highly efficient surface and structural-modified pyroxene (PY) nanoparticles. Low temperature hydrothermal synthesis route was used to prepare PY nanoparticles, which were characterized by different characterization techniques like XRD, BET, TGA,

TPD-CO_2, TPD-NH_3, and XPS. The prepared nanoparticles were employed for the adsorptive removal of visbroken residue asphaltenes (VR-C_5), and the solid-liquid-equilibrium (SLE) model was used to describe the adsorption isotherms. The catalytic thermo-oxidative decomposition of adsorbed VR-C_5 asphaltenes was investigated using TGA coupled with mass spectrometry. The catalytic behavior of the nanoparticles was compared based on the significant shift of the oxidation temperature, CO_2 evolution profiles, and the kinetic triplets of the oxidation reaction (i.e., $E\alpha$, $A\alpha$ and $f(\alpha)$). The isoconversional corrected method of Ozawa, Flynn, and Wall (OFW) was used to describe the reaction mechanism and estimate the kinetic triplets. After that the transitional thermodynamic parameters, such as $\Delta S^{\ddagger}$, $\Delta H^{\ddagger}$, and $\Delta G^{\ddagger}$ were investigated to get rise to the meaningful kinetic prediction.

458219 Gold Nanoplate-Based 3D Hierarchical Microparticles with High SERS Enhancement

Abstract:

Formation of intended nano- and micro-structures with regular building blocks has attracted much attention due to their potential applications in the fields of optic, electronic and catalysis. Herein, we report a novel strategy to spontaneously grow individual Au nanoplates into three-dimensional (3D) hierarchical cabbage-like microparticles (CLMPs). By reducing gold precursor to gold atoms, N-(3-amidino)-aniline (NAAN) itself was oxidized to form poly (N-(3-amidino)-aniline) (PNAAN), which specifically binds on Au (111) facet as capping agent and led to the formation of gold nanoplates. Due to the incomplete coverage of Au (111) facet, new gold nanoplate growth sites were spontaneously generated from the crystal plane of existing Au nanoplates for the growth of other nanoplates. This process continued until the nanoplate density reached its maximum range, eventually resulting in CLMPs with well-controlled structures. It opens a new avenue to utilize the imperfection during nanoparticle (NP) growth for the construction of microstructures. The individual CLMP shows excellent surface-enhance Raman scattering (SERS) performance with high enhancement factor (EF) and good reproducibility as it integrates the SERS enhancement effects of individual Au nanoplate and the nanogaps formed by the uniform and hierarchical structures.

472220 Single-Molecule Localization and Orientation for Super-Resolution Microscopy

Abstract:

Structural characterization by super-resolution microscopy has become increasingly widespread, particularly in the biological community. The technique is powerful because it can produce real-space images with resolutions of tens of nanometers, while sample preparation is relatively non-invasive. Previous studies have applied these techniques to important scientific problems in the life sciences, but relatively little work has explored the attainable limit of resolution using samples of known structure. In this work, we apply photo-activated localization microscopy (PALM) to polymer films that have been nanopatterned using electron-beam lithography. Trace amounts of a rhodamine spiroamide dye are dispersed into nanostructured poly(methyl methacrylate), and UV-induced switching of the fluorophores enables nanoscale localization of single molecules to generate a final composite super-resolution image. Features as small as 25 nm half-pitch are clearly resolvable.

We also explore the effect of molecular orientation of fluorophores, which can be used to report nanoscale chemical environment and dynamics in materials. Fluorophore orientation alters the single-molecule point-spread function, which can introduce systematic mislocalizations in imaging. Here, we avoid these phenomena by simultaneously determining single-molecule position and orientation. Distributions of orientations are then collected in response to illumination polarization, as well as mechanical stress and deformation.

448986 Characterization and Performance Evaluation of PVA Nanofiltration Membrane Coupled with TiO2 Nanoparticles

Abstract:

Abstract:

A novel PVA/TiO_2 composite polymer membrane was prepared by dip-coating of polysulfone(PS) UF membrane in an aqueous solution. Glutaraldehyde (GA) was used as a cross-linker for the composite polymer membrane in order to enhance the chemical, thermal and mechanical stabilities. The morphological study was investigated by atomic force microscopy (AFM).

Key words: nanofiltration, PVA, TiO_2 polysulfone

Introduction:

Polysulfone (PS) is one of the most extensively applied ultrafiltration (UF) membrane materials in industry area.But its hydrophobic nature is also well known, which results in the adsorption and deposition of hydrophobic solute on the membrane surface [2,3]. Titanium dioxide (TiO_2) has been the focus of numerous investigations for innocuity , resisting and decomposing bacterias, UV-proof and super hydrophilicity in recent years. Therefore, it has been applied to a variety of problems of environmental interest in addition to water and air purification [4]. In this work, we attempted to disperse the nano-sized TiO_2 particles on the surface of the PVA-coated polysulfone membrane in a reason of enhancing chemical, thermal and mechanical stabilities of the membrane.

2.Experimental:

PVA(1.5 wt%) dissolved in distillated water, was coated on PES barrier layer with dip coating technique. Therefore, GA(5%wt) was crosslinked on the surface of membrane with so called method. TiO_2 nanoparticles were dispersed in distillated water by sonication for 1hr and then stirred vigorously for 30 min .finally the result membrane was immersed immediately in TiO2 resultant bath. the composition of prepared membranes is shown in Table.1.

3.Results and Discussion:

THE GRAPHIC ON THIS PAGE WAS NOT AVAILABLE ON THE E-MEDIA VERSION OF THIS PAPER.

Table 1.Compositions of prepared membrane

THE GRAPHIC ON THIS PAGE WAS NOT AVAILABLE ON THE E-MEDIA VERSION OF THIS PAPER.

Fig.1. Three-dimensional AFM surface image of the PVA/TiO2 (0.2% wt) membrane

Since TiO2 has high affinity to water, the membrane performance of solute transport can be improved. The association between PVA and TiO2 also changed the surface potential of PVA. The obtained results clearly reveal that the antifouling property of TiO2-entrapped membrane is significantly improved .TiO2 nanoparticles plug membrane pores at high concentration .thus, the concentration of TiO2 should be optimized. according to flux and salt retention, the membrane containing 0.2%wt TiO2 nanoparticles had the best performance between so called membranes. In fact, the increment in TiO2 concentration has apparent influence on membrane behavior.but rising to 0.5%wt TiO2, does not significantly affect the membrane yield. Fig. 1 is the three dimensional AFM surface imageof the optimal PVA/TiO2 composite membrane (0.2%wt TiO_2). All roughness parameters of optimal membrane are shown in Table.2.as it can be seen, TiO2 nanoparticles are well-coated on the membrane surface and

THE GRAPHIC ON THIS PAGE WAS NOT AVAILABLE ON THE E-MEDIA VERSION OF THIS PAPER.

Table 2.Surface parameters of the PVA/TiO_2 (0.2 %wt) membrane

461085 Parametric Study on Gold Nanoparticles Flow Synthesis in a Microwave-Assisted Reactor

Abstract:

Gold nanoparticles (AuNPs) present unique electronic, optical and chemical properties which allow them to be used in various sectors such as electronics, catalysis, sensors, biotechnology and biomedical applications. Of key importance in the synthesis is the control of size, shape and particle size distribution (PSD). Narrow PSD of specific shape and size ensures realising the desired properties for the intended application.[1] Microwave heating has been used for nanoparticle synthesis in both batch and continuous flow reactors. The major advantage of microwave over conventional heating is the rapid volumetric and selective heating of the medium. Such heating mode can mediate particle formation on the inner reactor walls and lead to a â€œgreenâ€ technology due to low energy losses.[2]

The current study addresses the synthesis of gold nanoparticles in a flow reactor in a single-mode microwave applicator. Nanoparticles were synthesised using the citrate reduction method (Turkevich) and experiments showed repeatable results. The parameters studied were *(i)* the applied microwave power, *(ii)* the citrate/gold ratio and *(iii)* the flow rate of the synthesis mixture. UV-vis absorption spectra of the samples were taken immediately after collection from the reactor (stage I) and in regular intervals, until the surface plasmon resonance (SPR) band was stable (stage II). Analysis of the samples in stage II showed that increasing microwave power decreased polydispersity of synthesised particles, while SPR bands of the particles were in the range 525-535nm and final particle size was 20-30nm. The aspect ratio of the formed particles decreased with time from 1.5 (stage I) to 1.2 (stage II), indicating particles shifting from ellipsoidal to spherical shape.

Increasing the citrate/gold ratio of the inlet mixture from 1/1 to 60/1, while keeping constant flow rate, decreased the final size and dispersity of synthesized particles in stage II, from ~68 to 10nm, as citrate acts both as reducing and stabilizing agent.[3] Interestingly, stage I samples showed a minimum in particle size at citrate/gold ratio ~10. Increasing the flow rate of the synthesis solution from 4 to 10ml/min resulted in lower outlet reactor temperature, as microwave power is volumetric and affected by the flow rate of the medium.[4] Temperature affected the nucleation and growth stage of nanoparticle synthesis and UV-vis absorption and differential centrifugal sedimentation (DCS) analysis of the samples in stage II showed that increasing the flow rate resulted in larger and polydispersed PSD. A comparative analysis was conducted between control samples and samples in stage II. Control samples were prepared in batch by mixing citrate and gold precursor and letting them react at room temperature, until the surface plasmon resonance (SPR) band was stable. For the same citrate/gold ratio, control samples showed larger, more disperse AuNPs than samples prepared under microwave power in stage II.

References

[1] Seol, S.K., et al., *Microwave synthesis of gold nanoparticles: Effect of applied microwave power and solution pH.* Materials Chemistry and Physics, 2011. **131**(1-2): p. 331-335.

[2] Sturm, G.S.J., et al., *Microwaves and microreactors: Design challenges and remedies.* Chemical Engineering Journal, 2014. **243**: p. 147-158.

[3] Kumar, S., K.S. Gandhi, and R. Kumar, *Modeling of Formation of Gold Nanoparticles by Citrate Method.* Industrial & Engineering Chemistry Research, 2007. **46**(10): p. 3128-3136.

[4] Meredith, R., *Engineersâ€™ Handbook of Industrial Microwave Heating*. The Institution of Engineering and Technology, 1998.

467759 Liquid-Liquid Electrospray: A High-Throughput Nanomanufacturing Platform for the Synthesis of Micellar Nanocomposites

Abstract:

Scalable nanomanufacturing has emerged as a national need as researchers struggle to overcome the challenges associated with high throughput synthesis of high quality nanocomposites for numerous applications. The primary goal of this work is to establish a semi-continuous, tunable, and scalable platform for producing high quality micellar nanocomposites to be employed in biomedical applications, particularly cancer diagnostics and treatment. Liquid-Liquid Electrospray (LLE) is an emerging technology that incorporates electrohydrodynamic atomization (EHDA) via electrospray, an aerosol technique used to atomize fluids in the presence of electrical field, to emulsify organic solvent in water. Unlike traditional electrospray, in LLE, emulsion droplets are formed directly in the aqueous environment.

In our approach, amphiphilic block copolymers (i.e. poly(styrene-b-ethyleneoxide), PS-PEO), are added to the organic phase and processed via LLE, resulting in emulsion droplets that can undergo the 'interfacial instability' method of block copolymer self-assembly. Hydrophobic encapsulants (i.e. quantum dots, superparamagnetic iron oxide nanoparticles) in the organic phase can be encapsulated in the hydrophobic core of the micelle, yielding micellar composites with potential applications in bio-imaging and cell manipulation/separation. In this work, we demonstrate one specific advantage of the LLE-enabled, interfacial instability approach, namely that this process does not require the addition of surfactants to preserve stability; reducing downstream purification process complexity.

Here, we report EHDA of a dielectric solvent (i.e., chloroform) successfully performed via LLE. A high voltage was applied to an insulated electrospray needle placed inside of an aqueous solvent (i.e. ultrapure water), which sprayed the dielectric solvent in the form of a cloud of emulsion droplets. Using this technique, we synthesized empty micelles, as well as micellar composites encapsulating nanoparticles. TEM images of nanocomposites were analyzed, and the results show that the products obtained were monodisperse, between 80 to 90% spheres with a characteristic length of 50 – 75 nm. Ellipse and worm-like micelles were also observed with characteristic lengths of ~ 70 – 135 nm. In our ongoing studies, we are examining the morphology shift from sphere to worm-like micelles and other complex structures in response to increasing the nanoparticle to polymer ratio. In addition, the functionalities of composites are being tested, including fluorescent and magnetic responses of encapsulated nanoparticles. In addition, we will report on a modified version of LLE (LLE-FNP) that promotes the nucleation and growth of polymer amphiphiles, similar to the kinetically driven Flash NanoPrecipitation (FNP) process developed by Prud'homme and colleagues. In summary, by incorporating the electrospray technique into the interfacial instability process, LLE enables the semi-continuous generation of oil/water emulsions, enhancing throughput and eliminating the need for surfactant removal to achieve high concentration purified nanocomposites.

472539 Pushing the Frontiers of Tribology Via Newly Discovered Nanoscale Hybrid Lubricants

Abstract:

The power of modern supercomputers together with the availability of highly scalable atomistic simulation codes have begun to revolutionize the modeling and computational analysis of materials. An enticing illustration of the same comes from our recent work where large-scale molecular dynamics (MD) simulations performed on Argonneâ€™s supercomputer Mira closely coupled with experimentation led to a breakthrough in the field of tribology.[1] An elusive phenomenon termed â€œMacroscale superlubricityâ€ which refers to an extremely low friction state was shown to originate from an intriguing nanomechanical phenomenon: graphene patches wrap around nm sized diamond particles to form nanoscrolls that have significantly reduced contact area to an underlying surface. Atomistic simulations elucidated the overall mechanism and explained the mesoscopic link between the nanoscale mechanics and macroscale experimental observations. Given our recent success with nano-lubricants[1], we will focus on accurately capturing ensemble effects across tribological interfaces using some representative lubricant materials (2-D materials, nanoparticles) and their combinations for different types of environments and operating conditions. This talk also will include brief discussion on our ongoing efforts towards development of a new class of force fields to enable high-fidelity large scale dynamical and statistical simulations of tribo-interfaces, properties and functionalities of new lubricants, as well as pathways and mechanisms of their *in operando* synthesis and assembly.

References:

1. Berman (Sankaranarayanan) et al. Science, 348, 6239, 1118 (2015)

471065 High Dimension Biological Analysis of Carbon Nanotube Toxicity

Abstract:

The toxic effects of multi-walled carbon nanotubes (MWCNTs) have been studied for several years and adverse responses have been identified in several tissues, including cytotoxicity in keratinocyte cells and inflammatory and fibrogenic responces in pulmonary tissues. It has been also found that MWCNTs produce a time- and dose-dependent toxic response upon reaching the lungs in sufficient quantity. Beyond time and dose, chemical functionalization affecting water solubility, dispersibility and agglomeration tendency, as well as impurities, amorphous carbon, surface charge, shape, length, and layer numbers have been recognized as significant modulators of toxicity. MWCNTs have unique physical and chemical properties; they are characterized by high specific surface area and penetration ability, as well as higher activity than micro-sized particles. The fibrous structure of carbon nanotubes determines their pathogenicity, including mechanical damage in cell membranes and other sub-cellular structural elements. The toxic effects of carbon nanotubes are viewed often as the result of an explosive activation of oxidative processes related to generation of reactive oxygen species (ROS). Thus, we surmise that biological effects of MWCNTs and their potentially harmful health implications should be studied in advance prior to widespread deployment in consumer goods and pharmaceuticals. It is believed that an important role in the toxicity of MWCNTs is played by metallic impurities, which cannot always be removed completely by acids. Significant amounts of metal particles can be mobilized from the carbon nanotubes to the surrounding biological microenvironment. The final effect of this interaction will depend on the components and properties of the cellular and intracellular microenvironment and the physicochemical and toxicological characteristics of the metal particles. Based on the above, the current study aimed at identifying the effects of exposure to purified and unpurified MWCNTs at different levels of biological organization, and at shedding light on the potential mechanism(s) of action. For this purpose, (a) immunological, (b) biochemical, (c) gene expression and (d) biological pathway analyses were carried out, combining human biosamples and in vitro testing, focusing on the mechanisms through which MWCNT exposure induces immunological responses. The study involved 12 healthy volunteers with an age range from 25 to 30 years (mean age 27±2.1). Venous blood samples were collected in the early morning following overnight fasting in vacuum EDTA tubes (Vacutainer). The doses of MCWNTs were selected according to the recommended exposure level of 1 $\mu g/m^3$, according to NIOSH.

The immunological study included the evaluation of levels of CD3+CD95+, CD3+CD25+, CD3+CD126+ cells using the BD FACS CANTO II flow cytometer (BD Biosciences, USA). Mitochondrial membrane potential (MMP) was measured by the MitoProbe™ JC-1 Assay Kit using the BD FACS CANTO II flow cytometer. MMP was measured in leukocytes (MMPL) and mature CD3+ T lymphocytes (MMP CD3+). The percentage of cells with low MMP was reckoned on the basis of the measurement results.

Biochemical analyses were used to assess the levels of primary products of lipid peroxidation (LPO) – lipid hydroperoxides (LHP) and the intermediate products of LPO – diene conjugates (DC), ketodienes (KD), conjugated trienes (CT), compounds containing isolated double bonds (CCIDB). The LPO products have different wavelength absorption maxima (λLHP = 233 nm, λDC = 232 nm, λKD+CT = 278 nm, λCCIDB =220 nm), and they are differently extracted by heptane and isopropanol. The heptane phase contains the products of neutral lipid peroxidation; the isopropanol phase includes the products of phospholipid peroxidation. We used different ratios of heptane/isopropanol mixtures for the extraction. For LHP the ratio 2:1, v/v was used; for DC, KD, CT and CCIDB two mixtures in sequence were used at different heptane/isopropanol ratios, namely 1:1 and 3:7, v/v respectively.

A549 cells from a human Caucasian lung adenocarcinoma with the alveolar type II phenotype were obtained from ECACC (Sigma–Aldrich, Milan, Italy). The cells were cultured in Dulbecco's modified Eagle medium (DMEM), 2 mM l-glutamine, 50 IU/ml penicillin, and 50 µg/ml streptomycin, in a humidified atmosphere containing 5% CO2 at 37 0C and grown to 80% confluence. Exposure to purified and unpurified MWCNTs was done on sub-confluent cells. HaCaT cells were cultured in calcium-free DMEM, with 10% chelexed FBS, 4 mM L-Glutamine, and supplemented with calcium chloride at 0.03 mM or 2.8 mM final concentration. FBS was calcium-depleted by incubation with Chelex 100 resin for 1 h at 4 °C according to the BioRad protocol. The Chelex was subsequently removed using a 50 mL Millipore 0.22 um filter unit system. Microarray data analysis identified differentially expressed genes, applying different statistical filters to allow for only the statistically and biologically significant probes to be evaluated. The results of microarray analysis were validated using quantitative real-time polymerase chain reaction (PCR).

All analyses were done in triplicate. Data were presented as median and 25% to 75% interquartile range. Statistica 10.0 (Statsoft) software was used for the statistical analysis. The differences between the groups were analyzed using the Mann-Whitney test. A value of $p \leq 0.05$ was considered significant. Significant probe sets were evaluated for relevance to canonical pathway, molecular function and biological function using the Protein ANalysis THrough Evolutionary Relationships (PANTHER) Classification System (http://www.pantherdb.org). Statistically significant over- and under-represented annotation categories were determined by binomial statistics, using the observed number of expressed genes versus the numbers expected by chance within a certain annotation group. Categories with p-values > 10-2 were rejected.

Our study showed that the MWCNTs with lower content of metallic impurities (Type 1) affected lipid peroxidation parameters inducing pre-inflammatory effects. Oxidative disorders were manifested in the form of increase in the content of the primary products of lipid peroxidation (LPO) and redistribution of the intermediate products (CCIDB, DC, KD+CT) in neutral lipids and phospholipids of blood cell membranes. Type 1 MWCNTs increased the levels of LHP in blood by 1.8 times ($p<0.05$). The load test by the MWCNTs with the higher amount of metallic impurities (Type 2) on the whole blood caused a significant increase in LHP levels as compared to the purified MWCNTs (1.5 times, $p<0.05$) and 2.6 times as compared to the untreated blood. This may be due to the high content of metal ions in the unpurified MWCNTs that originate from Fe–Co catalyst and can be washed away by water. Peroxidation induction is consistent with other studies; it is presumably caused by the pro-oxidant action of iron ions (Fe2+), which generate hydroxyl and alcoxy lipid radicals. The initiation of new lipid peroxidation chain reactions leads to an increase in the levels of ketodienes and conjugated trienes. Exposure to MWCNTs with a minimal content of impurities caused an increase in the percentage of immune cells with reduced mitochondrial membrane potential. The number of mature T lymphocytes with reduced potential underwent a more than three-fold increase. At the same time, upregulation of the pre-apoptotic CD95 marker on CD3+ cells was observed.

Purified MWCNTs activated the early phase of T-cell response and caused an increase in the percentage of the functionally active T lymphocytes expressing interleukin-6 (IL-6 receptor (CD3+CD126+). The damaging effect of unpurified MWCNTs was manifested in the significant increase in the percentage of lymphocytes and CD3+ cells with reduced mitochondrial membrane potential (MMPL and MMP CD3+ respectively). The increased levels of CD3+CD95+, CD3+CD25+ cells were not significantly different from the levels corresponding to exposure to the purified MWCNTs. Whole genome transcriptomics in the lung epithelium (A549) and keratinocytes (HaCaT), which were selected as in vitro models in this study showed that gene expression significantly differs after exposure to purified and unpurified MWCNTs. 48 hours exposure resulted in a significant number of genes differentially expressed compared to the control; approximately 500 genes were expressed differentially in the case of MWCNTs with different levels of impurities. Similar results were found when we analyzed the data isolating the 25 genes involved in the molecular pathway associated to oxidative stress. The picture changes slightly in terms of modulated gene expression, when focusing on the genes involved in inflammation mediated by chemokine and cytokine signaling.

Analysis of the effects of different types of MWCNTs on gene expression showed that impurities influence significantly the induction of key toxicity pathways such as inflammation mediated by chemokine and cytokine signaling. Pathway analysis showed the modulation of genes related to the NFkB pathway, which is more evident for unpurified MWCNTs.

Our study showed that the main mechanism of effect for both types of MWCNTs is oxidative stress induction. Its intensity is directly related to the amount of metallic impurities in the MWCNTs. The subsequent development of inflammatory responses and enhanced peroxidation cause damage to mitochondria in immunocompetent cells. Unpurified MWCNTs generate more oxidation reactions than the purified CNTs, inducing significant mitochondrial dysfunction. Transcriptome analysis showed that activation of the NFkB pathway is the key biological process initiating a cascade of effects. This may cause a perturbation of the IL-6 pathway that aims to regulate the inflammatory processes and compensate apoptotic changes. Overall the immunological responses related to MWCNT exposure are considered as the result of the synergistic effect of systemic (mediated by cells of the routes of exposure) and local inflammation (blood cells).

454757 Toward Roll-to-Roll Production of Nanomaterials Using Microwave

Abstract:

Nanomaterials, such as carbon nanotubes (CNT), metal oxides, and polymers, possess superior mechanical, thermal and electrical properties, lead to broad applications in composite materials, smart structures, chemical sensors, energy storage and nano-electronic devices. However, the high cost and difficulty in getting large scale, high quality nanomaterials remain challenges. We demonstrate for the first time an affordable and scalable microwave approach for the direct growth of CNT, nanostructured metal oxides on a wide range of substrates, including carbon fibers, glass fibers, Kevlar, and Basalt fibers. The microwave initiated nanomaterial growth will take only 20-30 seconds under the microwave irradiation at room temperature in air, no need of any inert gas protection, and additional feed stock gases, usually required in CVD approach. The as-produced CNTs possess hollow centers, with the diameter ranging from 10-100 nanometers, and the length varies from several to tens of micrometers, which is the characteristics of multiwalled carbon nanotubes (MWCNT). Nanomaterials, including CNTs, meal oxides and metal sulfides, were manufactured using this microwave approach.

472319 Invited Presentation: Carbonaceous Nanoparticles and Their Interactions with Biological Cells

Abstract:

Particles originating from human activities have existed for millennia, e.g., smoke from combustion, but the recent development of industry and combustion-based engine transportation has increased anthropogenic particles pollution. At the same time, technological advancement has also changed the character of these particles, increasing the proportion of nanometer-sized particles --"nanoparticles"-- and expanding the variety of chemical compositions. Indeed, the manipulation of matter at the scale of atoms, "nanotechnology," is creating many new materials with characteristics not always easily predicted from current knowledge.

In this work we report on the interactions of carbon-based nanomaterials (both from combustion and synthetic sources) with biological cells, using atomistic simulations. As engineered nanomaterials, we will focus on carbonaceous quantum dots, which have recently emerged and ignited tremendous research interest. Their favorable characteristics include size- and wavelength-dependent luminescence, resistance to photobleaching, bio-conjugation, and functionalization to produce chiral nanostructures. Carbon-based quantum dots show promise in areas such optoelectronics, catalysis, bioanalysis and drug delivery.

463630 Nanomaterials: Occupational Exposure and Toxicity in Context

Abstract:

At the forefront of particulate health concerns are nanomaterials. This talk will propose some considerations for evaluating nanomaterials and their health effects.

1. Is the material as prepared truly â€œnanoâ€? For example, even though SWNTs satisfy the accepted definition of being nanoscale in 2-dimensions, in practice and certainly as synthesized, they are highly coalesced into aggregates of micron in size.

2. What distinguishing characteristics apply to the particular nanomaterial? High surface area is a universal hallmark of a nanomaterial. Yet fumed silica with a surface area of > 200 m^2/g is rated biologically safe and even used a food additive.

3. What are the synthesis methods? Even if synthesis is conducted within a closed system, subsequent processing steps of harvesting and processing can create far greater exposure risk.

3. Do the current markets and/or applications warrant concern? For example quantum dots contain heavy metals, yet they are a commercial product used in limited quantities for biological imaging.

4. What are the exposure routes and levels likely to be encountered in the manufacture, product development and consumer use of nanomaterial based products? Examples in which the nanomaterial, manufacturing process and product are generally considered safe include nanoclays used in various commercial polymers, TiO_2 in sunscreen and carbon black in car tires.

5. Is there any relevant information on related materials and/or nanoparticles and if so, is it applicable?

Examples of nanomaterials and synthesis conditions will be used to address these questions.

The talk will balance these considerations with those nanomaterials derived accidentally such from combustion-based stationary power generation or transportation. For the general population these sources overwhelm their potential exposures to nanomaterials, even after decades of study and emission controls. In fact recent studies have linked soot with asthma, heart arrhythmias and even death.

465578 Dose-Dependent Intracellular Reactive Oxygen and Nitrogen Species Production from Particulate Matter Exposure: Results from Ambient Samples and Chamber Experiments

Abstract:

Epidemiological studies have found associations between elevated particulate matter (PM) concentrations and increased incidences of cardiovascular and respiratory diseases. PM-induced oxidative stress has been suggests as a possible mechanism leading to these health effects. Here we present chemical and cellular oxidant measurements from ambient PM samples and controlled chamber experiments. We investigated reactive oxygen and nitrogen species (ROS/RNS) production from ambient PM samples and laboratory-generated secondary organic aerosols (SOA). Ambient samples ($n = 104$) were collected during multiple seasons from rural and urban sites around the greater Atlanta area as part of the Southeastern Center for Air Pollution and Epidemiology (SCAPE). SOA were generated from the photooxidation of six commonly emitted volatile organic compounds (VOCs). These experiments were conducted in the Georgia Tech Environmental (GTEC) facility under both dry and humid conditions in the presence of ammonium sulfate seed using H_2O_2 as an OH radical precursor. Murine alveolar macrophages were used to measure intracellular ROS/RNS production, while dithiothreitol (DTT) was used to quantify the concentration of redox-active species in each sample. For each sample, intracellular ROS/RNS was characterized using response parameters obtained by fitting dose-response data using the Hill equation. These response parameters (maximum response, EC_{50}, Hill slope, threshold, and area under the dose-response curve) were compared with DTT activity to determine whether chemical assays represent cellular ROS/RNS responses and with water-soluble PM components to elucidate species associated with PM-induced ROS/RNS production. From both ambient and chamber filters, we found that ROS/RNS production was highly dose-dependent, non-linear, and could not be represented by a single concentration measurement. We demonstrate that there is no simple correlation between DTT activity/PM composition and cellular ROS/RNS responses. For summer ambient samples, we found that DTT activity and organic species were significantly correlated with cellular ROS/RNS production, which motivates further studies involving PM-induced ROS/RNS production from summertime photochemically driven SOA. Finally, from chamber filters, we found that SOA generated from different precursor VOCs contained different concentrations of redox-active species and induced different levels of cellular ROS/RNS.

462217 Biofuel and Reference Diesel Particles: Differences in Inflammatory and Oxidative Effects

Abstract:

Particulate matter (PM) from engine exhaust has become an increasing concern due to its links to numerous adverse effects on human health and the environment. Recent efforts to reduce PM emissions and the associated climate change impacts from combustion include new technologies, such as the installation of diesel particulate filters and the use of biofuels. However, studies comparing the biological effects of PM from biodiesel and petroleum-based diesel provide conflicting results, and interpretation of these results is difficult because of different methods for particle generation and for biological assays.

This study aims to systematically evaluate the effect of particles generated from two biofuel blends, one biodiesel (BD) and one alcohol diesel (AD) and one reference standard diesel particle sample (NIST SRM 2975) on selected biological outcomes associated with responses to diesel particles in lung cells and in animals. These include: (a) CYP1A1 and 1B1 enzymes that can both detoxify and bioactivate PAHs in human airway epithelial cells, human bronchial epithelial cells and human macrophage cells lines, and (b) IL-8 and TNF-a that play critical roles in pulmonary host-defense and inflammatory responses. We also examined the role of activation of the transient receptor potential ankyrin-1 (TRPA1) cation channel, which is a target for diesel exhaust and similar combustion-derived particles and mediator of pro-inflammatory and irritant responses. Because PM deposits in small airways and the alveolar space, cell lines derived from small airway epithelial cells (H441) and modeling alveolar macrophages (THP-1) were selected for this study. PM was generated from simple surrogate fuels and characterized under well-controlled combustion conditions.

The results show that particle extracts caused similar responses compared to whole particles in both upper airway epithelial and macrophage cells. This suggests that the soluble organic components adsorbed on the particle surface are highly associated with the cellular responses we examined. When comparing the particles from the different fuel sources, BD particles are capable of activating key enzymes involved in PAH metabolism to very high levels compared to RD and AD. BD particles also had the greatest surface area, soluble organic fraction and oxidative potential. In contrast, RD was more pro-inflammatory than BD or AD, an effect that involved TRPA1 activation. This study adds to the growing body of evidence that biodiesel PM results in adverse lung health effects, potentially via different pathways than conventional diesel PM.

470397 Real Life PM Emissions from Traffic and Human Exposure Implications

Abstract:

The contribution of vehicle emissions to urban air pollution has been extensively investigated. Such emissions have been found to account for almost 90% of the total amount of substances released into urban ambient air. The ability of solid microparticles to penetrate deep into the human respiratory tract results in increased morbidity and mortality. There is also evidence that car exhaust gases are directly involved in the pathogenesis of several health endpoints, including respiratory, cardiovascular, neurological, and allergic diseases, as well as asthma. With regard to smaller particles that penetrate deeply across human respiratory tract, we need to highlight that particle translocation into systemic circulation is one of the major mechanisms related to air pollution attributed cardiovascular disease. Taking into account all the above, regulatory frameworks have been developed worldwide. Obviously, during the operation of a car, the emission control technologies efficiency declines with mileage, resulting in increased exhaust emissions; however, emissions of conditionally new cars that have a mileage of up to 1,000 km are not studied and unfortunately, not regulated. Considering all the above, the aim of this study is to characterize particulate (segregated by size and active surface) released by cars with low mileage and compare the results with those obtained from in-use vehicles with a mileage of 60,000-280,000 km, as well as to investigate the potential implications to human exposure.

The cars with high mileage and the ones with low mileage were kindly provided by the author colleagues. Conditional codes were introduced for the cars so as to avoid specifying the car manufacturers and models, covering various individual vehicle characteristics including manufacturing year, engine capacity and fuel type. For the aims of the study, the exhaust gas suspension (EGS) method was selected. The measurement methods were patented and used according to the following plan:

1. The test car was started and ran for 1-3 minutes before all extraneous dust and soot particles that had

settled there during the downtime and brought from the outside in the tailpipe were removed.

2. Then, the engine was shut off and the hose was connected to the tailpipe of the test car. After that, the hose was dipped into the plastic container, which was filled with 10 liters of deionized water.

3. After that, the car was started and worked at the neutral speed for 20 minutes. This time interval was chosen considering that the engine had been warmed up for approximately ten minutes (according to a coolant temperature sensor). Then, the warmed up engine idled for the remaining ten minutes.

Once the measurements had been completed, the distilled water container through which the exhaust gases had passed, was sealed with a lid and then sent to the laboratory. A significant advantage of this method is the lack of transformation of aerosols and solid particles. To measure particle size distribution, a laser particle size analyzer supplied with Fritch MaS software was employed. PM size distribution of the wet or dry dispersion units can be measured separately or simultaneously, with automated switching features. Optimal dispersion was accomplished in the NanoTec integral wet dispersion unit by using a combination of a robust, variable speed centrifugal pump with powerful ultrasonification.

To translate the different emission profiles identified in the study into actual exposure and uptake, taking also into account the deposition across HRT, a comprehensive methodological scheme has been developed, taking into account:

a. The amount of PM emitted per km travelled by the vehicles investigated, using the COPERT 4 methodology. PM emission factors (in mass of PM per km traveled) for the different vehicles is necessary so as to account of the overall mass of PM emitted by each individual vehicle type.

b. The PM size distribution of the emitted PM, as identified by the laser particle size analyzer.

c. The intake fraction (IF) of the emitted PM. Intake fraction is a metric of the emission-to-inhalation relationship, facilitating comparisons among sources (in our case the individual vehicles) in terms of their exposure potential. For a given emission source and pollutant, intake fraction is the cumulative mass inhaled by the exposed population divided by the cumulative emissions. Considering that IF depends on several parameters affecting the emission-to-intake process, (e.g. prevalent wind, emissions strength, population density), it is expected that it would vary with location and time. In a recent global intraurban intake fraction study, it was found that population-weighted mean intra-urban intake fraction for cities is equal to 32.

d. The deposition of PM across the HRT. To better understand the exposure implications of size-segregated particles distribution emitted by the investigated vehicles, HRT particle deposition modeling was applied to reckon the PM fraction deposited (DF) to the three parts of HRT. Major mechanisms of PM deposition across HRT include diffusion, sedimentation and impaction. Secondary mechanisms involve interception and electrostatic deposition. Different HRT regions involve different deposition mechanism as a function of PM size as follows:

- Naso-pharyngeal region (or upper respiratory tract – URT): impaction, sedimentation, electrostatic (particles > 1 μm)

- Tracheo-bronchial (TB) region: impaction, sedimentation, diffusion (particles < 1 μm)

- Pulmonary (P) region: sedimentation, diffusion (particles < 0.1 μm)

Several parameters that affect HRT deposition have been taken into account, including PM properties (concentration and size distribution), air flow parameters (lung capacity and breathing frequency) and HRT physiology (structure and morphology). HRT deposition was carried out using the Multiple Path Particle Deposition (MPPD) v. 2.1 model. Software inputs include morphological parameters of the pulmonary system – functional residual volume (FRC), tidal volume (TV), upper respiratory tract (URT) volume, as well as breathing frequency (BF) for each age group.

The results of the study indicated that the mean diameter of particulate emitted from high mileage cars ranges in hundreds of microns. Smaller particles were found in cars that were manufactured over 10 years ago and have considerable mileage. Thus, particles were detected in five of the 17 cars. The same percentage of cars (30%, 5 out of 17) is the source of PM50 class. Particles with a mean diameter of more than 100 µm were found in most samples of the cars without mileage (58%, 10 out of 17), but 42% of the cars without mileage are a source of particles of up to 20 µm. It appears that these aerosols correspond to different agglomeration stages. Data on the surface area of the exhaust gas particulate showed that entirely new cars emit extremely small particles with a large surface area (up to ca. 90,000 cm^2/cm^3).

The qualitative analysis of the exhaust gas particulates showed that there are limited differences in their composition between cars with high and low mileage respectively.

In terms of implications for population exposure and public health, age-dependent physiological differences are also of special importance. Taking into account the vehicle emission profile (mass of PM emitted per km travelled and PM size distribution), the respective intake fraction and the HRT deposition based on the size distribution, the amount of PM retained by all HRT regions was estimated. Among the newest vehicles, the highest amount of PM retained is associated to the emissions of the diesel vehicle VT 2012; the very low mean diameter of the emitted PM implies that its vast majority is within the respirable size spectrum. In general, among the newer vehicles, 0.0096 µg were deposited per km travelled for the diesel vehicles, while the respective amount for the gasoline cars was 0.0074 µg. Regarding older cars, gasoline cars contribution was the same to the newer ones (0.0074 µg per km traveled), while for diesel vehicles the amount (0.031 µg per km traveled) was the highest among all investigated vehicle categories. This observation is attributed to the higher amount of overall PM emitted (per km travelled) by older vehicles. Considering that newer diesel vehicles emit PM of lower aerodynamic diameter compared to older diesel vehicles, a higher fraction of the PM taken up is expected to deposit to the lower HRT.

Among the investigated vehicles, the diesel vehicle encoded VT 2012 combines all characteristics that may be potentially harmful for public health; the lowest average PM diameter (3.84 µm) that results in significant PM deposition across HRT (and especially the lower part), the highest active surface (89,871 cm^2/cm^3), and the highest adsorption potential of organic (e.g. PAHs) and metals.

A key finding of the study is that car mileage has virtually no effect on the size of the solid particles released into the environment, and that even cars with very low mileage (up to 1,000 km) are not "clean" in terms of emissions. On the contrary, special attention has to be paid to the lower aerodynamic diameter related to newer diesel vehicles, their higher specific surface (as well as the higher capacity of toxic substances adsorption) and how this is translated into actual increased human exposure and, consequently, health risk.

467482 Disinfection of Water Using Silver and Copper Nanoparticle Impregnated Activated Carbon

Abstract:

Access to safe drinking water is one of the biggest challenges across the globe. In this context, activated carbon is widely used as a water purification media for producing drinking water, which although removing most of the contaminants (responsible for taste, colour, odour, organics, heavy metals etc.), however is incapable of killing pathogens. To mitigate this, antibacterial properties of silver and copper nanoparticles (Ag-NPs and Cu-NPs) have been utilized. Therefore, the objective of this research is to combine both purification and disinfection effects of activated carbon and nanoparticles together and thereby produce drinking water in a continuous, flow-column set-up.

To this end, first we synthesized Ag-NPs by UV reduction of aqueous silver nitrate solution, in presence of tri-sodium citrate. The latter acts both as a reducing and a stabilizing agent. Following this, Ag-NPs were selectively impregnated on the external surface of plasma treated AC granules. This hybrid is referred to as Ag-AC hybrid, having 0.8 wt% of Ag-NPs of 28 nm mean size. Subsequently, microbial cell-killing experiments were performed in two set-ups: (i) shake flask (called batch experiments) and flow-column (called continuous-mode). We have used *E. coli* K12 (MTCC 1302) as a model organism (as it is typical of contaminated drinking water), always at a concentration of 10^4 CFU cells/ml in the inlet water. Batch results using 8 mg Ag-AC hybrid/ml of cell suspension showed that, all cells were killed within 25 minutes of contact time. Maintaining same *E. coli* concentration and Ag-NP loading percentage in Ag-AC (as in the batch experiment), a glass column (8 cm diameter and 25 cm in height) packed with Ag-AC was used for monitoring cell-killing performance over a long duration. Again, inlet water having 10^4 CFU of *E. coli*/ml could be completely disinfected to produce treated, outlet water having zero cell count, with a contact time of 23 minutes, very close to the designed value based on the previous batch cell-killing kinetics. Therefore, a continuous throughput of treating 1.62 L of contaminated water per hour could be maintained for at least up to 16 days, beyond which experiments were not conducted presently. Moreover, the Ag concentration in the outlet water reached only up to a maximum of 29.8 μg/L at steady state, which is well within the recommended maximum Ag concentration limit of 100 μg/L for drinking water (EPA). Thus these results demonstrate the viability of such a column-design to produce decontaminated water at a reasonable rate per hour, over a long period and with minimal loss of Ag in the treated water.

Cu-NPs being reported for its antibacterial activity too, are actually cheaper to make than Ag-NPs. For some of the Ag-resistant microorganisms, Cu-NP also shows better activity compared to Ag-NP. Thus, Cu-NPs have been synthesized using the thermal reduction method where L-ascorbic acid reacts with cupric chloride dihydrate to form Cu-NPs. In this green route, L-ascorbic acid works as a reducing agent. Cu-NPs of mean diameter 5.17 nm were synthesized and subsequently impregnated on plasma treated AC by wet impregnation. The Cu-AC showed a copper loading of 2.1 wt. %. Disinfection studies were performed using these Cu-AC, again in both batch and continuous modes, using the same wild type model microorganism *E. coli* K12 (MTCC 1302), starting with the same initial cell concentration of 10^4 CFU/ml. Similar to Ag-AC, on using 8 mg/ml of Cu-AC in the batch experiment, zero cell count was achieved in only 10 minutes. Based on this batch mode cell-killing data, the same glass column (but packed with only Cu-AC now) was used for continuous, flow experiments. When 10^4 CFU/ml of *E. coli*cells were introduced in the inlet water to this column, zero cell count in the outlet water was achieved in a contact time of 9 minutes, giving a maximum treated water flow rate of 3.77 L/h. Therefore, compared to 1.62 L/h of water in case of Ag-AC, a much higher amount of 3.77 L/h of water could be produced in the same column with Cu-AC. Simultaneously, in steady state continuous flow-column experiments, the maximum copper concentration in the outlet water was around 83.4 μg/L, which is far below the upper recommended limit of 1000 μg/L (EPA). Thus, Cu-AC shows better cell-killing performance compared to Ag-AC. This is primarily because the Cu loading (2.1 wt%) is higher in Cu-AC, compared to Ag loading of 0.8 wt.% in Ag-AC. Moreover, the steady state release concentration of Cu was much higher (83.4 μg/L) compared to Ag release concentration (29.8 μg/L) from Ag-AC. Therefore, Cu-AC shows higher antibacterial activity compared to Ag-AC and reaches zero cell count within a shorter contact time of 9 minutes in the continuous flow-column, in contrast to 23 minutes required for Ag-AC column to achieve zero cell count.

As Cu-AC shows better antibacterial activity compared to Ag-AC, further experiments are being pursued now with a mixture of Ag-AC and Cu-AC. Both these samples are mixed at 1:1 wt. ratio, that is, Ag-AC having 0.8 wt% Ag loading and Cu-AC having 2.1 wt% Cu loading. These have also been further used in both batch and continuous, flow-column experiments for cell-killing. Starting with the same 10^4 CFU/ml *E. coli*, zero cell count has been achieved in 8 minutes, using 8 mg/ml of the hybrid mixture. This implies that Ag-AC and Cu-AC combination potentially gives a synergistic effect in terms of *E. coli*cell-killing and is hence superior to the usage of only Ag-AC or only Cu-AC. Currently, further continuous column experiments are in progress, using the combination of Ag-AC and Cu-AC to optimize its performance towards the fastest possible cell-killing kinetics, which will maximize the production rate of drinking water.

Hence, water disinfection for potable quality (drinking) water (zero *E. coli* count with < 100 μg/L Ag and < 1000 μg/L Cu) could be achieved in a continuous manner at a large flow rate of nearly 4 L/h over a long duration, by using mixed packing of Ag-AC and Cu-AC.

457443 Metal Organic Framework Composites for Chemical Protection

Abstract:

Several MOFs have shown remarkable capabilities of degrading chemical warfare agents (CWAs) and their simulants. However, the exploration of MOFs in engineered materials such as textiles and clothing for chemical protection has not been prominently explored. It is envisioned that the next generation of protective clothing can shield environmental contaminants by integrating reactive materials, such as MOFs, into the fabric, where in addition to providing a physical barrier, the protective clothing can also undergo self-decontamination. Toward that end, we have engineered a nonwoven fiber composite containing reactive Zr-based MOFs that show the ability to degrade G- and V-type nerve agents. Our approach allows us to load a high weight content of MOFs on the fabric and offers access to reactive sites on the MOFs. We demonstrate through simulant and chemical agent reactions that MOFs in the fabric matrix remain highly reactive and mimic reactions of native MOF powders. These newly developed textiles are further evaluated for breathability, moisture vapor transport, and toxic gas permeation rates across the composite. We use DRIFTS to quantify the interactions between the substrate and reactive sites on the MOF to gain understanding of their reactivity. Our approach will serve as a platform for engineering MOFs into usable textiles without compromising MOF reactivity and performance.

473970 Atomistic, Coarse-Grained, and Statistical Mechanical Modeling of Dynamic DNA Nanostructures

Abstract:

DNA nanotechnology takes advantage of specific base-pairing interactions to self-assemble DNA strands into 2D and 3D nanostructures with precise shapes of increasing complexity, ranging from smiley faces to gear parts. However, most of these structures are structurally rigid and their mechanical functionality has not been explored. Recently, a new class of DNA-based nanostructures have been developed that combine the structural rigidity of double-stranded (ds) DNA components with the flexibility of single-stranded (ss) DNA connections to yield dynamic structures with tunable mechanical properties. Here we present our recent efforts in modeling several of such dynamic DNA nanostructures at various scales relevant to their design and function. In particular, we discuss the use of atomistic, coarse-grained, and statistical-mechanics models in providing new insights into the dynamical behavior of such nanostructures and in developing simple and useful design rules for achieving structures with targeted sensing and actuating properties.

456130 Self-Assembled Collagen-Mimetic Triple Helices with Antimicrobial Peptide Amphiphiles As Novel Antibacterial Agents

Abstract:

1. Introduction

The emergence of bacterial drug-resistance to antibiotics has caused significant difficulty for the treatment against infectious diseases. Bacteria can develop resistance within a few years after a new antibiotic is introduced. For instance, methicillin was first introduced into clinical use in 1959, which was rapidly followed by the emergence of methicillin-resistant *Staphylococcus aureus* (MRSA) reported for the first time in 1960. Moreover, MRSA are able to continuously develop a resistance to a broad range of antibiotics. It was reported that MRSA caused at least 80,000 invasive infections in the United States alone, and 53 million people worldwide may be colonized with MRSA. Therefore, the objective of this study is to design and develop novel self-assembling antimicrobial peptide amphiphiles to combat the antibiotic-resistant *S. aureus*.

Cationic antimicrobial peptides (AMPs) are naturally occurring short peptides that can penetrate across bacterial cell membranes and inhibit the development of bacterial drug-resistance. The antibacterial property of AMPs can be facilitated by their charge and secondary conformations. First, unlike the commonly used antibiotics, cationic AMPs can disrupt the bacterial cell membrane by electrostatic attachment and insertion, and these non-specific interactions can thus reduce the development of drug-resistance as the entire bacterial cell membrane is the main target site of most AMPs. Secondly, the amphiphilicity of AMPs, which represents the peptide, contains both hydrophilic charged domains and hydrophobic non-polar domains, can organize the AMPs into secondary conformations (e.g. β-sheet and α-helix) and facilitate the insertion of AMPs into bacterial cell membranes. Furthermore, since the collagen adhesion domains on the membrane of *S. aureus* have been shown to have a high affinity to the collagen-like triple helical structure, cooperation of a self-coiling collagen-mimetic peptide sequence with AMPs can be a novel and promising approach to promote the accumulation of AMPs on *S. aureus*, further enhancing the antibacterial efficacy of the AMPs.

2. Materials and Methods

Two types of self-assembled supramolecular nanostructures were successfully prepared in aqueous solution in this study: the liquid extracellular matrix-like (LEM) nanofibers and the amphiphilic Cardin antimicrobial (ACA) nanorods. Each of these two nanostructures was constituted by rationally designed amphiphilic peptide building blocks, which was driven by hydrophobic interactions and β-sheet stacking. The hydrophobic interactions were attributed to the aliphatic tail group conjugated to the N-terminus of the peptide amphiphiles, and the β-sheet stacking could direct the self-assembled structures into cylindrical shapes that facilitate the insertion into bacterial membranes.

In water, the self-assembled LEM nanofibers and ACA nanorods presented two different bioactive properties on the surface. The $(Gly\text{-}Pro\text{-}Hyp)_6$ repeating sequences of LEM nanofibers formed a collagen-mimetic triple helix, whereas the Cardin antimicrobial motifs $(Ala\text{-}Lys\text{-}Lys\text{-}Arg\text{-}Ala)_2$ were exposed on the outer surface of the ACA nanorods. More importantly, as the peptide amphiphiles of LEM nanofibers and ACA nanorods contained the same hydrophobic domain and β-sheet forming domain, these two peptide amphiphiles were combined into hybrid nanofibers by mixing LEM nanofibers and ACA nanorods at equal molars. These hybrid nanofibers then co-assembled via β-sheet stacking in the process of self-assembly, carrying both collagen-mimetic groups and Cardin antimicrobial motifs. Followed by the preparation of the self-assembled supramolecules, the morphology and size of the self-assembled structures were characterized by a transmission electron microscope (TEM) and dynamic light scattering (DLS) spectroscopy.

In bacterial assays, Gram-positive *S. aureus* (ATCC#25923) and MRSA (ATCC#43360) were incubated in 0.3% tryptic soy broth (TSB) media containing the self-assembled nanoscale supramolecules of peptide amphiphiles at concentrations from 20 to 100 μM at 37°C, and the growth curves were monitored by the optical density of a wavelength at 562 nm (O.D. at 562 nm) for 20 h. Also, to evaluate bactericidal effects, the self-assembled supramolecules of peptide amphiphiles at concentrations from 20 to 100 μM in 0.3% TSB were used to treat *S. aureus* and MRSA for 4 h in 37°C, and the viability of bacteria were quantified by colony forming units per ml (CFU/ml) after treatment. The viability of bacteria was evaluated by: [(CFU of treated sample)/ (CFU of untreated control sample)]*100%. All experiments were run in triplicate and repeated at least two times (N≥2) to demonstrate significance.

3. Results and Discussion

The TEM characterization showed successful self-assembly of the peptide amphiphiles discussed above. The self-assembled LEM nanofibers were about 10 nm in diameter and microns in length; the ACA nanorods had similar diameters but were shorter in length to about 100 nm, which probably resulted from the high cationic charge of the Cardin antimicrobial motifs. The hybrid nanofibers, which were combined by the LEM nanofibers and ACA nanorods, also exhibited diameters to about 10 nm and had a highly aggregated network structure, and the length of these nanofibers ranged from about 200 nm to 500 nm. The results of DLS measurements showed that although the hydrodynamic diameter of the LEM nanofibers (603 nm) was much larger than that of ACA nanorods (17 nm), the hybrid nanofibers had uniform hydrodynamic diameters (26 nm), indicating the successful co-assembly of the two peptide amphiphiles.

In bacterial assays, the ACA nanorods significantly inhibited the growth of both *S. aureus* and MRSA, as well as had high bactericidal effects towards these two bacterial strains. The 20 h bacterial growth curve of bacteria showed that the ACA nanorods at a concentration of 100 μM delayed the lag phase of *S. aureus* and MRSA to 18 h, whereas the untreated bacteria entered the exponential growth phase after the lag phase within 3 h. Also, the ACA nanorods at concentrations higher than 80 μM reduced the survival rate of both *S. aureus* and MRSA to as low as 10% compared with the untreated control sample. The results demonstrated for the first time that the self-assembled ACA nanorods had promising antibacterial properties against Gram-positive *S. aureus* as well as MRSA, suggesting that these self-assembling peptide amphiphiles could combat bacterial drug-resistance. In our ongoing studies, the antibacterial effects of the hybrid nanofibers in cooperation with a collagen-mimetic group, will be further elucidated.

459460 Design of 3-Helix Micelles with Tailorable Sizes and Shapes

Abstract:

Hierarchical self-assembled micelles from amphiphilic hybrid 3 helix peptide-polymer-lipid conjugates have shown great promise as nanocarriers in delivering therapeutic compounds in a targeted fashion. In order to realize the vast therapeutic potential of these micellar nanocarriers as well as the purpose of penetrating deep tissue or other biological barriers, there is a need to understand how PEG conjugation affects the micelle shape and size as well as self-assembly mechanisms. Atomistic molecular dynamics (MD) simulations of the self-assembly behavior in such length scales is challenging, which necessitates the development of a simpler coarse-grained (CG) model. First, we use a CGMD model based on the dissipative particle dynamics (DPD) technique to reveal the internal structure of 3-helix micelle and the behavior of conjugated PEG chains [1]. Next, we study the self-assembly patterns of the amphiphiles with different PEG molecular weights and conjugation positions on peptide bundles by considering the micelles aggregation number in the simulation box. We discover that the micelle size and stability is dictated by a competition between the entropy of confined PEG chains in the conjugation state, as well as intermolecular interactions among PEG chains that promote cohesion between neighboring amphiphiles. Furthermore, a computational phase diagram that can be used to design 3-helix micelles is conducted. This work opens pathways to control the morphology of self-assembled micelles with expectable and tailorable size, shape and stability as nanocarriers in drug delivery and other applications.

REFERENCES

[1] Ang, J.; Ma, D.; Lund, R.; Keten, S.; Xu, T. Internal structure of 3-helix micelle revealed by small-angle neutron scattering and coarse-grained MD simulation (In submission).

464817 Bioactive DNA-Peptide Nanotubes As Artificial Extracellular Matrices for Bone Tissue Engineering

Abstract:

Introduction: The extracellular matrix (ECM) is a complex network composed of collagen, fibronectin and other macromolecules that not only provides structural support but also relays crucial biochemical and biomechanical signals[1]. It dictates how cells attach, behave and interact with one another, thus, it plays a key role in tissue morphogenesis, differentiation and repair. In this context, the construction of artificial matrices that mimic the structure of the naturally occurring ECM and display biological signals might prove to be highly promising in controlling tissue regeneration[2]. Although electrospun fibers, polymeric hydrogels and self-assembled peptide amphiphiles have been widely investigated for cell adhesion and proliferation, thereâ€™s a need for scaffolds that are structurally programmable and easy to modify. DNA has emerged as one of the most promising building blocks for such nanoconstruction due to the high predictability of Watson-Crick base pairing, enabling the generation of one, two, three-dimensional assemblies for templated growth of nanowires, the organization of nanoparticles and as drug delivery carriers[3]. In the present study, DNA nanotubes were constructed by thermal annealing based on double-crossover motifs[4] and were further covalently functionalized with a BMP-7 derived peptide. The structure of the nanotubes was inspired by collagen with similar size and stiffness, and the BMP-7 derived peptide was selected because of its role in regulating the proliferation, differentiation and apoptosis of bone cells. The objective of the present study was to develop an extracellular microenvironment that can mimic the structural and regulatory characteristics of natural ECMs. Results showed that the DNA nanotube coated glass coverslips revealed a tangled network of fiber-like materials that resemble the morphology of natural ECM, suggesting DNA-peptide nanotubes could be a promising candidate as artificial extracellular matrices. Further experiments are still need to fully investigate their effects on osteoblast cell adhesion and proliferation.

Materials and Methods: Preparation of oligonucleotides stocks: All DNA strands were purchased with polyacrylamide gel electrophoresis (PAGE) purification from Integrated DNA Technology (IDT), dissolved at 500 Âµ M in IDTE buffer (IDT) and stored at -20 Â°C prior to use. For generating fluorescent tubes, the S3 sequence was ordered from IDT with a 3â€™-fluorescein (FAM) modification. Preparation of DNA-peptide hybrid conjugates: The BMP-7 derived peptide Ser-Asn-Val-Ile-Leu-Lys-Lys-Tyr-Arg-Asn (SNVILKKYRN) was conjugated to a DNA strand by Biosynthesis using a previously reported[2] Huisgen 1,3-dipolar cycloaddition reaction between a strained cyclooctyne and an azide. The resulting conjugate was purified by multiple high-performance liquid chromatography (HPLC) and characterized by mass spectrometry. DNA nanotube annealing: To form the DNA nanotubes, oligonucleotide stocks were further diluted to 50 µM in a TAE/Mg buffer (40 mM Tris, 20 mM acetic acid, 1 mM EDTA and 12.5 mM $Mg(OAc)_2$) and equal amounts of all constituent strands were mixed. The mixture was annealed from 95 °C to 25 °C over ~6 hours in a thermal cycler. After annealing, the solution was stored at 4 °C. Preparation of APTES-treated glass coverslips: In order to bind negatively charged DNA nanotubes, glass coverslips were treated with (3-Aminopropyl) triethoxysilane (APTES). The resulting coverslips had a positive charge at physiological pH. Briefly, to coverslips (12 mm in diameter) in a 12-well tissue culture plate, 100 µl of APTES were added into the spaces between wells (600 µl total). The plate was sealed with two layers of tape, wrapped in double layer of aluminum foil and then placed in oven at 90 °C for 18 hours. The coverslips were washed by ethanol and then DI water. Excess water was blown off by nitrogen gas and the coverslips were dried in a vacuum oven at 115 °C. Surface coating: To coat the surface of glass coverslips, 400 µl of the nanotube stock solutions were added on APTES-treated glass coverslips in a 24 well-plate and were incubated at 37 °C overnight to allow adhesion on the surface. Following incubation, the solution was removed and the coverslips were rinsed twice with sterile PBS before seeding cells. Transmission electron microscopy: To characterize size and morphology of DNA nanotubes, they were visualized using transmission electron microscopy operated at 80 kV. Briefly, 10 µl of the stock solution was applied on 300-mesh copper-coated carbon grids for 5 min. Following wicking and rinsing with deionized water, the grids were then stained with 10 µl of a 1.5% uranyl acetate solution twice for 10 s and dried for 15 min prior to imaging. Fluorescent imaging of the surface coating: Following annealing, the DNA nanotube solution was coated on APTES-treated coverslips according to the method described in the surface coating section. Images of the coatings were taken using a fluorescence microscope under 40X objective. A Z-stack was acquired to estimate the approximate thickness of the coating and a 3D reconstruction of the z stack was obtained using a 3D viewer plugin in ImageJ. Cell viability assay: To determine cell viability, human osteoblasts (HOB, Promocell) were seeded at a density of 20,000 cells/cm^2 in 96-well plate and subsequently co-cultured with the DNA nanotubes at various concentrations (0/1.25/2.5/3.75/5 µM) for 24 hours under standard cell culture conditions (37 °C, humidified, 5% CO_2/95% air) . MTS assays were used to determine cell density after incubation. Briefly, 20 µl of MTS dye solution was added per 100 µl of solution, and the absorbance readings were taken at 490 nm after 4 hours of incubation. Cell density was then determined with correlation to a standard curve. Statistical analysis: All experiments were conducted in triplicate and repeated at least three different times. Statistical differences were determined using analysis of variance (ANOVA) where $p< 0.01$ was considered statistically significant.

Results and Discussion: TEM images confirmed that the rationally designed DNA tiles can self-assemble and form long tubelike structures about 10-12 nm in diameter and often many micrometers in length with both walls and central pore clearly visible. Incorporating hairpin into one of the strands yields an almost indistinguishable structure, indicating that modifications at selected positions are not likely to affect tube formation. This offers the largest advantage of this system over other alternatives in that there is potential to vary the biological epitope without affecting the core structure. Stability of the nanotubes and nanotube-peptide hybrids against temperature, cation depletion and nuclease degradation were shown to be better than DNA double helices using TEM.The fibrillar morphology of the coating resembles that of extracellular matrices, indicating the potential of these DNA nanotubes to act as mimics of a native extracellular matrix. Finally, these nanotubes were shown to have no toxic effect on osteoblast cells with a concentration of up to 5 µM. Conversely, at low concentrations, specifically 1.25 and 2.5 µM, increases in cell density were observed.

Conclusions: Through above experiments, DNA nanotubes self-assembled from DNA tiles were successfully designed, constructed and characterized. The ECM-like fibrillar morphology and the minimal toxicity to osteoblast cells indicated their potential to be used as improved artificial extracellular matrices.

References:

1. Frantz C, Stewart KM, Weaver VM. The extracellular matrix at a glance. *J Cell Sci*. 2010;123(24):4195-4200.

2. Stephanopoulos N, Freeman R, North HA, et al. Bioactive DNA-peptide nanotubes enhance the differentiation of neural stem cells into neurons. *Nano Lett*. 2015;15(1):603-609.

3. Aldaye FA, Senapedis WT, Silver PA, Way JC. A Structurally Tunable DNA-Based Extracellular Matrix. *J Am Chem Soc*. 2010;132(42):14727-14729.

4. Rothemund PWK, Ekani-Nkodo A, Papadakis N, Kumar A, Fygenson DK, Winfree E. Design and Characterization of Programmable DNA Nanotubes. *J Am Chem Soc*. 2004;126(50):16344-16352.

464853 Colloidal Directed Assembly of Pi-Conjugated Oligopeptides for Supramolecular Electronics

Abstract:

The directed mesoscale engineering of nanoscale building blocks holds enormous promise to catalyze a revolution in new functional materials for advanced electronics. Bio-inspired systems can play a key role in this effort due to their inherent "programmable" function. In nature, biological polymers form chemically complex and functional heterostructures that underlie life processes. From this perspective, the ability to efficiently harness the assembly of synthetic peptides could offer effective pathways to produce functional materials. In this work, we study the self-assembly and directed assembly of non-natural peptides containing conjugated synthetic cores. In these materials, pi-conjugated units are flanked by symmetric oligopeptides with defined sequences, thereby yielding materials that can assemble into hierarchical structures via straightforward triggering with acidic solutions. We employ a general and facile strategy to induce self-assembly of synthetic peptides by simple evaporation of polar aprotic solvents. Notably, macroscopic peptide fiber arrays were crafted by confining the evaporation of peptide monomer solutions within preformed colloidal microchannels. Following assembly, we characterize the optical properties of peptide fiber arrays containing pi-conjugated cores using a variety of techniques. We further interrogate the electronic properties of these materials by integrating synthetic peptide assemblies as the active layer in a field-effect transistor (FET). In this way, the present study offers new insights into the design and fabrication of supramolecular electronics.

464997 A Self-Assembled pH-Responsive Multi-Component Platform for Oral Vaccination

Abstract:

Next-generation vaccine research must achieve safe, cost-effective and easily administered vaccines with 1) needle-free delivery mechanism, and 2) immunization with subunit vaccines. Oral vaccines improve ease of administration and confer protection at mucosal surfaces for more robust immune responses. Subunit vaccines (e.g. protein antigens) are desirable to overcome safety issues associated with current attenuated or inactivated oral vaccines, but are poorly immunogenic and absorbed, requiring the use of adjuvants. We have previously demonstrated success using anionic hydrogel carriers to deliver protein therapeutics, such as insulin and calcitonin, in a pH-dependent manner for targeted intestinal delivery. Successful vaccination strategies require direct antigen internalization and elicitation of immunogenic responses.

We have developed a system that combines two biomaterial platforms. In its core, polyanhydride nanoparticles (PNPs) based on 1,6-bis-(*p*-carboxyphenoxy)hexane (CPH) and sebacic acid (SA) serve simultaneously as adjuvant and delivery vehicle of subunit antigens. These are protected by pH-responsive polymers based on poly(ethylene glycol) (PEG) and poly(methacrylic acid) (PMAA), shielding a payload of PNPs through the harsh conditions of the gastrointestinal tract for selective release in the small intestine, proximal to antigen-sampling M cells. This platform is referred to as Polyanhdyride-Releasing MicroParticle Technology, or PROMPT.

PROMPT formulations were synthesized by pH-mediated anti-solvent precipitation to encapsulate PNPs. Briefly, the pH-responsive copolymer, denoted P(MAA-g-EG), was synthesized by UV initiated free radical polymerization of MAA and polyethylene glycol methacrylate and characterized by ^{13}C NMR, FT-IR, and GPC. PNP formulations were synthesized by anti-solvent nanoprecipitation to encapsulate model protein antigen, ovalbumin. P(MAA-g-EG) dispersions were flocculated in the presence of PNPs by desolvating the polymer through pH reduction. PROMPT

particle morphology and PNP encapsulation were evaluated by scanning electron microscopy and confocal microscopy, respectively. *In vitro* release kinetics were evaluated at physiological conditions and quantified using the micro bicinchoninic acid (microBCA) assay. Enzyme-linked immunosorbent assay (ELISA) was used to evaluate that antigenicity of ovalbumin released from PROMPT formulations. Cytocompatibility of self-assembled PROMPT formulations was evaluated in two cell lines: Caco-2 intestinal epithelial cells and RAW264.7 macrophages. RAW264.7 cells were also used to evaluate PNP internalization. Results indicated that PROMPT formulations can achieve targeted release of PNPs in intestinal conditions. Formulations preserve protein antigenicity, and nanoparticles released from PROMPT are readily internalized by macrophages, indicating PROMPT has potential as an adaptable platform for oral subunit vaccine administration.

Acknowledgements: This work was supported by a grant from the National Institutes of Health (5-R01-EB-000246-20) and a NSF GRFP. We thank the research group of Prof. Balaji Narasimhan from Iowa State University for their collaboration on this project.

468606 Filomicelles Self-Assembled from Degradable Di-Block Copolymers Deliver Retinoids and Chemotherapeutics in Durable Control of Carcinoma Cell Fate

Abstract:

Introduction: Conventional injection of semi-soluble drugs hit both healthy and tumor cells, causing side effects that limit dose. While this approach might be broadened with nanocarriers (such as flexible cylindrical micelles or filomicelles that have been previously shown to circulate longer and accumulate better *in vivo* [1]), the lack of aromatic groups in the core of the nanoparticles may hamper it from reaching peak loading of common aromatic chemotherapeutics.

Results 1: In order to improve the delivery of aromatic drugs in filomicelles, aromatic groups were integrated into the hydrophobic block of a degradable di-block copolymer. Aromatic filomicelles were formed by self-assembly of biodegradable amphiphilic di-block copolymer polyethylene glycol- poly-Î±-benzyl caprolactone (PEG-PBCL) with suitable block ratios, and loaded with a common chemotherapeutic, Paclitaxel. Aromatic filomicelles loaded more Paclitaxel than analogous aliphatic systems. Cell death and the aneuploidy in surviving cells (which indicates toxicity) were higher in both liver and lung carcinoma lines *in vitro* compared to aliphatic filomicelles. Importantly, the filomicelles themselves were bio-compatible and led to minimal cell death. Initial tests *in vivo* also suggested more potent tumor shrinkage, with minimal effects on thrombocrit and hematocrit.

Thus, aromatic filomicelles demonstrated a marked an improvement over the corresponding aliphatic nano-carriers [2]. However, despite inducing higher levels of aneuploidy in surviving cancer cells, the cells reverted back to the original proliferative state over several periods of the normal cell cycle. This is consistent with relapse commonly seen clinic. Retinoic acid (RA) and other retinoids regulate RA receptor transcription factors that induce differentiation and arrest

proliferation of many cell types, including cancer cells [3]. Combining RA with select chemotherapeutics has produced durable cures of select cancers, notably pro-myeloblastic leukemia (PML) where RA differentiates cells while chemotherapeutic kills the cancer stem cell [4].

Results 2: The above mentioned filomicelles were loaded with Paclitaxel, as well as RA, for testing effects on cancer cell lines *in vitro* and *in vivo*. Across multiple liver and lung carcinoma lines, we find dual treatment with RA plus Paclitaxel increases, aneuploidy, Lamin-A (a marker of differentiation) levels and cell death beyond those achieved by either drug single-handedly, with effects being durable. A month after treatment, relapse rates are low for RA-Paclitaxel treated cells (15%), compared to almost all (92%) for cells treated with Paclitaxel alone. Decreases in levels of key cell cycle factor Cyclin-D1 and particularly proliferation marker Ki-67 help clarify the basis for drug synergy. These effects are greatly enhanced by loading the drugs into filomicelles. Co-loading the drugs into filomicelles lead to a more potent system compared to separate loading, with no loss in the integration efficiency of drugs. Notably, relapse rates were ~2% three months after treatment, highlighting the improvement offered by nano-carriers. Ki-67 was the key affected protein, with levels half of those in free RA-Paclitaxel treated cells. Preliminary tests *in vivo* demonstrate sustained delivery for days as well as efficacy in shrinking tumors. Injections of RA-Paclitaxel combination are able to shrink tumors and arrest growth for over 15 days, a feat not achieved by Paclitaxel alone [2].

Conclusion: The above results highlight the irreversible synergy of killing cancerous cells while driving differentiation. Filomicelles accumulate better in tumors, thus increasing the efficacy of delivery and hence the treatment. The overall result is a system that efficiently delivers drugs to the tumor that leads to higher cell death, while arresting proliferation in surviving ones.

References:

1. Geng Y, Dalhaimer P, Cai S, Tsai R, Tewari M, Minko T, Discher DE. Shape effects of filaments versus spherical particles in flow and drug delivery. *Nature Nanotechnology. 2(4)*, 249-255 (2007).
2. D. Christian, S. Cai, O. Garbuzenko, T. Harada, A. Zajac, T. Minko, and D. Discher Mol. Pharmaceutics, 2009 (6) 1343-1352
3. R. Connolly, N. Nguyen, and S. Sukumar. Clinical Cancer Research 2013 (19) 1651-1659.
4. Y. Jing, L. Wang, L. Xia, G. Chen, Z. Chen, and W. Miller. Blood 2001 (97) 264-269.

455879 Construction of Graphene Oxide (GO) Framework on Hollow Fiber Membranes Via Layer-By-Layer Approach for Heavy Metal Removal

Abstract:

A layer-by-layer (LbL) approach was employed to deposit an ultrathin graphene oxide (GO) framework layer on a modified Torlon® hollow fiber support, enabling the composite membrane with superior nanofiltration (NF) performance. The GO framework was molecularly designed following the 2-step method: repeated GO and ethylenediamine deposition and an amine-enrichment modification by hyperbranched polyethylenimine (HPEI). Surface characterizations by SEM, XPS and zeta potential confirmed the formation of the GO framework structure. A 5 to 10 deposition cycle was proven efficient to seal the surface defects, reduce the pore size and narrow down the pore size distribution of the composite membranes. The synergistic combination of the GO framework layer and the Torlon® support can effectively reduce the polymer consumption for the fabrication of traditional integrally skinned NF membranes. The GO/Torlon® composite membrane has rejections higher than 95% towards Pb^{2+}, Ni^{2+}, and Zn^{2+} with a superior water permeability of 4.7 L $m^{â''2}$ $h^{â''1}$ $bar^{â''1}$. The membrane also exhibits excellent long-term performance stability during a 150-h NF test. Thus, the newly developed membrane has great potential for heavy metal removal and may provide useful insights on the fabrication of new generation 2-dimensional (2D) NF membranes.

461795 Analysis of Graphene Membranes and Time-Varying, Stochastic Gas Transport

Abstract:

Single layer graphene is a promising system as a separation membrane. We consider mechanisms and methods of analysis for gas transport through graphene. We also present a detailed analysis of experimental gas permeation data through single layer graphene membranes under batch depletion conditions parametric in starting pressure for He, H_2, Ne, and CO_2 between 100 and 670 kPa. We show mathematically that the observed intersections of the membrane deflection curves parametric in starting pressure are indicative of a time dependent membrane permeance (pressure normalized molecular flow). Analyzing these time dependent permeance data for He, Ne, H_2, and CO_2 shows remarkably that the latter three gases exhibit discretized permeance values that are temporally repeated. Such quantized fluctuations (called â€œgatingâ€ for liquid phase nanopore and ion channel systems) are a hallmark of isolated nanopores, since small, but rapid changes in the transport pathway necessarily influence a single detectable flux. We analyze the fluctuations using a Hidden Markov model to fit to discrete states and estimate the activation barrier for switching at 1.0 eV. This barrier is and the relative fluxes are consistent with a chemical bond rearrangement of an 8â €“10 atom vacancy pore. Furthermore, we use the relations between the states given by the Markov network for few pores to determine that three pores, each exhibiting two state switching, are responsible for the observed fluctuations; and we compare simulated control data sets with and without the Markov network for comparison and to establish confidence in our evaluation of the limited experimental data set.

467795 Graphene and Graphene Oxide Nanoplatelets As Fillers to Reduce Ageing and Improve Permselectivity of PTMSP Membranes

Abstract:

We have determined the effect of physical dispersion of small amounts (1wt%) of nanoplatelets of graphene and graphene oxide to the gas permeation and ageing behavior of poly(1-trimethylsilyl-1-propyne) (PTMSP). PTMSP has extremely large permeability but is subject to ageing, that causes loss of free volume and permeability with time.[1]

The membranes formed with an optimized protocol based on solvent casting with an organic solvent are stable and mechanically robust, although the repeatability of membrane properties is still not optimal, a main factor to be controlled being the solvent evaporation rate.

To analyze the ageing, that is a thickness-dependent mechanism, we fabricated both thin (1-10 microns) supported over a microporous support, and self-standing thick membranes (50-180 microns).

Although the quantitative extent may vary from one sample to another, it has been found that the addition of Graphene Oxide to PTMSP enhances its gas permeability detectably in all cases, despite the small amount added (1 wt%). On the other hand, the effect of the addition of graphene nanoplatelets depends strongly on the preparation protocol and the final sample behavior cannot be controlled.

The ageing was studied with gas permeability tests over time (up to 9 months) and it was concluded that both graphene and graphene oxide nanoplatelets reduce ageing, for both thin and thick samples. The effect of temperature was studied in the range 30-60°C. In the case of He/CO_2 mixture, the samples show a switchable behaviour (He or CO_2 selective) that can be tuned varying the temperature and the filler.

[1]Olivieri, Ligi, De Angelis, Cucca, Pettinau, *Effect of Graphene and Graphene Oxide Nanoplatelets on the Gas Permselectivity and Aging Behavior of Poly(trimethylsilyl propyne) (PTMSP)*, «INDUSTRIAL & ENGINEERING CHEMISTRY RESEARCH», 2015, 54, 11199 - 11211

468193 Chitosan-Graphene Oxide Composite Membranes for Water Filtration

Abstract:

Chitosan (CS) is a natural, biologically-derived polymer in which deacetylated and acetylated units are randomly distributed. CS can be produced by treating crustacean shells such as shrimp crab, and squid shell with alkali sodium hydroxide. The use of CS as a polymer film, membrane, or scaffold has been explored for different applications, including drug delivery, wound healing, tissue engineering, biosensors, and water treatment. Due to its biocompatibility, biodegradability, low toxicity, and antibacterial and hemostatic properties, CS continues to be explored as a key biopolymer and potential low-cost, naturally-occurring alternative to petroleum-based synthetic polymers. Moreover, CS contains amino and hydroxyl functional groups. The functional groups make CS hydrophilic and amenable to adsorption-based dye and heavy metal removal from water. Two critical challenges in CS utilization as a membrane material in water treatment applications are the weak mechanical properties and the solubility of CS in acidic aqueous environments. To overcome these drawbacks, modification methods, including cross-linking strategies and the use of mechanical reinforcement agents (e.g., graphene oxide, hydroxyapatite, cellulose whiskers, clay), can address these two challenges and result in a more robust membrane material. The use of nanoscale additives has shown particular promise. The greater available surface area per unit mass or volume and the unique properties of materials at the nanoscale allow greater mechanical stability and benefit membrane performance, as well as add additional functionality (e.g., reactivity).

Due to its hydrophobic nature, pristine graphene has a tendency to aggregate, which can reduce the available surface area and its usefulness as a nanofiller. To address this challenge, hydrophilic graphene oxide (GO) is produced by chemical modification of graphene, where oxidation causes the addition of hydroxyl, carboxyl, and epoxide functional groups to the basal planes and edges of the graphene sheets. Low cost, high surface area, water solubility, and high negative charge density make GO a notable adsorbent for removal of heavy metal ions and cationic dyes from water.

The oxidative surface modification of GO also enables GO to be a useful, dispersible nanofiller for water filtration membranes, due to the strong interactions between hydrophilic polymer functional groups and GO.

In this talk, we will discuss characterization and performance results for a set of chitosan-graphene oxide composite membranes, where two types of graphene oxide, a nanoscale GO and a granular GO, were evaluated and compared to results for chitosan-only and GO-only membranes. Electron microscopy of membrane surfaces and cross-sections demonstrate significant differences in structural organization and GO layering within the composite membrane materials. Elemental analysis by energy dispersive x-ray spectroscopy identified the presence of aluminum in GO-only membranes and no aluminum in chitosan-GO composites, as is expected from differences in membrane synthesis. X-ray diffraction measurements show that a GO-only membrane has a high level of order with a calculated interlayer spacing of 8.38 Å, which increases slightly to 8.52 Å when wetted with pure water. In comparison, chitosan-GO composite membranes result in broad diffraction peaks and the loss of apparent ordered interlayer spacing upon wetting. X-ray photoelectron spectroscopy results confirmed the presence of aluminum in GO membrane samples, as well as a decrease in amine and an increase in protonated amine and amide contributions in the nitrogen N1s spectrum when GO was added to chitosan. This increase in both amide and protonated amine species suggests oxidation and protonation of the amine groups on the chitosan polymer backbone through interactions with the oxygen-based functional groups on the surface of GO. Membranes were tested for pure water flux and rejection with model water contaminant methylene blue; performance results will be discussed for methylene blue concentrations of 1 â€“ 100 mg/L.

469087 Reduced Graphene Oxide (rGO) Membranes for Water Treatment Applications

Abstract:

Graphene oxide (GO) membranes have recently been studied for separations of molecules in nanoscales. With an interlayer spacing of 1.3-1.5 nm, GO nanochannels can reject dyes (rejection> 95%, size: 1.3 nm) depending on the size of dye and can have moderate rejections for divalent ions. It provides a robust active layer for separations as compared to polymeric membranes. However, GO is readily dispersible in water due to the presence of the different oxygen-containing functionalities like carboxylic, hydroxyl, and epoxy on the basal plane and along the edge. GO layer can be peeled off the porous substrate easily under the shear force exerted by water, thus limiting its practical applications. This study focusses on the reduction of graphene oxide membranes and understands its impact on stability in aqueous environment. Catalytic properties of these membranes in advance oxidation are also explored.

GO membranes were synthesized using vacuum filtration for varying loading of GO on porous microfiltration membrane (Nanostone PVDF 200). GO membranes were further chemically reduced using either $NaBH_4$, or ascorbic acid. Reduced graphene oxide membranes (rGO) were more hydrophobic as compared to GO membranes. Characterization of rGO membranes was carried out to understand reduction of oxygen containing functionalities, and to quantify the mechanical strength of the film.

The rGO membrane was observed to be stable in a crossflow membrane setup. The membrane had a rejection of 95% for neutral red dye (1.3 nm diameter) and 35% rejection for $CaSO_4$. GO is known to possess antibacterial properties and thus makes it suitable for applications where biological fouling is severe. We investigated the fouling behavior of these membranes in the presence of Humic acid (HA), and bovine serum albumin (BSA), and compared it with commercial nanofiltration membranes (Nanostone Inc. NF8 and DOW NF270).

GO and rGO also have been reported to have catalytic properties in advanced oxidation processes. This makes these membranes interesting as no transition metal catalyst are involved. We observed that rGO membranes had better efficiency in the decomposition of sodium persulfate (to generate sulfate free radicals). This project is supported by, NSF KY EPSCoR program, and by NIH-NIEHS-SRC, and by Australian Research Council (through Monash University).

469316 Dynamic Microstructure and Compaction of Graphene Oxide Thin Film Membranes in Nanofiltration

Abstract:

Graphene-based materials has attracted a lot of interest for developing ultra-thin and high performance membranes due to their unique two dimensional structure, one atomic thickness and strong mechanical strength. Graphene oxide (GO) membranes made by assembling small GO flakes have shown excellent performance in nanofiltration due to their well-defined interlayer space. However, besides the interlayer distance, the microstructures of GO membranes such as the stacking of GO flakes and the defects within membranes may also affect the membrane performance [1-2]. In our study on GO membranes in nanofiltration, serious membrane compaction was observed, where water permeation flux decreased considerably over time. The compaction of GO membranes could be due to the change in microstructure, where highly order GO laminar microstructure was induced when pressure was applied during the permeation test. However, the disorder of microstructure was reversible and could be restored when GO membranes were dried in air, and compaction reoccurred every time when GO membranes were retested after drying. The change in microstructure was found in GO membranes prepared by most of the synthesis methods (pressure driven or surface tension driven) and can be affected by the conditions where the membranes were stored. Understanding of the dynamic of the microstructures is crucial as it could affect the performance of GO membranes/ thin films in applications where the microstructure is critical. The observation of the change in membrane microstructures also provided important information for the understanding of water transport phenomenon in GO membranes.Â Though the microstructure evolved over time, GO membranes still showed an excellent molecular cut off for nanofiltration, ~250 Dalton after repeated drying-wetting cycles.

[1] Aba, N.F.D., et al., Graphene oxide membranes on ceramic hollow fibers â€" Microstructural stability and nanofiltration performance. Journal of Membrane Science, 2015. **484**: p. 87-94.

[2] Chong, J. Y., et al., UV-enhanced sacrificial layer stabilised graphene oxide hollow fibre membranes for nanofiltration. Scientific Reports, 2015. **5**: 15799.

469607 Tuning the Gas Barrier Properties and Selectivity By Means of Graphene-Based Coatings on Polymeric Film

Abstract:

The peculiar properties of graphene and its derivatives are very attractive for several applications, and significant efforts are devoted to their implementation in various fields. Among the others, the very low gas permeability of the graphene sheet together with the characteristic 2-D structure are very attractive for the fabrication of barrier materials, as well as of gas separation membranes. However, the production of a large and defect-free graphene surface is not practically feasable, and consequently the assembling technique and the method used to couple graphene with e.g. polymeric materials are crucial. Clearly, the resulting material cannot be fully impermeable as a single graphene sheet, and gaseous species are forced to diffuse around graphene sheets. The organization can exploited for gas separation purposes, selectively blocking only some penetrants, as in molecular sieves.

In this work, two different assembly methods are employed to fabricate graphene oxide based coatings on polymer substrates.

First, a low oxygen graphene oxide (G, C/O â‰ˆ8) solution obtained by electrochemical exfoliation was used to produce graphene coatings by vacuum-assisted filtration; the obtained graphene cakes were then transferred, by roll-to-roll or simple press, onto various polymeric substrates, suitable e.g. for packaging, such as polyethylene terephthalate (PET), polyproylene (PP) or polylactic acid (PLA). The coating was firmly attached to the polymer matrix, and the adhesion proved to be very effective resisting also over time, while the distribution of the graphene sheets and their assembly have been observed by Scanning Electron Microscopy measurements. This simple production technique was successfully used to produce low oxygen permeability multilayer films, showing a reduction of the O_2 transfer rate of more than 70% with respect to the bare substrate in case of PET and more than 90% in case of PP and PLA.

A second assembly method involved the use of the layer-by-layer (LBL) technique, to produce alternating ultra-thin multilayer structures to coat the polymeric matrixes. After its proper activation, the substrate is alternatively dipped into polyelectrolyte or graphene oxide water solution, exploiting electrostatic forces to obtain a molecular-thin layer with a high order degree. Highly oxidized graphene oxide (GO, C/O â‰ˆ1) was used as a electronegative-behave species, together with the partially-positive polyethyleneimine (PEI) to built a well ordered, self assembling coating on top of various substrates. The SEM images revealed a quite flat top surface of the coating, in which the GO sheets can be clearly identified, while the analysis of the sample cross section suggested the existence of an alternated structure.

The highly ordered GO structure produced by this method was able to provide very effective barrier systems, leading to a 96% reduction of the O_2 transfer rate on PET, by means of a 25 bilayer translucent coating (whose thickness can be estimated as 200nm). The estimated O_2 permeability of the coating only resulted remarkably low, comparable to the best already-available gas barrier solutions.

The sieving ability of the obtained GO coating was also investigated by means of direct permeability test of CO_2 and smaller gas molecules, such as H_2 and He, allowing the determination of selectivity values. The peculiar assembly produced by the LbL technique, indeed, conferred to the coating a sieving ability based on the probe different molecular sizes, which is suitable for H_2/CO_2 separation, relevant in pre-combustion CO_2 capture applications. Indeed, the evaluation of the coating selectivity returned an astonishing value of 70, proving the validity of such approach. The same method is also applied to another polymer matrix of relevance in membrane field, Matrimid polyimide, characterized by moderate H_2/CO_2 separation performances, obtaining the same results.

Therefore, this technique allows to conveniently tailor the gas transport properties through very thin coating layers characterized by large sieving selectivity values, and moderate resistance towards the diffusion of small probes such as H_2.

470109 Multiscale Graphene Topographies Programmed By Sequential Mechanical Deformation

Abstract:

Multiscale Graphene Topographies Programmed by Sequential Mechanical Deformation

Po-Yen Chen[1,2]*, Robert H. Hurt[2], Ian Y. Wong[2]

[1] *Department of Chemical Engineering, Massachusetts Institute of Technology, Cambridge, MA, 02139*

[2] *School of Engineering, Brown University, Providence, RI 02912*

Email: pychen@mit.edu

Complex surface topographies emerge in ultrathin-layered materials undergoing large mechanical deformations. Yet, the ability to independently engineer feature size and orientational order across multiple length scales remains a challenge. In this study, we demonstrate hierarchical graphene surface architectures generated using various sequences and combinations of extreme mechanical deformation. The method involves multiple cycles of compression driven by thermal actuation of pre-stretched polymer substrates followed by polymer dissolution, graphene film transfer, and recompression. In each generation, the compression can be biaxial (2D) or unidirectional (1D), and the 1D contraction step can be aligned parallel or perpendicular to the previous step, leading to a family of different hierarchical wrinkle/crumple textures (**Figure 1,** the scale bars are 4 micrometers). Analysis of the three-generational genealogy of these graphene films shows that both directionality and sequence play a role in final texture, and that the characteristic length scale of the features increases with subsequent generations. This behavior can be explained through the increase in effective film thickness (which is directly related to feature size) as complex, out-of-plane textures are added in each successive generation. These films show systematic increases in hydrophobicity and electrochemical current density with each successive generation, and the final product of three-generational extreme compression shows superhydrophobicity (static contact angle >160 degree) and high electrochemical activity (20-fold increase in comparison with planar GO film). We believe this texturing concept can be extended to other 2D material films and has potential applications in anti-fouling substrates, stretchable electronics and advanced electrode architectures.

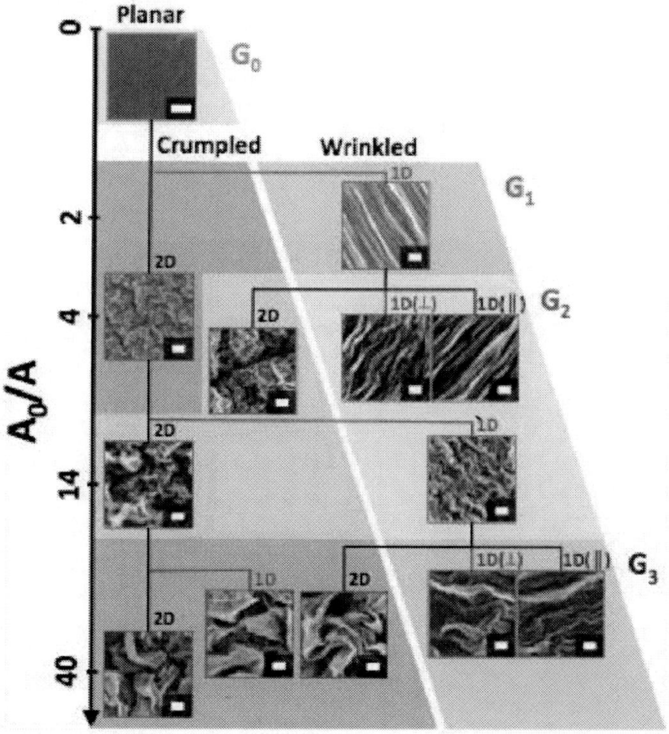

Figure 1. The genealogy of GO multigenerational structures from planar generation 0 (G_0) coatings to multiscale generation three (G_3) structures. A_0 indicates the area of initial planar film; A is the area of multigenerational GO film. All of the scale bars are 4 micrometers.

458151 In-Vitro Dosimetry Model for Toxicity Ranking of Metal Oxide Nanoparticles

Abstract:

There is a debate in the nanosafety community regarding whether cellular toxicity of engineered nanoparticles (NPs) should be ranked on the basis of delivered dose (i.e., NP mass settled per suspension volume) or administered dose (i.e., initial NP mass concentration). Calculation of delivered dose with sufficient confidence for in-vitro toxicity ranking of NPs requires quantification of NP sedimentation considering fundamental suspension properties of NPs in cell culture media. Accordingly, we developed a first principles â€œparticles in a boxâ€ sedimentation model, which accounts for the particle size distribution, fractal dimension and permeability of agglomerated NPs. The developed sedimentation model was evaluated against both external and our own experimental sedimentation data for metal-oxide NPs, demonstrating significantly improved estimates of settled NP mass over other simplified sedimentation calculations with the assumption of impermeable and/or monodispersed NP agglomerates. The model was then applied to calculate the delivered NP doses in different cell culture media for the dose-response analyses of seven metal oxide NPs that were identified toxic from a library of twenty-four NPs in our previous studies. Hierarchical hazard ranking of the identified seven toxic NPs, according to the their EC_{50}, was not measurably different when determined on the basis of either delivered or administered doses. In contrast, simplified sedimentation calculations based on the assumption of monodispersed impermeable NP agglomerates led to significant underestimation of the settled NPs mass and thus distorted toxicity ranking.

472232 Amorphous Silicon Dioxide Nanoparticle Interactions with Pulmonary Epithelial Cells with and without a Pre-Existing Protein Corona

Abstract:

Silicon dioxide nanoparticles are increasingly abundant in consumer products and industrial applications, which increases their potential exposure to humans. In addition, these nanoparticles are under investigation for biomedical applications within the body such as targeted drug delivery. In either case, upon entering the body, nanoparticles interact with biological fluids and are immediately coated with proteins. Ultimately, it is this protein layer that interacts with the bodyâ€™s cells and tissues, and thus dictates the subsequent fate of the nanoparticles in the body. In this work, we investigated the adsorption of an abundant protein in biological fluids, albumin, to silicon dioxide nanoparticle surfaces and the impact of this protein layer on subsequent nanoparticle uptake within human cell cultures.

A previous mechanistic study conducted in our group has indicated that pH plays a strong role in the protein affinity for the nanoparticle surface (*publication in preparation*). Using attenuated total reflectance Fourier transform infrared (ATR-FTIR) spectroscopy, we obtained a real-time adsorption profile for BSA on the nanoparticle surface. Protein coverage on the nanoparticle surface was quantified using thermogravimetric analysis. Results from these studies indicated that at the isoelectric point of the BSA-silicon dioxide nanoparticle complex, pH 3.7, protein coverage was highest. Whereas at physiological pH, pH 7.4, much lower adsorption was observed over the same time period.

Cellular responses to uncoated particles and particles coated at these two pHs, to model a low-coating and a high-coating of protein on the nanoparticle surface, were conducted .Two pulmonary epithelial cell lines, 16HBE14o- and A549, were chosen for these studies as models of the different regions of the lung (bronchioles and alveoli) where nanoparticles deposit. Cell viability was measured via the MTT assay, particle uptake through TEM, and inflammatory markers through ELISA. Results from these studies indicated that coated particles behaved differently than uncoated particles in every parameter examined and that uptake depended on the cell system studied.

Overall, exposure to amorphous silicon dioxide nanoparticles induced a cellular response in pulmonary epithelial cells. Since these particles impacted the two cell lines differently, the deposition location of particle in the lungs after inhalation is important for understanding their toxicity. Surface functionalization prior to cellular exposure also played a key role in the cellular response, further supporting the idea that the protein corona affects cellular recognition and responses.

466807 Physiochemical Properties of Nanoparticles Determine Their *in Vitro* Cytotoxicity

Abstract:

Solid nanoparticles have proven useful in overcoming many of the barriers to drug delivery. Additional development has seen the attachment of a myriad of surface coatings such as polyethylene glycol (PEG). This commonly used PEG layer masks particles from innate host defence mechanisms within the body and increases circulation half-life of the therapeutic upon injection. Although this approach is widely accepted, the underlying mechanism is not well understood. This study examined the effect of hydrophilic surface coatings on solid silica nanoparticles with their interactions with Chinese Hamster Ovary (CHO) cells. Specifically, we examined the effect of differing lengths of PEG on intracellular uptake and viability.

In order to elucidate the effects of particle surface characteristics, studies were conducted with PEG 2k, 5k and 20k, as well as aminated silica nanoparticles. Two nanoparticle sizes, 60 and 120 nm, were chosen to draw size effect comparisons between the particles. These particles are generally recognized as a safe material by the Food and Drug Administration and provided a uniform foundation via size distribution, surface characteristics and mild reaction chemistry that precluded any nanoparticle-mediated effects.

In vitro distribution of the silica nanoparticles was found to directly correlate with their surface coating. Pegylated particles were found to remain interspersed among the CHO cells, whereas amine coated particles were found to produce punctate regions of fluorescence on and around the cells. Confocal microscopy confirmed uptake over simple surface association of the nanoparticles. Similar trends were correlated with flow cytometry. It was found that all PEG coatings decreased particle uptake by the CHO cells in comparison to those with an amine coating. The aminated particles had the greatest effect on cell growth. The 60 nm particles yielded greater half maximal inhibitory concentrations (IC_{50}) values over the 120 nm particles. The IC_{50} values for 60 nm aminated, PEG 2k, 5k and 20k particles were found to be 0.42 ± 0.18, 2.58 ± 0.09, 1.75 ± 0.11 and 3.535 ± 0.15 mg/mL, respectively. The IC_{50} values for the 120 nm aminated, PEG 2k, 5k and 20k particles were found to be 0.59 ± 0.05, 1.435 ± 0.13, 1.57 ± 0.125 and 1.333 ± 0.17 mg/mL. Growth inhibition directly correlated with particle concentration regardless of surface character. Pegylated 60 nm particles showed the greatest differences among their PEG coatings whereas the 120 nm pegylated particles showed similar results among all PEG sizes.

This study found that size and surface coatings have a dramatic influence on the cell viability and uptake *in vitro*. In addition, these studies show there may be an interplay between both of these variable sets. This work has provided a basis of comparison that could be extended to various nanoparticle systems. Understanding the impact of widely used solid nanoparticle compositions, sizes, coatings and their combinations on mammalian systems is an important step in developing robust drug delivery vehicles.

471757 The Impact of Titanium Dioxide Nanoparticles on the Lysosome-Autophagy System and Cellular Clearance

Abstract:

Titanium dioxide (TiO_2) is "generally regarded as safe" by the US Food and Drug administration and is found in a wide variety of consumer products, from cosmetics and pharmaceuticals to food additives. Recent studies suggest that the cellular uptake of nano-sized TiO_2 may cause oxidative stress, lysosomal impairment, and inflammation. However, the impact of nano-sized TiO_2 on cell physiology remains unclear. To address this knowledge gap, we evaluated the cellular response to the uptake of anatase TiO_2 nanoparticles (15, 50, and 100 nm in diameter), with particular emphasis on the lysosome-autophagy system, which is the cell's first response to the uptake of a material perceived as foreign and potentially harmful. Following characterization of the TiO_2 nanoparticles, cytotoxicity analyses were conducted and each of the three TiO_2 nanoparticles were found to induce minimal toxicity at concentrations in the cell culture medium up to 1000 μg/mL. Induction of the lysosome-autophagy system was then tested by monitoring the subcellular location of the transcription factor EB, a master regulator of lysosome- and autophagy-related genes. We found TiO_2 nanoparticles to induce activation of the transcription factor EB at concentrations as low as 10 μg/mL and to induce autophagic vesicle formation. However, uptake of high concentrations of TiO_2 nanoparticles (100 μg/mL and above) was also found to prevent vesicle turnover. We hypothesized that this observed effect on vesicle turnover was due to lysosomal impairment. Therefore, we analyzed lysosomal hydrolase function and membrane stability following TiO_2 nanoparticle uptake and observed a correlation between lysosomal function and integrity and vesicle turnover. In conclusion, we established that cellular uptake of TiO_2 nanoparticles induces activation of the lysosome-autophagy system, and thus upregulation of cellular clearance, but results in a blockage of autophagic flux and degradation *via* impairment of the lysosome.

458173 Potential Impact of Sublethal Levels of Nanomaterials on Interactive Behavior of Environmental Bacteria

Abstract:

Although bactericidal effects of a wide range of nanomaterials on environmental bacteria have been extensively reported, little is known about the impact of nanomaterials at sublethal levels on environmental bacteria and their interactive behaviors, which is a critical gap toward a comprehensive understanding of the impacts of nanomaterials on microbial ecosystems. In this study, using model environmental bacteria, we report the impact of single-wall carbon nanotubes (SWCNTs) and nanoSilver (nAg) on cell-to-cell interaction through influencing the production of â €œpublic goodsâ€ or the production of quorum sensing (QS) signals. Sublethal level of SWCNTs (40 or 80 Âµg/ml) inhibited the production of pyoverdine (PVD) (an important metabolite that involves in interactive behavior of microbial communities, especially in iron-limiting environments) in planktonic cultures of *Pseudomonas aeruginosa*. The inhibitory effect has been confirmed at the mRNA level by transcriptional assay and quantitative polymerase chain reaction (qPCR) analysis. We further explored the effect of SWCNTs on mature biofilms of *P. aeruginosa*. Our results showed that the exposure to SWCNTs for up to 48 h did not significantly influence PVD production in cells encased in mature biofilms; however, the exposure to SWCNTs promoted cell detachment from the biofilms and decreased PVD production of the detached cells. To explore the influence of nanomaterials on quorum sensing, we used *P. syringae* and *Pantoea stewartii* as model organisms and quantified the influence of SWCNTs and nAg on the production of different acylated homoserine lactones (AHL) signals. Our results showed for the first time that differential effects could be exhibited by nanomaterials on different QS signals. Our findings advance current knowledge toward a comprehensive understanding of the ecotoxicity of nanomaterials.

467722 Dissipative Particle Dynamics (DPD) Simulations of Star Polymer Microdroplets with Tunable Hollowness and Porosity Controlled By Nanoscale Block Architecture

Abstract:

Recently we have reported the ability to simultaneously control the nano-scale and micro-scale assembly of droplets, made from star polymers, which are used as scaffolds for tissue regeneration.[1] A variety of hollow, non-hollow, and porous structures were obtained via experiment, and subsequently corroborated by simulation. Importantly, the droplet microstructure was tunable via a single control parameter - the relative density of unreacted star polymer end groups (hydroxyls) within the block. Using dissipative particle dynamics simulations (DPD) containing millions particles we describe how the hydroxyl density controls the droplet structure; by varying the star polymer block architecture (arm length and number) we stabilize different droplet morphologies that can be predicted via the hydroxyl density. Finally, we also highlight the role of solvent selectivity between the hydroxyl groups and the polymer itself in determining the resultant mesostructure.

1. Zhang, Z.; Marson, R. L.; Ge, Z.; Glotzer, S. C.; Ma, P. X. â€œSimultaneous Nano- and Microscale Control of Nanofibrous Microspheres Self-Assembled from Star-Shaped Polymers.â€ Advanced Materials (2015). doi: 10.1002/adma.201501329.Â

469888 Theoretical and Experimental Investigation of Microphase Separation in Mixed Thiol Monolayers on Silver Nanoparticles

Abstract:

Controlling spatial patterning in nanoparticle monolayers is key to tuning nanoparticle interfacial properties. One method of inducing phase separation in nanoparticle monolayers is using ligands with chain length differences in the monolayer. This chain length mismatch is thought to create an entropic driving force for phase separation into stripy monolayers which maximize free volume for the longer ligand. Our work focuses on quantifying this effect using a combination of experimental and computational techniques. To do this we look at phase separation in alkanethiol monolayers on silver nanoparticles. The monolayers consist of two different alkanethiol ligands, one long ligand, dodecanethiol, and a shorter alkanethiol ligand, undecanethiol to butanethiol. In our work we use a MALDI-TOF technique, developed in a previous paper[1], which quantifies degree of phase separation in the nanoparticle monolayer experimentally. Additionally, we run Self-Consistent Field Theory (SCFT) and Monte Carlo atomistic simulations for each alkanethiol pairing to determine the exact spatial patterning that we expect to see with each system. We then calculate a predicted MALDI-MS spectrum from each simulation using a technique established in our previous paper[1]. This allows us to directly compare the experimental and computational results both quantitatively and qualitatively. With this comparison we get greater assurance that the simulation results accurately reflect the experimental system.

References

[1]Â Farrell, Z.; Merz, S.; Seager, J.; Dunn, C.; Egorov, S.; Green, D. L. Angew. Chemie Int. Ed. 2015, 54, 6479â €"6482.

470055 The Response of Anisotropic Colloidal Particles to a Nearby Electrode

Abstract:

Recent work has shown that colloidal crystals with intricate and complex microstructure can be fabricated from anisotropic particles responding to an electric field. Anisotropic particles, such as doublets with lobes of dissimilar zeta potential, will tend to form assemblies with more complex microstructure than those fabricated from isotropic particles because anisotropic particles will interact via directionally dependent forces. This presentation will summarize recent work in our group focused on the response of non-spherical particles to a nearby electrode. The force, torque, and resulting motion of homo- and heterogeneous doublets was calculated in response to a DC electric field. A non-zero force on the doublet in the direction perpendicular to the applied electric field was obtained as a consequence of the unbalanced equilibrium charge electroosmotic (ECEO) flow generated along the particleâ€™s surface; the strength of the flow field depended on the sign and magnitude of the lobe zeta potential and weakly on the shape of the particle. Moreover, the ECEO flow field and the accompanying interparticle interactions were anisotropic. Experiments measuring the response of doublets and ellipsoids to DC and AC electric fields will also be presented. These results suggest a strategy by which one can design 2D interparticle interactions to achieve a specific ensemble microstructure.

467500 Locally Glassy Dynamics in Colloidal Systems with Competing Interactions

Abstract:

Colloidal systems with a combination of short-range attraction (SA) and long-range repulsion (LR) have received significant interest due to their wide variety of solution states. Of these, cluster fluids and kinetically trapped states reminiscent of purely SA systems have received particular attention. In particular, these states have been observed in concentrated formulations of monoclonal antibodies, a powerful and prevalent oncology therapeutic. While dynamic arrest at intermediate volume fractions (i.e., gelation) driven purely by attraction has been heavily studied (and remains under debate), the influence of an additional repulsive force is poorly understood. To study these systems, we use highly purified lysozyme, which is well-known to interact via an SALR potential at low ionic strengths. Over a broad range of protein concentrations, the solution structure is quantified using small angle neutron scattering while the dynamics over short times is directly measured using neutron spin echo and over long times is inferred from viscosity measurements. While the viscosity displays Newtonian behavior under all conditions, the short time dynamics become sub-diffusive at elevated protein concentrations and low temperature. This disparity in short and long time dynamics is correlated with the formation of intermediate range order, but only occurs at concentrations well beyond the structural percolation transition. These heterogeneous percolated networks contain localized regions of high density that reduce protein motion, resulting in glassy-like behavior over short times, yet maintain macroscopic liquid-like behavior. The structural and dynamic characteristics of SALR systems found here are distinct from macroscopically heterogeneous attractive driven gels, but may provide insight into the mechanism of kinetic arrest in those systems.

472514 Development and Self-Assembly of Coarse-Grained Skin Lipid Models Derived Via Multistate Iterative Boltzmann Inversion

Abstract:

The barrier function of the skin is known to be localized to its outermost layer, the stratum corneum (SC). Specifically, the multilamellar lipid matrix surrounding the dead, flat corneocytes in the SC is thought to be the main barrier to chemical penetration through the skin. Although the chemical makeup of the SC is known, the molecular level details of its structure are not, with several different models proposed for the arrangement of this complex lipid mixture. Molecular simulation presents a suitable means to study such systems, allowing a clear molecular level perspective of the ordering and precise control over system composition. However, to date, simulations of SC lipids have focused primarily on preassembled structures. While such studies have offered much insight into lipid-lipid and lipid-water interactions, it is likely that the preassembled structures are unduly influenced by their initial configuration, due to the low diffusivity of the lipids. As such, self-assembled structures may present a more realistic view of SC lipid systems, although the long time scales needed due to the rough free energy landscape make atomistic level modeling impractical, necessitating the use of more computationally efficient coarse-grained models.

Here we develop coarse-grained (CG) models and investigate the behavior of the key lipid classes present in human SC, namely ceramide (CER), cholesterol (CHOL), and free fatty acids (FFA). CG potentials are derived from atomistic simulations via multistate iterative Boltzmann inversion [1], which yields potentials with increased levels of transferability between states and thus are suited to study self-assembly. Particular emphasis is placed on the hydrophobic interactions, where wetting properties are used to tune the relative attraction between CG beads in order to accurately capture phase segregation at the CG level. We investigate the behavior of CER-rich bilayers and multilayers and find that the CG CERs self-assemble into assemblies that closely match several key structural properties of atomistic models and experimental CER systems.[3] Self-assembly of mixed lipid systems is then investigated to probe structural conformations of more realistic SC mixtures and explore the relevance of several experimentally proposed models of SC lamellar organization.

References

[1] Moore, T. C., Iacovella, C. R. & McCabe, C. Derivation of coarse-grained potentials via multistate iterative Boltzmann inversion. J. Chem. Phys. 140, (2014).

[2] Bouwstra, J. A., Gooris, G. & Ponec, M. The lipid organisation of the skin barrier: Liquid and crystalline domains coexist in lamellar phases. J. Biol. Phys. 28, 211â€"223 (2002).

[3] Moore, T. C., Iacovella, C. R., Hartkamp, R. & McCabe, C. A Coarse-grained Model of Stratum Corneum Lipids: Free Fatty Acids And Ceramide NS. Submitted.

470827 Designing Interactions That Stabilize Assemblies to Changes in Density or Temperature: Application to Square, Snub Square and Kagome Lattices

Abstract:

Manufacture of materials with specific morphologies for novel applications (e.g. photonics) remains a formidable challenge. While some design control is possible at nanometer length scales using top-down fabrication approaches, bottom-up self-assembly of a targeted material structure presents an alternative that, in some cases, may be less expensive and more robust. Design methods for determining the interactions required for targeted assembly fall under the category of 'forward' methods, whereby a systematic discovery of equilibrium structures is made from a given interaction form; or 'inverse methods', whereby a desired structure is targeted as the thermodynamically stable structure, and the tunable interaction parameters required to establish it found via a constrained optimization. Here, we demonstrate that the optimization of potential parameters for ground state assembly can be formulated analytically as a non-linear problem, which can then be solved numerically with, e.g., the mathematical modeling software GAMS (Generalized Algebraic Modeling System). This formulation allowed us to recently test how the interaction required for assembly might be constrained to favor structures that are unusually stable to changes in temperature or density. We discuss the implications of our approach for the design of soft repulsive interactions that assemble into square [1], snub square, and kagome lattices [2].

[1] W. D. Piñeros, M. Baldea, T. M.Truskett, J. Chem. Phys. 144, 084502 (2016)

[2] W. D. Piñeros, M. Baldea, T. M. Truskett, submitted (2016)

471997 Small Angle X-Ray Scattering Determination of PS-b-PMMA Bulk Phase Diagram

Abstract:

As the demand for increased feature density grows in the semiconductor and microfabrication industry, the interest for thin film materials that can self-assemble into periodic features with a small spacing grows as well. Block copolymers (BCPs) have shown the ability to phase separate into line/space and cylindrical patterns with a feature spacing as low as 5 nm. The ordering of these features can be controlled by a chemical and/or topographic contrast in the substrate the films are cast on. Computer simulations are very useful in quickly weeding through many possible patterned substrates to determine their effect on the phase separated features. However to do this the model being used in the simulation must accurately represent the block copolymers. In this work the bulk phase diagram of a commonly used block copolymer (PS-b-PMMA) is generated in order to properly parameterize a coarse-grained molecular dynamics model such that the model can accurately reproduce the experimental phase diagram and then be used to explore the block copolymer. While PS-b-PMMA has been used as a standard for most BCP experiments, to the best of our knowledge, its phase diagram has never been published. The importance of the actual phase diagram cannot be understated, as it acts as a guide for the necessary properties a BCP requires to form the desired features. The bulk phase diagram for low dispersity block copolymers will be mapped out using small angle x-ray scattering.

449661 Polymer-Mediated Polymorphic Control over Open Colloidal Crystals

Abstract:

Globally ordered colloidal crystal lattices have broad utility in a wide range of optical and catalytic devices, for example, as photonic bandgap materials. However, the self-assembly of stereospecific structures is often confounded by polymorphism. Small free energy differences often characterize ensembles of different structures, making it difficult to produce a single morphology at will. Current techniques to handle this problem adopt one of two approaches: that of the "top-down," or "bottom-up" methodology, whereby structures are engineered starting from the largest or smallest relevant length scales, respectively. However, recently a third approach for directing high fidelity assembly of colloidal crystals has been suggested which relies on the introduction of polymer co-solutes into the crystal phase. [1] By tuning the polymer's morphology to interact uniquely with the void symmetry of a single desired crystal, the entropy loss associated with polymer confinement has been shown to strongly bias the formation of that phase. [2-3] However, previously this approach has only been demonstrated in the limiting case of close-packed crystals. Here we show how this approach may be generalized and extended to complex open crystals, illustrating the utility of this third "structure-directing agent" paradigm in engineering the nanoscale structure of ordered colloidal materials. [4] The high degree of transferability of this paradigm's basic principles between relatively simple crystals [1-4] and much more complex ones indicates this represents a fundamental addition to presently known self-assembly techniques, and opens new avenues to better engineer the structure of these materials.

References

[1] Mahynski, Panagiotopoulos, Meng, and Kumar, Nature Communications 5, 4472 (2014).
[2] Mahynski, Kumar, and Panagiotopoulos, Soft Matter 11, 280-289 (2015).
[3] Mahynski, Kumar, and Panagiotopoulos, Soft Matter 11, 5146-5153 (2015).

[4] Mahynski, submitted (2016).

451870 The Design of Micelles for Molecular Diagnostics

Abstract:

Imaging modalities such as magnetic resonance imaging (MRI) are routine tools to characterize disease progression in the clinic. However, MRI has limitations as it aims to assign physical traits to assess the molecular state of the disease. In an effort to image and quantify molecular markers in vivo to accurately determine disease progression, we report on gadolinium (Gd)-incorporated, fibrin-targeting micelles for cardiovascular diseases, specifically targeting atherosclerosis. Micelles were self-assembled with DSPE-PEG2000-CREKA, DSPE-PEG2000-Cy7, and one of three different Gd-chelators: DSPE-PEG2000-DTPA(Gd), BSA-DTPA(Gd), or DSPE-DTPA (Gd). While the latter two molecules are commercially available and commonly used in nanomedicine and MRI, DSPE-PEG2000-DTPA(Gd) was synthesized for the first time to develop a chelating molecule with the same hydrophobic tail as the other micelle components. The particles were characterized and their targeting efficacy assessed in vitro and in vivo. Overall, we sought to gain additional insights regarding how molecular and micelle design can affect targeting and diagnostic capabilities and the outcome is discussed.

Specifically, CREKA micelles consisting of DSPE-PEG2000-DTPA(Gd) were spherical and approximately 10 nm confirmed by transmission electron microscopy (TEM) and dynamic light scattering (DLS). CREKA micelles incorporating either DSPE-DTPA(Gd) or BSA-DTPA(Gd) formed a mixed populations of spherical and cylindrical nanoparticles. Upon addition to fibrin clots in vitro, CREKA micelles with DSPE-PEG2000-DTPA(Gd) showed statistically significant binding capabilities compared to its non-targeting counterpart (no peptide). In comparison, CREKA micelles with DSPE-DTPA(Gd) or BSA-DTPA(Gd) showed no difference compared to its non-targeting counterparts. From these results, CREKA micelles incorporating DSPE-PEG2000-DTPA(Gd), or CREKA-Gd micelles, were used for further studies. T1 values decreased with increasing Gd content (2278-897 ms for 100% DSPE-PEG2000-CREKA micelles-25/75% DSPE-PEG2000-CREKA/DSPE-PEG2000-DTPA(Gd)) confirming efficacy as contrast agents. Upon injecting into late-stage atherosclerotic mice, enhanced signal was found at the aortic arch via MRI by both types of micelles, further confirmed via ex vivo fluorescence imaging. Clearance of micelles was found to be mainly through the liver and spleen, or the reticuloendothelial system, and H and E results of organ sections showed no morphological abnormalities.

In sum, novel nanoparticles for MRI were designed using various gadolinium chelators. Our data supports the importance of molecular design of self-assembling particles as it relates to functional and biological efficacy for disease diagnostics. Future studies will incorporate therapeutic components and dual MRI/positron electron tomography capabilities to enhance signal sensitivity.

472495 Controlled, Self-Directed Assembly of Novel in-Situ Forming Biodegradable Nanostructures for the Delivery of Ocular Therapeutics

Abstract:

The physiological and anatomical complexity of the eye make it a highly protected organ, substantially limiting drug delivery despite the ability to treat topically. Designing effective drug delivery systems to target specific areas of the eye is a major research challenge. Currently, topically applied drugs, including eye drops, make up over 90% of the ophthalmic drug market. Eye drops are grossly inefficient delivery vehicles, since less than 5% of the administered dose reaches the intraocular tissues. Thus, there is a considerable unmet need for more efficacious and convenient delivery systems of ocular therapeutics. Our work concentrates on engineering novel physically crosslinked nanocarriers, that are easily formulated using FDA-approved biodegradable polymers. These tailorable carriers provide solubility and prolonged bioavailability and residence time of the ocular therapeutic.

We have developed, designed, and synthesized novel in-situ forming nanocarriers, consisting of amphiphilic PLGA-PEG-PLGA. These nanostructures self-assemble in aqueous environments via reverse thermal gelation. By varying the lactic acid (LA) to glycolid acid (GA) ratio of the hydrophobic PLGA block, the molecular weight of PEG, the PLGA to PEG ratio and the polymer solution concentration, we produce injectable, in-situ forming nanogels that are optically clear at physiological temperatures. Characterization studies include optical property assessments, critical gelation temperature (CGT), rheology, critical micelle concentration (CMC), in-vitro degradation, and controlled and sustained drug release kinetics.

We demonstrate that carriers with a LA/GA ratio between 1 and 15, at compositions between 10%(w/v) and 25%(w/v), allow for over 90% visible light transmittance, gel formation between 34 and 37°C, and sustained and controlled in-vitro polymer degradation and drug release kinetics for over one week. Below the CGT, the triblock copolymer of interest forms a homogeneous micelle therapeutic injectable solution, held together by hydrogen bonding interactions.

Above the CGT, the solution swells forming an ordered packing, with hydrophobic forces dominating the 3D nano-network of bridged micelles with the drug embedded within the PLGA units. The controlled self-directed assembly of the amphiphilic molecules allows visible spectrum wavelengths to pass through the core-corona structure, making the gel optically transparent. The self-assembled polyester/polyether matrix slowly degrades via the cleavage of ester bonds, allowing for continuous and controlled drug release.

Changes in the physical and morphological states of the hydrogel are rapid and controllable, making this novel thermosensitive gel attractive for the delivery of ocular pharmaceutics. Our technology offers a more effective and efficient method of ocular therapy by providing a controlled and extended release of medication and has wide and far reaching potential to improve a number of ocular therapies.

465886 Molecular Gel Formation As a First Order Phase Transition

Abstract:

Self assembling systems based on dipeptide molecules that form ordered fibrous structures are seen as promising new materials for applications in tissue engineering, drug delivery, biosensors, photoelectronics, etc. However, the gelation mechanism in these types of systems has been largely overlooked and remains poorly understood, limiting the development of comprehensive design guidelines for specific applications. We explore the gel transition of the aromatic dipeptide molecule fluorenylmethoxycarbonyl diphenylalanine (Fmoc FF). The addition of water to a solution of Fmoc FF in dimethyl sulfoxide (DMSO) results in the self assembly of Fmoc FF molecules into a space filling fibrous network. We provide evidence that gel formation is associated with a first order phase transition resulting in nucleation and growth of strongly anisotropic crystals with high aspect ratios. The strength of attraction between Fmoc-FF molecules as a function of water concentration is estimated from long time self-diffusion measurements using ^{1}H NMR diffusion ordered spectroscopy (DOSY). The resulting phase behavior follows that observed for a wide range of other crystallizing nanoparticles and small molecules, consistent with the short-range nature of the intermolecular attractions. Furthermore, a lag time in the formation of the new phase is observed, indicating the existence of a free energy barrier to the formation of a crystal nucleus of critical size. The application of classical nucleation theory for a cylindrical nucleus indicates that one-dimensional crystal growth is driven by an imbalance of the surface energies of the ends and sides of the fiber. We believe that the discussed concepts can be extended and applied to other similar molecular gel systems, thereby providing guidelines that would facilitate the design of novel molecular gelators.

468014 DNA-Programmable Assembly of Enzyme Superlattices

Abstract:

Enzymes are one of Natureâ€™s most diverse and powerful nanoparticles. Assembling them into multicomponent crystals has been of great interest and can be accomplished through a variety of means, typically based on protein surface chemistry. The programmability of DNA makes it an attractive structure-directing ligand for assembling nanoparticles into well-defined materials. Recently, DNA has emerged as an effective ligand by which catalytically active proteins can be assembled into extended crystalline superlattices of CsCl symmetry. In these systems, protein surfaces are covalently functionalized with nucleic acid ligands, which direct interparticle interactions through Watson-Crick base pairing. In this study, we predicted that under certain conditions the anisotropic shape and surface chemistry of the enzyme could induce directional bonding interactions resulting in the assembly of a different lattice symmetry. To probe this, the enzyme catalase was functionalized with DNA ligands of varying flexibility and co-assembled with gold nanoparticles possessing different numbers of DNA bonding ligands. A phase transformation was observed with decreasing DNA flexibility and increasing gold nanoparticle ligand number from CsCl to Th_3P_4 symmetry, with mixtures of the two lattice symmetries present at intermediate states. Through fitting of the SAXS diffraction patterns, the composition of the mixtures was quantified. Coarse grained molecular dynamics simulations were used to further understand the energetics of the two phases. This is the first instance in these systems that phase transformation has been observed by tuning ligand flexibility, expanding on the means by which well-defined functional materials can be assembled from proteins using DNA.

468197 Microreactors As a Tool for Producing Polymer Nanoparticles By a Self-Assembled Process

Abstract:

Microreactors as a tool for producing polymer nanoparticles by a self-assembled process.

A. Tabernero, A. González-Garcinuño, Miguel A. Galán, Eva M. Martín del Valle

Department of Chemical Engineering; University of Salamanca, Salamanca, Spain

Microreactors have been used for these last years for improving nanomaterials synthesis because several reaction parameters can be easily controlled. This type of reactors provides a continuous shear stress on nanoparticles dispersion, allowing a uniform outcome and controlling the nanoparticles formation kinetics. In this work, we propose the use of a microreactor to avoid nanoparticles aggregation of a fructose biopolymer (levan). This polymer is amphiphilic and is able to produce nanoparticles by a self-assembled process and shows a great potential for biomedical applications. Furthermore, the effect of several experimental parameters on biopolymer particles aggregation and Z-potential will be studied in order to determine the best conditions to avoid nanoparticles aggregation. Microreactors nanoparticle results will be in addition compared with the results obtained with a conventional methodology to produce levan nanoparticles.Results indicate that particle size decreases from 400 nm. to 100 nm by using microreactors, whereas the Z-potential remains more or less the same (between -25 and -30 mV). Moreover, we studied the use of microreactors with the carboxymethylated form of the polymer (carboxymethyllevan), obtaining the same tendency. Therefore, microreactors can be used as a tool for producing nanoparticles of materials that can self-assemble in nanoparticles due to its amphiphilic nature.

471763 Polymer-Induced Liposome Aggregation: Toward the Application of Naked-Eye Bio-Detection

Abstract:

Signal amplification has been one of the important issues for detecting proteins or pathogens at low concentration regime. Here, we investigate a liposome clustering process induced by an amphiphilic triblock copolymer through hydrophobic interactions. The clustering is visible and thus can be used as an amplification platform for naked-eye detection. In this study, the liposomes are composed of dipalmitoylphosphatidylcholine (DPPC, di-C_{16}) and dihexanoylphosphatidylcholine (DHPC, di-C_6) induced by triblock copolymer linkers, (propylene glycol)$_m$ – (ethylene glycol)$_n$ – (propylene glycol)$_m$ ($PO_m - EO_n - PO_m$, reverse pluronic copolymers). The aggregation occurs when the two PPO hydrophobic blocks insert in the bilayers of different liposome resulting in a liposomal clusters. The clustering is presumably facilitated by the mismatch of acyl chain between DPPC and DHPC, as evident in the rapid increase of UV absorption and hydrodynamic radius (R_H). In contrast, the clustering of DPPC liposomes is hardly observed in the presence of polymer linkers. The experimental results are further corroborated by the simulation work, indicating that less energy is required for the polymer insertion into the interfaces between DPPC and DHPC compared to the pure DPPC regime. Moreover, we observed distinct aggregation kinetics between polymer linkers with different hydrophobicity. Specifically, increase of the polymer concentration leads to more obvious clusters for the relatively hydrophilic polymer linker ($PO_{14} - EO_{24} - PO_{14}$), but inhibits the aggregation for hydrophobic polymer linkers (i.e. $PO_{25} - EO_7 - PO_{25}$ and $PO_{26} - EO_{20} - PO_{26}$). This phenomenon is explained by the mechanism that the hydrophobic polymers strongly anchor at and fast occupy the "active" (DHPC-rich) sites of bilayer compared to the hydrophilic ones, whose hydrophobic PPO may frequently associate with and dissociate from the "active" sites, enhancing the stringing process between liposomes.

472283 Information-Directed Assembly of Dynamic Covalent Molecular Ladders

Abstract:

Nucleic acids present the most versatile class of materials for producing nanostructures to date and, through careful consideration of their residue sequence, can be designed to self-assemble via the hybridization of complementary strands into arbitrary structures with nanometer precision. Unfortunately, the versatility of nucleic acid assemblies is tempered by their thermal and mechanical instability, attributable to the weakness of the hydrogen bonds that hold the strands together. The substitution of the canonical nucleic acid base pairs with covalent inter-strand interactions precludes this instability and yields a unique and powerful nanofabrication strategy wherein the complex, information-driven assembly of nucleic acids is combined with the strength of covalent bonds. Notably, the creation of exquisite nanostructures by self-assembly and the toughness imparted by covalent bonds are generally perceived as mutually exclusive owing to the prevalent irreversibility of covalent bond-generating reactions. Fortunately, several covalent interactions are known to be reversible under particular reaction conditions, enabling the error correction mechanism that is essential for the selective fabrication of supramolecular structures. As a result of the enormously greater strength and directionality offered by covalent bonds in comparison to the weaker interactions observed in biology, these 'dynamic' covalent chemistries offer an elegant approach to nanostructure assembly that combines complexity AND toughness. Here, we demonstrate the sequence-dependent self-assembly of peptoid-based oligomers to afford molecular ladders with up to sixteen rungs via scandium(III)-catalyzed imine metathesis. Moreover, we examine the ability of this approach to generate branched ladders that can assemble into larger, robust, multi-dimensional nanostructures.

464838 Microwell Arrays for Screening Interactions Between Root-Associated Microbes

Abstract:

Plant growth promoting rhizobacteria (PGPR) associate with root surfaces in order to provide benefits to the plant host, some of which include improved nutrient uptake, pathogen protection, and plant hormone regulation. An improved understanding of the relationships between PGPR and root hosts has spurred recent interest in developing constructed, artificial communities of PGPR for use as a low-energy, sustainable alternative to chemical fertilizers. However, wide-spread application of constructed communities has been limited due to the fact that the vast majority of microbe-microbe interactions occurring between PGPR in the rhizosphere are unknown or poorly characterized. This limitation stems from the fact that traditional microbiological assays designed to characterize microbial interactions do so using co-culturing methods that are qualitative and low-throughput. Additionally, these assays fail to re-create the physical and chemical complexities of natural ecosystems, such as soil or rhizospere, which are spatially confined microenvironments. To address these types of limitations, we have developed a microfabricated cell-screening platform termed the â€œmicrobe arrayâ€ which uses a novel biofabrication method to randomly assemble thousands of independent bacterial meta-populations into microscale wells with controlled physicochemical properties. It was recently shown that by tuning well dimensions, interacting bacteria can be combined into wells at low dispersity, allowing for the assembly of replicate initial populations; or at high dispersity, allowing for thousands of unique meta-populations to be assembled across a single array of wells. After this, communities can be simultaneously screened for growth or decay to identify combinations of bacteria and environmental parameters that promote or inhibit colonization. In development of this platform, we investigated the effect of spatial confinement on the growth of *Pseudomonas* bacteria. Here, populations were assembled into wells at high dispersity, and growth screening experiments revealed that successful colonization of microbial communities in a spatially confined, diffusion limited environment is highly dependent on inoculum levels. Our current work focuses on screening rhizobacterium populations in this platform to identify pathogenic, competitive or symbiotic interactions that inform the development of constructed communities for biofertilizer and biocontrol applications.

EFFECT OF SIZE AND CHARGE OF METAL IONS ON HYDROGEN PEROXIDE STABILITY IN SILICA HYDROGELS

Ezgi Melis Dogan, Fulya Sudur Zalluhoglu, Nese Orbey, University of Massachusetts Lowell, Lowell, MA, USA

Abstract

Currently, the most efficient method to increase the stability of hydrogen peroxide (H_2O_2) is microencapsulation in silica hydrogels and xerogels via environmentally-friendly sol-gel chemistry using sodium silicate as the starting precursor. The high stability of H_2O_2 in these systems arises from strong hydrogen bonding between silanol groups (Si-OH) and peroxide molecules. Our previous study showed that the amount of Na^+ ions have a strong effect on the structure and morphology of hydrogels. Therefore in the present work, the effect of the size and charge of the ions other than Na^+ on the properties of the hydrogel and stability of peroxide was studied. Monovalent (K^+) and divalent (Mg^{+2}) ions were introduced into the sodium silicate precursor in the form of their respective salts. The pH of the sol was changed by using an ion exchange resin to replace a portion of the ions in the solution with H^+ ions. Gelation time and structure, pH of the sol and stability of H_2O_2 in hydrogels were investigated. Results showed that the stability and gel time can be adjusted to the desired level by changing the ion content and pH of the sol. The increase in total ion concentration in the sol resulted in a decrease in gel time and the best H_2O_2 stability was obtained with the addition of the magnesium salt: seventy percent of the peroxide was retained in hydrogels containing Mg^{2+} ions, as compared to 61 % retention with hydrogels containing Na^+ ions at the end of 70 days. Hydrogels were characterized by using scanning electron microscopy (SEM) and Brunauer-Emmett-Teller (BET) analysis.

Introduction

Hydrogen peroxide (H_2O_2) is an environmentally friendly oxidant which has a limited stability and tend to decompose into oxygen and water according to following reaction.[1]

$$H_2O_{2(aq)} \rightarrow H_2O_{(l)} + \frac{1}{2}\, O_{2(g)} \qquad (1)$$

The concentration of the peroxide, temperature, pH of the aqueous solution, and presence of impurities or stabilizers significantly affects the decomposition rate of H_2O_2. Storage at low temperatures is required for H_2O_2 solutions due to this instability.[2] However, H_2O_2 stability at ambient temperatures is desired by many industrial applications such as sanitation,[3] disinfection,[4] and tooth whitening.[5]

Various methods have been experimented to decrease the decomposition rate and increase the stability of H_2O_2.[6-10]Among these methods, microencapsulation of H_2O_2 in silica hydrogels and xerogels has been the most efficient methods to enhance peroxide stability because of formation strong hydrogen bonds of H_2O_2 molecules with the silica surface.[11]

The synthesis of hydrogels and xerogels is performed using sol-gel chemistry which is a simple and low cost method.[12] The most common precursor of the gels is sodium silicate solution. A colloidal suspension known as sol is first formed by introducing silanol groups into aqueous sodium silicate and removing sodium ions from the solution. For this purpose, a cation exchange resin is used to adjust the sodium content via ion exchange in the solution.[11] This exchange occurs according to following reaction where Rz is the styrene-divinylbenzene copolymer and SO_3H is the sulfonic acid group.

$$Na^+ + RzSO_3^-H^+ \rightarrow H^+ + (RzSO_3)Na \qquad (2)$$

A hydrogel is observed in the sol-gel transition point where the viscosity of the solution significantly increase. At this point, Si-O-Si bonds are formed via polymerization of silanol groups in the solution resulting a three-dimensional interconnected network (hydrogel). Xerogels are more contracted gel networks obtained by drying the hydrogels. Silica xerogels containing H_2O_2 has been studied by Żegliński et al.,[6] Bednarz et al.[7] and Wolanow et al..[8] Hydrogel and xerogel properties depend on the drying and storage temperatures, pH of silica sols, and initial precursors. Sudur et al.[9, 10] studied the transition conditions of hydrogel to xerogel and observed the effects of sodium ion on the pH of the sol and H_2O_2 (3.3 to 33.2wt %) stability at room temperature (RT). Significant effects of sodium ion content on the structure, the release rate of entrapped H_2O_2, and the thermal properties of hydrogels and xerogels were found.

The effects of size and charge of ions present in the sol on the structure of hydrogels and stability of entrapped H_2O_2 were questioned because of the observations in our previous work.[9] In the literature, the effect of ion type and concentration on the binding power of sodium silicate was studied for applications in petroleum drilling and construction.[13-15] However, H_2O_2 was not incorporated into the formulations used in these studies. To best of our knowledge, the effect of size and charge of ions on the hydrogel formation or hydrogen peroxide stability has not been studied.

In the present work, H_2O_2 was encapsulated in silica hydrogels using K^+ ions into the sodium silicate precursor in the form of potassium chloride (KCl) at three different molar ratios of K^+: Na^+ (0.14:1, 1:1, and 1.76:1). The effects of K^+ on the gelation time, pH of the sol, stability of H_2O_2 and gel structure were studied. Similar to Na^+, K^+ is monovalent but it has a larger atomic size. On the other hand, silica hydrogels were obtained using Mg^{2+} in the precursor solution to study the effect of charge of the ion on H_2O_2 stability.

Experimental Procedure

Materials

Sodium silicate solution (reagent grade, 10.6% Na_2O, 26.5% SiO_2), hydrogen peroxide (30wt % in H_2O, ACS reagent), Amberlite IR120 (hydrogen form), orthophosphoric acid (85% wt in H_2O); potassium permanganate (ACS reagent >99%) and sulfuric acid (99.999%) were purchased from Sigma Aldrich Chemical Co. Potassium chloride, magnesium chloride and calcium chloride (ACS reagent, >99%) were purchased from J.T. Baker.

Hydrogel synthesis

The procedure of hydrogel synthesis, previously available in the literature[9] was modified. Hydrogels containing Mg^{2+} and K^+ ions along with Na^+ ions were synthesized dissolving magnesium chloride ($MgCl_2$) and potassium chloride (KCl) in deionized water before the addition of sodium silicate. Sodium silicate solution was added dropwise to salt solutions at a volume ratio of 1:5. For hydrogels containing K^+, three different K^+: Na^+ molar ratios of 0.14:1 (low), 1:1 (equimolar), and 1.76:1 (excess) were formulated. In hydrogels containing Mg^{2+}, the molar ratio of Mg^{2+}: Na^+ was adjusted as 0.2:1. Then, all formulations are combined with 30wt % H_2O_2 solution and 7 µL of phosphoric acid (H_3PO_4). The pH of the sols were controlled by adding different amounts (0-3.2 g) of ion exchange resin (Amberlite IR 120, hydrogen form) to replace a portion of the ions in the solution with H^+ ions. The mixture was stirred with an overhead mixer for 4 min. The sol was transferred to a glass Petri dish (D = 100 mm, h = 15 mm) after observing the settlement of the resin. The petri dish was placed in a BlueM gravity oven at 40 °C until observing the gelation.

Characterization

The pH of the sol was changed using different amounts of ion exchange resin at a constant K^+: Na^+ ratio and an initial H_2O_2 concentration (19.9wt %). Gelation time was observed with respect to pH value of the solution for each K^+ formulations. H_2O_2 stability was observed in hydrogels containing K^+ as a function of pH and molar ratio of K^+: Na^+ ions. Stability of H_2O_2 in hydrogels containing Mg^+ was compared based on the presence of ion exchange resin in the starting sol. Hydrogels were stored under ambient conditions at room temperature, and H_2O_2 content was determined using titration method with potassium permanganate ($KMnO_4$) as the titrant. Measurements were made daily for the first 10 days and, then periodically until the concentration of H_2O_2 dropped to 2wt %. Two repeated titrations were performed for each sample and the reproducibility was found to be within 1%.

Quantachrome Autosorb-3B nitrogen sorption porosimeter and JEOL JSM 7401F scanning electron microscope were used to study the morphology and structure of the hydrogels. The surface area and pore size were measured through BET (Brunauer-Emmett-Teller) analysis in the porosimeter. Before these measurements and SEM imaging, aerogels were obtained from hydrogels using a critical-point dryer (Tousimis SAMDRI-795) to preserve the solid network and retain the pore structure. Aerogel samples were coated with gold/palladium target by using a vacuum sputter coater. SEM images of samples were obtained at an acceleration voltage of 10 to 15 kV.

Results and Discussion

Gel time as a function of pH and K^+: Na^+ ratio

Gelation time of low and equimolar of K^+: Na^+ ratio in the sol was studied for different pH values of the sol. Results are plotted in Figure 1. Gelation was most rapid between pH=4 and 6 because the sol passed through the isoelectric point, at which it did not carry a net electrical charge and rapid aggregation of the particles was observed.

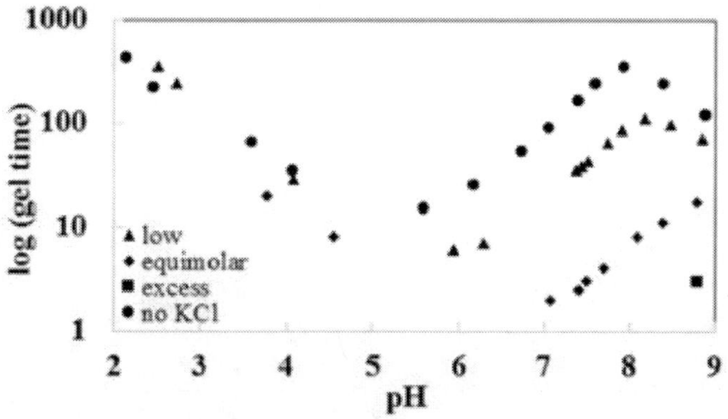

Figure 1: Relationship between gel time and pH for different molar ratios of K^+: Na^+ ions
▲: low (0.14) ♦: equimolar (1) ■: excess (1.76) ●: previous results obtained with unmodified sodium silicate precursor[9]

For all the pH values studied, addition of the ions resulted in faster gelation compared to the previous results obtained in the presence of Na^+ ions only.[6] At the pH range of 2-4, a slow gel formation was resulted because of low concentration of hydroxyl ions. However, longer gel times were obtained after the isoelectric point (pH=6-8) because the gel become negatively charged by high concentration of hydroxyl ions. In the presence of KCl, a shift toward the basic side was observed in Figure 1. This is because the K^+ and Cl^- ions have higher affinity for silicate ions rather than OH^- ions and, therefore more OH^- ions remain in the sol.

Stability of H_2O_2 in hydrogels containing K^+

H_2O_2 content was measured during 70 storage days to study the stability of H_2O_2 as a function of the K^+: Na^+ ratio and the pH of the sol. Stability results of hydrogels containing different K^+: Na^+ ratio at a constant pH of 8.8 are shown in Figure 2. H_2O_2 stability was observed to increase with the addition of K^+ ions as seen more clearly in short term stability (first 10 days during storage).

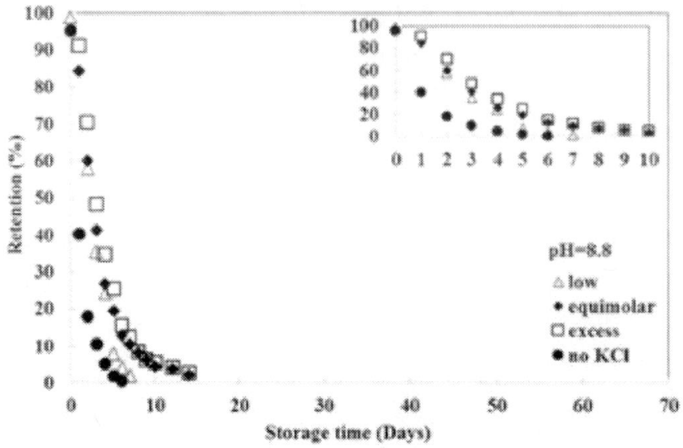

Figure 2: Room temperature stability of H_2O_2 as a function of storage time at constant pH=8.8 (for different molar ratios of K^+: Na^+ Δ: low ♦: equimolar □: excess ●: previous results obtained with unmodified sodium silicate precursor[9])

For example, 18.1wt % peroxide retention was obtained at the end of 2 days of storage at room temperature in our previous work when no K^+ ion was present in the system.[9] In this study, the peroxide retention of hydrogels containing low, equimolar and excess K^+:Na^+ molar ratios was 58.1, 60.2 and 70.6wt %, respectively, during the same period. Moreover, complete decomposition of H_2O_2 in hydrogels containing K^+ took longer (7 to 14 days) than that in previously studied hydrogels[9] (5 days).

Figure 3 a & b show the effects of pH on peroxide stability for low and equimolar amounts of K^+: Na^+, respectively. For each ratio, when the pH of the sol decreased, higher retention of H_2O_2 was observed. At lower pH values of the sol, higher amounts of H^+ ions increase the entrapment efficiency and retention by creating more available sites for peroxide in silica network.[16]

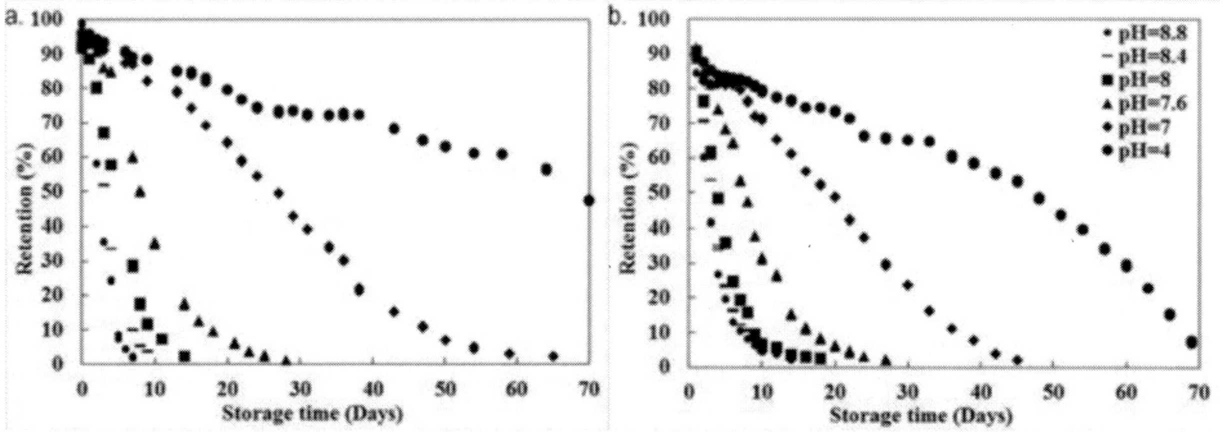

Figure 3. The comparison of H_2O_2 stability in hydrogels containing low (a) and equimolar (b) ratios of K^+: Na^+ at different pH values

Stability of H_2O_2 in hydrogels containing Mg^{2+}

H_2O_2 stability of hydrogels containing Mg^{2+} was studied to observe the effect of ion charge added into the starting silicate solution. Magnesium was selected to be partially exchanged with the sodium in the silicate at 1:0.2 molar ratio of Mg^{2+}: Na^+ because magnesium ion is divalent and smaller in size than Na^+. Magnesium ion could not be completely replaced with Na^+ because magnesium silicates are not water soluble unlike potassium silicates. The following reaction shows the precipitation reaction of magnesium silicates.

$$2(\equiv \mathrm{SiO^-Na^+}) + \mathrm{MgCl_2} \rightarrow 2 \equiv \mathrm{SiO^-Mg^{2+}} + 2\mathrm{NaCl} \quad (3)$$

The pH of the sol containing Mg^{2+} was decreased using ion exchange resin (Amberlite IR120) which replaces ions with H^+. Figure 4 shows H_2O_2 retention in hydrogels synthesized in the presence (pH=4.6) and absence (pH=8.8) of ion exchange. Previous stability results of hydrogels containing only Na^+ are also included in Figure 4 for comparison. The presence of Mg^{+2} showed improved peroxide stability regardless of the presence of ion exchange. We believe that the reason of this result is more sites available for hydrogen bonding between silanol groups and peroxide molecules due to the smaller size and higher charge of magnesium ions than Na^+.

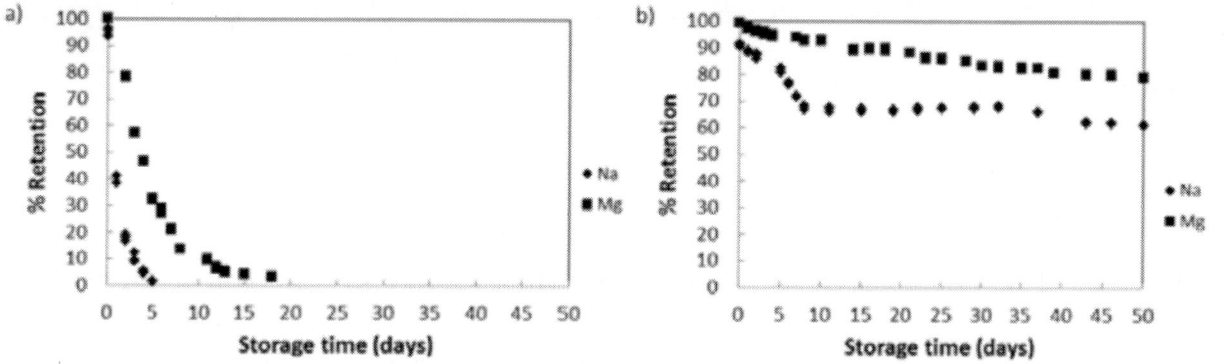

Figure 4. Room temperature stability of H_2O_2 as a function of storage time (a) without ion exchange (pH=8.8) and (b) with ion exchange (pH=4.6)

Morphology and structure of hydrogel

BET and SEM analyses showed that the presence of potassium ion resulted less compact gel structure at constant pH. This is because K^+ has larger atomic size than Na^+ and results larger pores in the hydrogel structure. Figure 5 shows SEM images of samples having different molar ratios of K^+: Na^+ at pH=8.8.

Figure 5. SEM images (at X200,000) of hydrogels for different molar ratios of K^+: Na^+ : a) low, b) equimolar, c) excess (at constant pH=8.8)

Nitrogen sorption studies were carried out to obtain quantitative results of the effect of potassium ion on pore size and surface area of hydrogels. Results are shown in Figure 6. When K^+: Na^+ molar ratio was increased, larger pore sizes were obtained because of the larger atomic size of K^+ (Figure 6a). For each case, small pore sizes were observed as pH decreased because hydrogen ions have capability to bring silica particles together by charge screening effect. In Figure 6b, a decrease in surface area was resulted in the presence of K^+. However, surface area of hydrogels increased at lower pH values as expected due to the decrease in pore size.

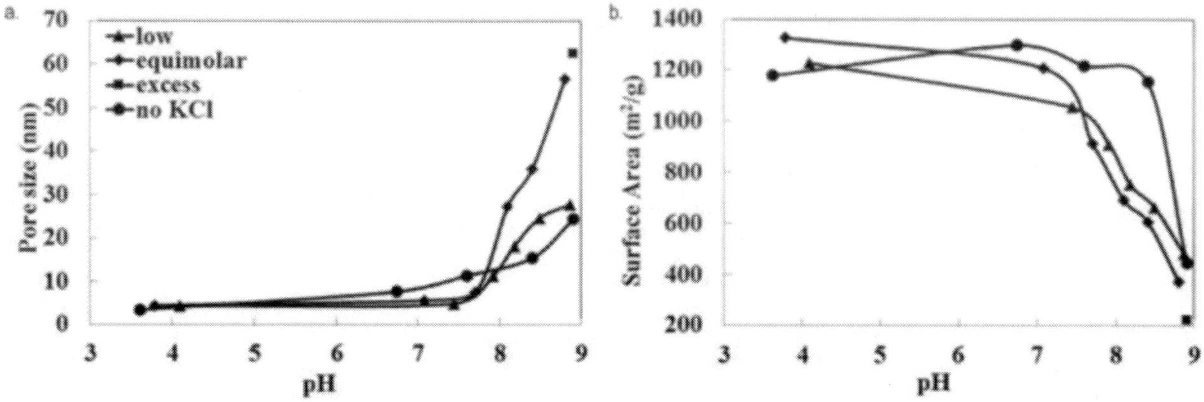

Figure 6. Structural properties of silica hydrogels with changing pH and K^+: Na^+ molar ratio, ▲: low ♦: equimolar ■: excess ●: previous results obtained with unmodified sodium silicate precursor[9]

Conclusions

A significant effect of the size and charge of ions on H_2O_2 stability and the hydrogel structure was observed. Gelation time decreased as K^+: Na^+ ratio were increased in the starting precursor. The stability of encapsulated H_2O_2 was improved with the addition of Mg^{2+} and K^+ ions and decreasing pH of the sol. With the increase in K^+: Na^+ ratio, average pore diameter was found to increase and surface area decreased at constant pH (8.8), therefore the gel became less compact. The higher retention and better stability of H_2O_2 at room temperature was resulted by controlling the ion content (size and charge) and pH. The properties of hydrogels can be tailored to control the gel time and H_2O_2 stability. A detailed study might be required for better understanding of the effect of size and charge of ions on peroxide stability and the hydrogel structure. For example, calcium ions can be included in hydrogel synthesis because the sizes of Ca^{2+} and Na^+ ions are similar whereas it has higher charge like Mg^{2+}.

References

1. Schumb WC, Satterfield CN, Wentworth RL. Hydrogen peroxide. ACS Monograph Series. New York: Reinhold Publishing Corporation, 1955.
2. Sigma Aldrich Product Information Sheet: Hydrogen Peroxide Solution. http://www.sigmaaldrich.com/
3. Colodette JL, Rothenberg S, Dence CW. Factors Effecting Hydrogen Peroxide Stability in the Brightening of Mechanical and Chemimechanical Pulps. J. Pulp. Pap. Sci. 1989; 15: 3−10.
4. Sulieman M, Addy M, MacDonald E, Rees JS. The Effect of Hydrogen Peroxide Concentration on the Outcome of Tooth Whitening: An In Vitro Study. J. Dent. 2004; 32: 295−299.
5. Kim JC, Lee KU, Shin WC, Lee HY, Kim JD, Kim YC, Tae G, Lee KY, Lee SJ, Kim JD. Monoolein Cubic Phases Containing Hydrogen Peroxide. Colloids and Surfaces B: Biointerfaces. 2004; 36: 161−166.

6. Żegliński J, Cabaj A, Strankowski, M, Czerniak J, Haponiuk JT. Silica Xerogel-Hydrogen Peroxide Composites: Their Morphology, Stability, and Antimicrobial Activity. Colloids and Surfaces B: Biointerfaces. 2007; 54: 165−172.
7. Bednarz S, Rys B, Bogdal D. Application of Hydrogen Peroxide Encapsulated in Silica Xerogels to Oxidation Reactions. Molecules. 2012; 17: 8068-8078.
8. Wolanov Y, Shurki A, Prikhodchenko PV, Tripolskaya TA, Novotortsev VM, Pedahzur R, Lev O. Aqueous Stability of Alumina and Silica Perhydrate Hydrogels: Experiments and Computations. Dalton Trans. 2014; 43: 16614.
9. Sudur F, Pleskowicz B, Orbey N. Hydrogen Peroxide Stability in Silica Hydrogels. Ind. Eng. Chem. Res. 2015; 54 (6):1930−1940.
10. Sudur F, Orbey N. Properties of Hydrogen Peroxide Encapsulated in Silica Hydrogels and Xerogels. Ind. Eng. Chem. Res. 2015; 54(44): 11251-11257.
11. Bergna HE, Roberts WO. Colloidal Silica: Fundamentals and Applications. Boca Raton, FL: CRC Press, 2005.
12. Baccile N, Babonneau F, Thomas B, Coradin T. Introducing Ecodesign in Silica Sol-Gel Materials. J. Mater. Chem. 2009; 19: 8537−8559.
13. Brady AP, Brown AG, Huff H. J. The polymerization of aqueous potassium silicate solutions. Colloid Sci.1953; 8(2):252-276.
14. Shroff AV, Shah DL. Grouting Technology in Tunneling and Dam Construction. Rotterdam, Brookfield: Oxford & IBH Publishing Company, 1992.
15. Chun-xi Z. Recent Advances in Water glass Sand Technologies. Shanghai, China: School of Material Science and Engineering, Shanghai Jiaotong University; 2007.
16. 27. Iler RK. The Chemistry of Silica: Solubility, Polymerization, Colloid and Surface Properties and Biochemistry. New York: Wiley, 1979.

472182 UV-Assisted Synthesis of Carbon Nanotube-TiO_2 Nanocomposites for Enhanced Photocatalytic Air Purification

Abstract:

Efficient photocatalysts consisting of carbon nanotubes (CNTs) decorated with TiO2 nanoparticles were synthesized by a simple two-step process that is free of aggressive acid treatments that often introduce defects to CNTs. The novel synthesis process involved UV-assisted oxidation of CNTs using H_2O_2, followed by the deposition of TiO2 via a wet chemical method using TiCl4 as a titanium source. The resulting CNT/TiO2 nanocomposites have been characterized by X-ray diffraction (XRD), transmission electron microscopy (TEM), X-ray photoelectron spectroscopy (XPS), and Raman spectroscopy. XRD results revealed that the crystalline structure of TiO2 is anatase and the TEM results showed the undamaged tubular morphology of CNTs decorated with TiO2 nanoparticles. The XPS data revealed the intimate chemical interactions between the CNTs and TiO2 in the nanocomposites. The expected synergistic effect and efficient separation of photogenerated electron-hole pairs in the synthesized CNT/TiO2 nanocomposites were verified in gas-phase photocatalytic degradation of VOCs.

470129 NOVEL Magnetic Nanocomposite Materials for RAPID Removal of Polychlorinated Biphenyls from Contaminated Water Sources

Abstract:

Magnetic nanocomposite systems able to capture polychlorinated biphenyls (PCBs) in aqueous solution were developed, providing a cost-effective water remediation technique. Two distinct methods were employed to develop these polyphenolic nanocomposite materials, where the polyphenolic moieties were incorporated to create high affinity binding sites for organic pollutants. The first method utilized a surface initiated polymerization of polyphenolic-based crosslinkers and co-monomers on the surface of iron oxide magnetic nanoparticles to create a core-shell nanocomposite. The second method utilized a bulk polymerization method to create macroscale films composed of iron oxide nanoparticles incorporated into a polyphenolic-based polymer matrix, which were then were processed into microparticles. Both methods produce nanocomposite materials that can bind chlorinated organics, can rapidly separate bound organics from contaminated water sources using magnetic decantation, and can use thermal destabilization of the polymer matrix for contaminant release and material regeneration. The polyphenol functionalities used to bind organic pollutants were quercetin multiacrylate (QMA) and curcumin multiacrylate (CMA), which are acrylated forms of the nutrient polyphenols with expected affinity for chlorinated organics. Particles were characterized using transmission electron microscopy (TEM), dynamic light scattering (DLS), Fourier transform infrared spectroscopy (FTIR), thermal gravimetric analysis (TGA) and UV-visible spectroscopy stability analysis. Pollutant binding studies were performed using PCB 126 as a model PCB and chlorinated organic pollutant to determine binding affinity and capacity, and this was quantified using gas chromatography coupled to electron capture detector (GC-ECD). It was demonstrated that the materials effectively bound PCBs in aqueous media.

472163 Influence of Silica-Based NanoparticlesÂ Embedded in Sand Bed Filtration for Cleaning-up Industrial Wastewater

Abstract:

Sand filtration technology is widely used to remove the suspended solids and colloids from the industrial wastewater effluent. However, such standalone technology cannot remove the soluble organic compounds from the industrial wastewater. In the current application, we are targeting to improve not only the removal of suspended solids but also increase the removal efficiency of total organic carbons (TOC) by enhancing the sand surface area via integrating it with functionalized silica-based nanoparticles. These kinds of nanoparticles are environmentally friendly and naturally occurring materials at the macro-level.

In this study, different types of functionalized silica-based nanoparticles were successfully prepared, characterized, and tested for TOC adsorptive removal within batch and continuous adsorption processes. Briefly, the Brunauerâ €"Emmettâ€"Teller (BET) surface area, X-ray diffraction (XRD), Fourier transform infrared (FTIR) spectroscopy and high resolution transmition electron microscopy (HRTEM) were used to confirm the surface area, crystalline domain size, functionality, and structural topology, respectively. The characterization results showed that the prepared nanoparticles are spherical in shape with a crystalline diameter of 10 nm and BET surface area around 180 m^2/g. Adsorption experiments were conducted in a batch mode to investigate the adsorption affinity and capacity of nanoparticles toward TOC removal from a locally supplied industrial wastewater sample. The results indicated that the TOC removal efficiency was improved significantly by 50% using the functionalized sand with nanoparticles at low mass fraction (<10 wt%). Afterwards, the filter bed column, continuous mode, was conducted using the functionalized and the un-functionalized sands. Interestingly enough, the functionalized sand with in-house prepared silica-based nanoparticles showed prolonger breakthrough time (>50 times) in comparison with the un-functionalized sand. These promising results are believed to help the current conventional wastewater treatment methods not only on a lab scale but also on an industrial level.

464632 Inactivation of *E. coli.* Using a Novel TiO_2 Nanotube Electrode

Abstract:

Electrochemical technology has gained increasing interests for water/wastewater treatment in recent years. In particular, it may find applications in decentralized treatment due to its simplicity, amenability to automation, compact instrumentation, and ease of operation. The performance of electrochemical processes highly depends on the material of electrodes. However, the high cost of electrodes, formation of toxic byproducts, and short electrode lifetime have impeded its wide application. Therefore, development of cost-effective anode materials are needed to support future adoption of electrochemical technologies for environmental applications. In our study, a novel TiO_2 nanotube anode has been fabricated in the lab via simple electrochemical anodization using metallic Ti as the substrate followed by electrochemical hydrogenation/cathodic polarization. The one-dimensional nanostructure is shown in FESEM images. The TiO_2 nanotube anodes with stainless steel as the cathodes were tested for inactivation of a model microbial contaminant *E. Coli.* using a filter-press type cell. Effects of current density applied (0.625 ~ 2.5 mA/cm^2), electrolyte concentration (0.3 mM~3 mM NaCl), and different electrolyte (NaCl, Na_2SO_4, Na_2HPO_4, $NaNO_3$, and $NaHCO_3$) on *E. Coli.* inactivation were investigated. Electrochemically generated oxidizing species were measured as free chlorine. Further, production of reactive oxygen species (•OH, O_2•-, SO_4^{2-}•) were examined using probe compounds. In addition, electro-disinfection experiments were conducted in real water matrices, i.e. samples of surface water, treated surface water before disinfection, and secondary effluent before disinfection collected from local water/ wastewater treatment plants. Formation of inorganic (e.g. ClO_3^-) and organic (THMs) disinfection byproducts were also monitored. Our results have demonstrated that the novel TiO_2 nanotube anode can effectively achieve 5~6 log reduction of E. Coli. with low byproducts formation. The low energy consumption also indicates that such disinfection system may be

operated using solar energy stored in a DC battery for decentralized treatment or as a point-of-use device.

451725 Metal Organic Framework Derived Nanoporous Carbon Â As a Novel Adsorbent for Water Treatment

Abstract:

Metal organic framework derived nanoporous carbon

as a novel adsorbent for water treatment

Zahra Abbasi[a], Ezzatollah Shamsaei[a], Soo Kwan Leong[a], Bradley Ladewig[a,b], Xiwang Zhang[a], Huanting Wang[a*]

[a]Department of Chemical Engineering, Monash University, Clayton, Victoria 3800, Australia

[b]Department of Chemical Engineering, Imperial College London, Exhibition Road, London SW7 2AZ,

United Kingdom

Phone: +61 3 9905 3449, Email: Huanting.wang@monash.edu

Nanoporous carbon materials have attracted a great deal of attention because of their potential application in many areas such as adsorption and catalysis. These materials have desirable properties such as high surface area, good chemical and thermal stabilities as well as fast kinetics, which make them suitable for adsorption applications. Porous metal organic frameworks (MOFs) consist of crystalline structures that provide high internal surface areas and tunable porosities which could be considered as starting materials for developing novel carbon based nano composite

adsorbents.

Different types of adsorbents have been studied for the removal of organic pollutants from water and wastewater such as activated carbon, fly ash waste materials and different kind of MOFs. However, their adsorption capacity is low in comparison with nanoporous carbons. Therefore, developing novel carbon based materials with simple synthesis methods for adsorption of pollutants from water is of great importance.

Zeolitic immidazolate frameworks (ZIFs) are a class of MOFs that exhibit remarkable features like high thermal and chemical stability and demonstrate similar pore topologies to zeolites. Among various ZIFs reported, ZIF-8 with the formula of $Zn(2\text{-methylimidazolate})_2$ and with a sodalite-related zeolitic structure has been used for many applications such as adsorption and catalysis. The surface and bulk properties of ZIF-8 can be modified to make it a more competent adsorbent for water treatment purposes. Carbonization of ZIF-8 could be the easiest and most practical modification approach to amend this metal organic frame work's characteristics.

The adsorption capabilities and characteristics of nanoporous carbon particles derived from ZIF-8 were investigated in this study. Direct carbonization of ZIF-8 was used to synthesize carbon nano particles. SEM and TEM images revealed that the ZIF-8 derived carbon carbonized at several temperatures retained the original structure and morphology of ZIF-8. These carbon nano-particles exhibited significantly greater adsorption capacities for the removal of dye from water compared to the ZIF-8. The high adsorption capacity is attributed to the change in surface charge and pore size. The surface functionality and hydrophobicity of the materials were investigated by Raman Spectroscopy, N_2 adsorption-desorption, FTIR, TGA and contact angle measurements. The change in the surface charge of the carbon particles was confirmed by zeta potential of the samples. The dye uptake followed pseudo-second order kinetics and the equilibrium data was best fit to the Langmuir model for the carbon nano-particles. Intra-particle diffusion studies showed that the adsorption process was governed by both the diffusion step and external mass transfer. The ZIF-8 derived nano porous carbons were found to be highly efficient adsorbents for water and waste water treatment.

474742 Interfacial Assembly and Engineering of Ordered Functional Mesoporous Materials for Applications

Abstract:

With recent progresses made in modern nanoscience and nanotechnology, ordered mesoporous materials have been one of the hottest research topics in scientific community spanned chemistry, materials science, physics and biology. The construction of mesoporous materials is mainly concerned with building monodispersed mesosized (2-50 nm) pore voids and arranging them in a long-range ordered array. Generally, two kinds of templates are used to produce the mesopores: supramolecular aggregates such as surfactant micelle arrays, and rigid preformed solids such as ordered mesoporous silica, carbon, and colloidal crystals. Noticeably, besides the templates, the interface also plays a central role in the synthetic process, because it provides a rich and crucial space for the assembly and construction of mesostructures. Generally, two kinds of interfaces involve in the synthetic system. The first one is at between surfactant templates and guest species, which has been extensively investigated. Another important interface is the two-phase (solid, liquid and gas) one, including liquid-solid, gas-liquid, liquid-liquid, gas-solid, and solid-solid interface, which has been well developed for the synthesis of ordered mesoporous materials. Compared with the one phase synthesis referring to homogeneous nucleation and growth, the introduction of a two-phase interface in the system can change the growth behaviors of mesoporous materials and lead to the formation of molding or multifunctional mesoporous materials. For example, mesoporous thin films or membranes have been widely fabricated on a substrate *via* an evaporation-induced self-assembly (EISA) method. Multifunctional core-shell structured mesoporous materials can be obtained by rationally depositing mesoporous shells on well-designed cores at the interface. Recently we have developed a novel facile approach i.e. a solvent evaporation-induced aggregating assembly (EIAA) to synthesize large pore mesoporous silica materials. In addition, the well-known hard-templating method for mesoporous materials is also a typical interface reaction.

448965 High-Energy Density Metal-Free Biobatteries Powered By Soft Drinks

Abstract:

More and more portable electronics urgently require better batteries featuring high-energy density, good safety, and low environmental impacts. Enzymatic fuel cells are a promising micro-power source for small electronics. However, incomplete oxidation of sugars mediated by few redox enzymes and low faraday efficiency result in low energy storage densities. Here we demonstrate the complete oxidization of sucrose, glucose, and fructose at the anodic compartment of biobatteries via a synthetic 15-enzyme pathway. The biobatteries generated maximum power densities from 0.65 to 0.98 mW cm^{-2}. Also, a nitrogen-doped activated carbon cloth was used as a flexible cathode to achieve oxygen reduction reaction faster than that catalyzed by platinum-coated membrane electrode assembly. Biobatteries comprised of 15 enzymes oxidize sucrose, fructose and glucose simultaneously, generated nearly 24 electrons per hexose with faraday efficiencies of approximately 95%. Metal-free biobatteries powered by commercial soft drinks whose energy-storage density was ~400 Ah kg^{-1} of drinks exhibited maximum power densities of up to 0.55 mW cm^{-2}, implying that soft drinks could be perfect fuels for bioelectricity generation.

474830 Lego-like Micropillar/Microwell Chip for High-Throughput Functional Analysis of Genes Encoding Pathogen-Specific Antimicrobial Enzymes

Abstract:

The overuse and misuse of antibiotics have resulted in the emergence of antibiotic-resistant pathogens along with the perturbation of healthy human microbiota via killing putatively beneficial commensal bacteria. Hence, targeting infectious bacterial pathogens is important for reducing the evolution of antibiotic-resistant bacteria and preserving the endogenous human microbiome. Cell lytic enzymes including bacteriophage endolysins, autolysins, and peptidoglycan targeting enzymes are useful antibiotic alternatives, and genetic information of numerous cell lytic enzymes is currently available. However, the identification of their antimicrobial function and specificity has been limited due to time-intensive protocols to identify their specific targets. Here, we developed Lego-like sandwich microchip for rapidly accessing the function of diverse genes that are suggestive of encoding cell lytic enzymes. This approach can be used to quantify antimicrobial activity in a high-throughput manner by detecting bacterial cell growth in 40 nL alginate gel spots on micropillar supports. In addition, cell-free protein synthesis could be achieved in a complementary microwell chip. The combination of the chip-based antimicrobial assay with parallel cell-free protein synthesis enabled the rapid identification and evaluation of species-specific antimicrobial function of 11 genes encoding known and putative cell lytic enzymes.

466405 Chaperonin-Inspired Enzyme Protection By Mesoporous Silica

Abstract:

Enzymes can catalyse bioorganic transformations with much higher specificity than traditional chemical catalysts, which makes them valuable for fine chemical and pharmaceutical manufacturing. However, enzymes have evolved to operate within their host cells: most often in neutral, aqueous solutions and at moderate temperatures. Therefore, they tend to have poor stability in chemical reactors that involve high temperatures, acidic or basic conditions, or organic solvents.

In vivo, enzymes are stabilised from high temperature or unfavourable solution conditions by chaperonins. Chaperonins, such as the GroEl/ES complex, help refold partially unfolded proteins within a narrow, cylindrical pore that becomes negatively charged. While chaperonins' complete mechanisms are complex, we are inspired by some of their fundamental properties when developing materials for enzyme immobilisation.

Mesoporous silica SBA-15 is used as a synthetic chaperonin analogue because of its controlled mesopore diameter and its negatively charged interior surface. Enzymes are immobilised on SBA-15 of different pore diameters, and their stabilities and biocatalytic activities are tested. SBA-15 is shown to be a valuable candidate for enzyme immobilisation, because it can protect enzymes from denaturing conditions. Selected findings include: immobilised myoglobin's protection from acidic reaction conditions, with activity improvement of up to 350%; immobilised myoglobin's complete protection from the protease pepsin; immobilised lysozyme's protection from basic reaction conditions, with activity improvement of up to 200%; and the improvement of immobilised lysozyme's activity by altering SBA-15's pore diameter.

More fundamentally, SBA-15 is a useful material for the systematic investigation of enzyme immobilisation parameters because of its highly tuneable pore diameter and its easily functionalised surface. The relative importance of hydropathy, electrostatic attraction, and steric confinement are studied in order to improve the future design of enzyme immobilisation materials.

464833 Intermediate Channeling Via Nanoscale Confinement

Abstract:

Multi-step, enzyme-catalyzed reaction cascades occur frequently in metabolic and catabolic pathways of living cells with high efficiency, selectivity, and yield. Understanding these metabolic pathways can lead to engineered synthetic pathways for the production energy and high-value molecules.

Transfer of intermediates between the active sites without diffusing into the bulk is known as "substrate channeling." Confinement by molecular tunnels, electrically charged pathways, spatial organization of catalysts and covalent bonding of intermediates are examples of substrate channeling mechanisms. Under most relevant reaction conditions, the rate of diffusion is faster than the rate of reaction.[1] Hence, confining the active sites and intermediate by creating a physical tunnel between active sites enables controlled transport without bulk diffusion.

In the present study, computational modeling has been performed to study the effect of geometric, kinetic, and transport parameters on intermediate channeling via confinement, using a continuum model. Efficiency of transport is quantified by reactant yield and flux control coefficients (FCC).[2] Retention of intermediates within the tunnel, with minimal access to bulk solution, is shown to be the key to efficient channeling.

Acknowledgement

We gratefully acknowledge support from Army Research Office MURI (#W911NF1410263) via The University of Utah.

References

1. I. Wheeldon, S. D. Minteer, S. Banta, S. Calabrese Barton, P. Atanassov and M. Sigman, "Substrate channelling as an approach to cascade reactions", *Nat. Chem.*, **8**, 299–309 (2016). doi:10.1038/nchem.2459.

2. D. A. Fell, "Metabolic control analysis: a survey of its theoretical and experimental development", *Biochem. J.*, **330**, 313–330 (1992). doi:10.1042/bj2860313.

474743 Engineering Ultrastable Protein Scaffolds for the Controlled Assembly of Multifunctional Nanobiomaterials

Abstract:

Self-assembling protein templates are of increasing interest in the field of nanoscale fabrication of biomaterials, where precise patterning of functional biomolecules, such as enzymes, is often desired. In particular, protein building blocks can be strategically chosen to exhibit desired functionality, while engineering their assembly allows for the controllable positioning of the subunits. The filamentous protein g-prefoldin (g-PFD) from the hyperthermophilic archaeon *Methanocaldococcus jannaschii* is an excellent candidate for such a tunable scaffold. Its remarkable stability, unique modularity, and self-assembly into filaments with chaperone activity render it an ideal candidate for the bottom-up construction of novel protein nanostructures.

We have recently reported the assembly of g-PFD monomers into geometrically defined templates of tunable dimensions and symmetry, as well as the controlled positioning of nanoparticles along the filaments. We aim to achieve more complex patterning, while expanding the applicability of g-PFD templates to diverse enzymatic systems. Fusing different enzymes to each subunit should enable periodic positioning of multiple enzymes along the filament to catalyze sequential reactions and metabolic pathways. In addition, we have demonstrated the ability to position cytochrome C proteins at high density along the filament as a first step to creating novel conductive nanowires. Finally, we are studying g-PFDâ€™s ability to form 3D crosslinked multifunctional protein hydrogels. Ultimately, the ability to position functional molecules in a controlled manner will considerably enhance our ability to fabricate advanced multifunctional nanobiomaterials.

460929 Concanavalin A Enabled High Performance Enzyme Cascades on Magnetic Nanoparticles

Abstract:

Concanavalin A (ConA) is a lectin that specifically binds to carbohydrates containing non-reducing terminal α-D-mannosyl or α-D-glucosyl moieties. Here we present a facile and aqueous method preparing multiple glycolenzyme catalysts using ConA coated magnetic nanoparticles (ConA-MNP). The first step is to from a ConA layer on the surface of MNP (ConA-MNP) by crosslinking with glutaraldehyde. Then glycoenzymes were agglutinated on the surface of ConA-MNP spontaneously, followed by crosslinking with glutaraldehyde.

The resulted GOx-HRP@ConA-MNP showed an over 80% residual activity compared to their native counterparts in solution. In contrast, merely no activity was observed when the enzymes were directly immobilized on MNP via crosslinking with glutaraldehyde. The GOx-HRP@ConA-MNP retained over 90% of the initial activity after 10 cycles. Above mentioned procedure was applied to form GOx-Cat@ConA-MNP for catalyzing the conversion of glucose to gluconic acid. It was evidenced by both structural characterization and reaction kinetics that the co-localization of the cascade enzymes enabled a substrate looping between two enzymes, i.e., oxygen produced by Cat was confined to GOx producing H_2O_2 as a substrate for Cat. As a result GOx-Cat@ConA-MNP achieves 91% glucose conversion in 75 minutes whereas the aqueous GOx and Cat reached a conversion of 86%. The GOx-Cat@ConA-MNP retained 99% after 8 cycles with a glucose conversions over 90% in each run.

The high catalytic performance and the ease of operation makes the multiple enzymes @ConA-MNP appealing for enzyme cascades for various applications.

467438 Design of a Heterogeneous Biocatalyst for Cofactor Regeneration and Improved Catalytic Characteristics

Abstract:

Enzyme immobilization generally allows for improved lifetime and stability of catalytically active protein. Typically, immobilization systems involve adsorption to surfaces, cross-linked enzyme aggregates or entrapment of enzyme in a polymer network. However, these systems often can introduce unwanted limiting effects, such as high resistance to mass transfer or distortion in the enzyme structure due to adsorption or crosslinking. To overcome these shortcomings, it is possible to use more porous materials which are imbued with immobilization domains. Supraparticles formed by a hierarchically structured self-assembly of protein-inorganic nanoflowers provide a porous and stable support for enzyme capture with a high surface area and low mass transfer resistance. In this work, we used hybrid protein-inorganic supraparticles made of calcium phosphate and protein with leucine zipper binding domains for high-affinity immobilization of fusion proteins containing the complementary binding domain and enzymes. Particularly, we showed the modularity of the supraparticle system by immobilizing two distinct enzymes on the particle in tandem.

As a model system, we used the amine dehydrogenase, cFL1AmDH, which is coupled for cofactor regeneration with formate dehydrogenase, cbFDH. The cFL1AmDH oxidizes NADH to convert 4-fluorophenylacetone and ammonia to 4-fluoroamphetamine, an important chiral intermediate. However, the enzyme lacks desirable kinetic characteristics and reaction condition, given a current optimum at 5M ammonium chloride and a pH of 9.6. The cbFDH regenerates NADH from NAD+ by oxidizing formate to carbon dioxide, reducing the cost associated with replacing an expensive cofactor. By co-immobilizing these enzymes, we seek to improve the activity of cFL1AmDH by supplying a constant source of NADH. To account for the effects of alkalinity and ionic strength from the optimal cFL1AmDH reaction, the supraparticles were synthesized with several concentrations of ammonia and at neutral and alkaline pHs. The particles synthesized under the reaction conditions proved to be comparable to the neutral, ammonia-free synthesis in size and number. We then generated a set of fusion proteins with four different linker peptides connecting the catalytic domain to the immobilization domain and tested their activity. After selecting the enzymes with the highest specific activity, we incubated them with the particles and quantified the enzyme immobilization and specific activity. The particle porosity, effectiveness factor, and Thiele modulus were quantified to determine the effects of the particles on mass transfer. This work demonstrates our ability to create an industrially relevant heterogeneous biocatalyst which incorporates cofactor regeneration, is stable in enzymatically relevant conditions, and allows for enzyme recycling.

1. Sheldon, R. A. Enzyme Immobilization: The Quest For Optimum Performance.ChemInform. 2007, 38, 1289–1307.
2. Park, W. M.; Yee, C. M.; Champion, J. A. Self-Assembled Hybrid Supraparticles That Proteolytically Degrade Tumor Necrosis Factor-α. J. Mater. Chem. B. 2015, 4, 1633–1639.
3. Bommarius, B. R.; Schürmann, M.; Bommarius, A. S. A Novel Chimeric Amine Dehydrogenase Shows Altered Substrate Specificity Compared to Its Parent Enzymes. Chem. Commun. 2014, 50, 14953–14955.

461059 Concanavalin a Coated Activated Carbon for Enzyme Immobilization

Abstract:

While the large pore size, good mechanical property, non-toxicity, as well as the availability of chemical or physical modification makes activated carbon (AC) an ideal support for enzyme immobilization, direct adsorption of enzymes onto AC often yields poor apparent activity and reusability. This might be attributed to the strong interaction that leads to undesired conformational changes of the adsorbed enzymes. Here we present a novel and facile method, as featured by coating AC with Concanavalin A (ConA), a lectin that specifically binds to glyosidic molecules, to prepare AC immobilized enzymes.

During a run, AC was loaded into ConA solution of appropriate concentration and incubated at room temperature for 12 hours. Then glycoenzyme such as laccase and HRP was loaded into the suspension of ConA coated AC and incubated for 12 hours. BSA, which was frequently used as a stabilizer of aqueous protein or during free drying was also used as a control to evaluate the stabilization effect of ConA.

It is shown that HRP immobilized on ConA coated AC remained 50% initial activity after 25 cycles. In contrast, the directly adsorbed HRP on AC lost all activity after 12 cycles. Replace of ConA with BSA resulted in 18% apparent activity remain after 25 cycles. The laccase immobilized on ConA coated AC appeared 96% of initial activity after 11 cycles. In contrast, the direct adsorption of laccase on AC led to a reduction of apparent activity by 74%. Moreover, the immobilized enzymes showed a 2.5-fold increase in activity compared to direct adsorption and an improved stability over a wider range of pH and at a higher temperature. The high yield of activity, enhanced reusability and stability of immobilized enzyme, in combination with the high adsorption ability of AC, make the enzyme immobilized on ConA coated AC promising for various applications of enzymatic catalysis.

474089 Tunable Synthesis of Au and Ag Hybrid Asymmetric Nanofingernails with Circular Dichroic Response

Abstract:

Single metal Au or Ag nanostructures are known to exhibit optical activity due to plasmonic effects and have been rationally designed for applications in photonics, sensing, bioimaging and especially surface enhanced Raman scattering (SERS) spectroscopy. Hybrid Au and Ag nanostructures possess additional functionalities due to the presence of both noble metals. However, the current methods used to achieve such hybrid nanostructures are either costly, require many fabrication steps, or are unreliable. Here we introduce a novel hybrid nanostructure composed of Au and Ag whose shape resembles that of a fingernail. The fabrication technique involves sequential thermal evaporation of metals onto track-etched hard templates to produce arrays of nanofingernails reliably, quickly and inexpensively. Moreover, the method is tunable with respect to nanofingernail height and thickness. The optical activity of the fabricated arrays is confirmed via circular dichroism (CD) spectroscopy. The CD spectroscopy demonstrates that the reduced symmetry and hybrid nature of the Au-Ag nanofingernails, which produce localized hot-spots of high intensity that change with incident circular polarization handedness, tailor the CD response. These novel structures have potential utility as platforms in surface-enhanced plasmonic sensing applications where the polarization-dependent hotspots may identify the location of adhered molecules.

450690 Non-Standard Amino Acid Labeling of the Extracellular Matrix of Probiotics for *In Vivo* Imaging of the Gut

Abstract:

Current methods for diagnosing and observing the healing or progression of diseases of the gastrointestinal tract are often invasive and do not allow for easy time monitoring. Alternative approaches would be beneficial if they allowed for continuous monitoring of the patientâ€™s condition non-invasively and for the use of molecular signatures or changes in the local gut environment to diagnose diseases. Probiotics are tools of choice for colonizing the gut, promoting health and even delivering therapeutics. By engineering probiotics to produce a detectable signal, they could also be used to image and probe diseases.Â In order to accomplish specific probing of probiotics, bio-orthogonal functional groups can be incorporated in proteins produced by probiotic bacteria via genetic engineering and use of non-standard amino acids (NSAA). Of particular interest are non-standard residues with side-chains containing azides, alkynes or other moieties that can participate in simple and efficient chemical reactions for labeling proteins with dyes, peptides or nanoparticles. Combining probiotics and NSAA could lead to the development of non-invasive methods for *in vivo*imaging in the gut.

Here we present the engineering of a probiotic strain, *Escherichia coli Nissle 1917* (ECN), to produce curli nanofibers containing NSAA that can participate in click chemistry reactions. Curli nanofibers are part of the extracellular matrix of ECN, and contribute to the bacterial adhesion in the gut and to biofilm formation. *In vitro* experiments demonstrate that ECN can effectively incorporate NSAA into proteins without recoding of the ECN genome, and only with minimal effects on the fitness of the bacteria. Subsequent dye binding via click chemistry allows for the selective labeling of curli nanofibers containing NSAA. During *in vivo* experiments, after colonization of the gut and expression of the curli nanofibers in the presence of NSAA, fluorescent labels can be used to detect the presence over time and the spatial distribution of ECN bacteria in the gastrointestinal tract. At a longer term, our technology could be optimized for triggered expression of NSAA-containing curli nanofibers at specific locations in the gut where disease markers are produced.

459775 Corona Phase Molecular Recognition for Protein Targets

Abstract:

Corona phase molecular recognition (CoPhMoRe) uses a heteropolymer adsorbed onto and templated by a nanoparticle surface to recognize a specific target analyte. This method has not yet been extended to macromolecular analytes, including proteins. Herein we develop a variant of a CoPhMoRe screening procedure of single-walled carbon nanotubes (SWCNT) and use it against a panel of human blood proteins, revealing a specific corona phase that recognizes fibrinogen with high selectivity. In response to fibrinogen binding, SWCNT fluorescence decreases by >80% at saturation. Sequential binding of the three fibrinogen nodules is suggested by selective fluorescence quenching by isolated sub-domains and validated by the quenching kinetics. The fibrinogen recognition also occurs in serum environment, at the clinically relevant fibrinogen concentrations in the human blood. These results open new avenues for synthetic, non-biological antibody analogues that recognize biological macromolecules, and hold great promise for medical and clinical applications.

460007 Probing Sub-Cellular Organelle Contents with Carbon Nanotube Optical Reporters

Abstract:

The optical properties of non-covalently dispersed single-walled carbon nanotube (SWCNTs) are well-suited for biological imaging and sensing applications due to intrinsic, photostable, and environmentally-sensitive photoluminescence in the near-infrared region. Appropriately functionalized nanotubes enter live cells and remain in the endolysosomal pathway. Here, we show that certain polymer-nanotube complexes can detect abnormal accumulations of lysosomal lipids in live cells, indicative of a family of diseases known as lysosomal storage disorders. Further, we show that the readout of the nanotube optical reporter is reversible in nature, and can be used to monitor the regression of disease upon pharmacological treatment.

468646 Selective Capture and Release of Biomolecules in Complex Solutions Using Gold Nanoparticles and Electromagnetic Fields

Abstract:

Functionalized gold nanoparticles have been studied extensively for various biological applications including drug delivery to cells and tissues, tumor therapy, and biosensing in complex biological systems. The most commonly studied particle shapes are spheres and rods of varying diameters and aspect ratios. In comparison to gold nanospheres, gold nanorods are particularly advantageous because they offer two-dimensions for tuning particle properties - the length of the nanorods determine the longitudinal surface plasmon resonance (LSPR), and the width of the nanorods determine the transverse or lateral surface plasmon resonance(TSPR). The former is particularly interesting for engineering complex functionalities into gold nanoparticles. One such example is the photothermal ablation effect studied extensively in tumor therapy. When gold nanorods are exposed to electromagnetic waves matching LSPR, the incident energy is converted into localized heat which can be deployed for killing tumor cells. We have extended the basic physics of this light-matter interaction to develop sophisticated biosensing approaches for detection of biomolecules in complex biological fluids. Our objective in this study is to demonstrate (i) the ability of peptide-functionalized gold nanorods to selectively bind to specific biomolecules such as proteins, enzymes and antibodies in a complex mixture, and (ii) the ability to trigger release of the biomolecules, on demand, under the influence of an electromagnetic field. To achieve this objective, we designed nanorods of aspect ratios varying from 2-4 that were coupled to 3 different cysteine-terminated peptides via the gold-thiol bond. The peptides employed in this study were chosen for their high affinity and high specificity of binding to three different biomolecules of varying size, charge, hydrophobicity and tertiary structures â€“ an IgG antibody, human cardiac troponin I, and streptavidin. Fetal bovine serum (FBS) was used as a background for the mixture of these proteins to simulate a complex biological sample. Peptide-functionalized nanoparticles were incubated simultaneously in a mixture of the target proteins, each labeled with a different fluorophore, in FBS and were washed and centrifuged to remove unbound material. On exposure to light of specific wavelengths from 550 nm, 800 nm and 1100 nm, we demonstrate the selective release of the target biomolecules in a wavelength-dependent fashion. The amount of material released from the nanorods was quantified using UV spectrophotometry. Selectivity was quantified as the ratio of the desired biomolecule to background material released from the nanorods. The amount of material released was a strong function of the incident wavelength, intensity and time of exposure. Capture and release capabilities in FBS were compared with similar measurements performed in a low complexity background containing bovine serum albumin (0%, 1% and 10% wt/vol) in phosphate buffered saline. The results from these studies can be further extended for the detection and quantification of a large number of proteins in biological samples of clinical relevance such as human serum. These studies are also promising for the detection of protein variants for developing analytical techniques in downstream bioprocessing applications.

468790 Theranostic Multibranched Gold Nanoantennas for Breast Cancer Diagnostics and Therapeutics

Abstract:

Breast cancer is one of the leading causes of non-communicable disease death in women worldwide. Especially dangerous and lethal are the breast cancers that have genetic mutations for surface receptors utilized in drug therapy. Specifically, breast cancer cells triple negative for the progesterone, estrogen, and Her2/neu receptors are impervious to most therapies and can even become resistant to some chemotherapy drugs. This resistance calls for new and innovative treatments that be optimized for cancers based on an individual patient basis/cancer phenotype.

Nanoparticle based diagnostics and therapeutics have recently emerged as a novel platform for management and mitigation of cancer at all stages. Gold nanostructures, specifically, have multiple characteristics that make them ideal for cancer theranostics including high biocompatibility, ease of bioconjugation, ability to tune their plasmon resonance to absorb tissue penetrating near infrared light, their use as contrast agents, and ability to convert light to heat when excited at the plasmon resonance for photothermal ablation of cancer cells. In this work we demonstrate the use of near-infrared light absorbing multibranched gold nanoantennas (MGNs) to simultaneously deliver diagnostic and therapeutic (theranostic) capabilities in triple negative breast cancer cells (MDA MB 231). MGNs consist of a core that absorbs light and protrusions that serve as emitters confining light into a localized area generating enhanced local fields via the nanoantenna effect and intense photothermal response.Â By exploiting these optical properties of MGNs we have utilized them for ultrasensitive surface enhanced Raman scattering (SERS) imaging to visualize MDA MB 231 cells with single-cell resolution *in vitro* and subsequently performed photothermal therapy to enable cell death using low laser powers of $< 5W/cm^2$.

By synthesizing various Raman reporter/receptor target combinations, multiplex imaging can be achieved, providing a cellular â€œtraffic mapâ€ demonstrating receptor expression *in vitro*. Further, due to the narrow peak width

fundamental to Raman spectroscopy, these multiple tags can be distinguished *in vivo*, even while amidst the many signals within tissue. We confirmed the feasibility of theranostic MGNs *in vivo* as multiplexing agents by observing the SERS signal over a longitudinal study and determined the max accumulation within human cancer xenografts. Lastly, these MGNs were utilized as contrast agents for photothermal optical coherence tomography (PTOCT) combining the high sensitivity of SERS and the high resolution and high tissue penetration depths of PTOCT within a single nanoagent to perform dual modal imaging.

468954 Colorimetric Detection of Therapeutic Levels of Ionizing Radiation Using Plasmonic Nanoparticle Gels

Abstract:

Radiotherapy has evolved into a highly complex and efficient treatment modality for ablation of malignant tumors. Treatments are typically administered in fractions of 2 Gy each (1 Gy = 1 Joule energy absorbed / kg), leading to a cumulative dose of 20-70 Gy depending on the disease. Despite several technological advances, determination of the dose delivered to the tumor and adjacent tissues remains a challenge due to complex fabrication, cumbersome operation, and high costs associated with current dosimeters. Advances in molecular and nanoscale systems can lead to novel platforms for the detection of therapeutic levels of ionizing radiation in radiotherapy. Here, we describe fundamental studies and development of a novel gel-based colorimetric nanosensor for detecting therapeutic levels of X-rays (1-10 Gy) administered in clinical radiotherapy. Following exposure to X-rays, surfactant micelles, incorporated together with gold salts in a gel, were able to convert the monovalent gold ions (Au1) to nanoparticles within the matrix, resulting in the formation of a maroon-colored plasmonic gel. Differences in color intensity of the gel following irradiation were used as a quantitative indicator of the radiation dose employed. The gel-based nanosensor was able to detect doses as low as 0.5 Gy, and demonstrated a linear detection range of 0.5 â€“ 5 Gy, which shows its application in the fractionated radiotherapy regime. Importantly, the gel-based dosimeter was able to report on the spatial distribution of the dose, indicating its use for identifying tissue regions that may receive higher-than-intended doses. The gel was also able to successfully detect therapeutic levels of radiation doses administered to anthropomorphic tissue phantoms. The range of detection, ease of fabrication, simplicity associated with colorimetric detection and relatively lower costs indicate that this technology can be potentially translated to different radiotherapy applications in the clinic.Â

471251 Carbon Nanotube-Based Microdevices for Tracking Single Macrophages By Raman Scattering

Abstract:

Carbon Nanotube-Based Microdevices for Tracking Single Macrophages by Raman Scattering

Zhibin Wang[1], Junfei Xia[1], Li Sun[2], Phong Tran[3], Sida Luo[3], Yi Ren[2], Tao Liu[3], Jingjiao Guan[1,4]

[1] Dept. of Chemical and Biomedical Engineering, FAMU-FSU College of Engineering, Florida State University.

[2] Dept. of Biomedical Sciences, College of Medicine, Florida State University.

[3] Dept. of Industrial & Manufacturing Engineering, FAMU-FSU College of Engineering, Florida State University.

[4] Integrative NanoScience Institute, Florida State University

Introduction: The ability to track therapeutic cells in vivo is useful for developing effective cell therapies. Existing cell-tracking techniques such as magnetic resonance imaging and positron emission tomography suffer from the need for ionizing radiation or limited capability for multiplex labelling. In contrast, Raman imaging does not involve ionizing radiation; are highly material dependent and can be excited by NIR, which permits a tissue-penetration depth at the centimeter scale. Carbon nanotubes exhibit strong resonance-enhanced Raman modes and have a unique Raman spectrum. On the other hand, we developed a low-cost, highly versatile technique for fabricating microdevices and preparing the complexes of the microdevices and live cells. We thus extended this method to produce carbon nanotube-based microdevices for tracking single macrophages by Raman scattering.

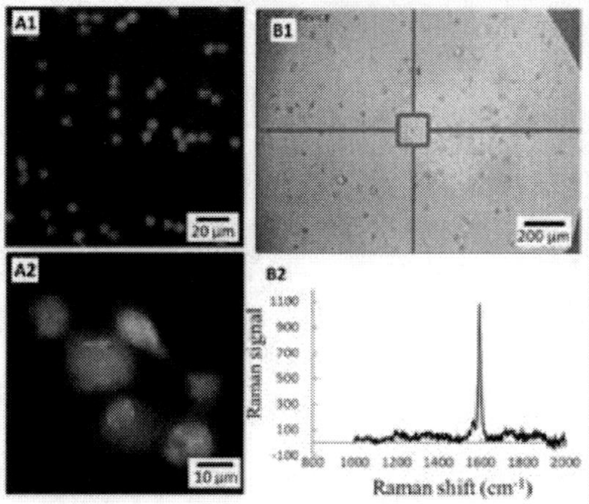

Fig. A1. Released microdevices. A1. Optical micrograph of cells imaged under a Raman microscope. A2. Macrophage-microdevice complexes. B1. Micrograph of cells and carbon nanotube-based microdevices B2. Raman scattering of the cell enclosed in the square in B1.

Materials and Methods: A polydimethylsiloxane stamp carrying an array of 7 ¦Ìm-diameter circular pillars was coated a trilayer of poly(allylamine hydrochloride) (PAH), and poly (sodium 4-styrenesulfonate) (PSS) and PAH. An aqueous solution of surfactant-stabilized single walled carbon nanotubes dispersed were sprayed on the stamp by using a self-made automatic spraying machine operated by a computer numerical control system. After spraying, poly(lactic-co-glycolic acid) (PLGA) containing a fluorescence dye octadecyl rhodamine B chloride (R18) was spin-coated on carbon nanotubes film surface. The microdevices were transferred onto a poly(vinyl alcohol) (PVA) -coated glass surface and released by dissolving PVA with water. Raman scattering of 785 nm laser by single microdevices was overserved. The microdevices were incubated with mouse bone marrow-derived macrophages to study interactions between the microdevices and the macrophages using fluorescence microscopy and flow cytometry. The viability of macrophages carrying microdevices were investigated by staining the macrophages with calcein AM. Raman scattering of single microdevices were characterized using a Raman microscope and a 5¡Á objective lens.

Results and Discussion: Fig. A1 shows released carbon nanotube-based microdevices loaded with R18 were d in. Fig. A2 shows microdevices were associated to the macrophages after 24 h incubation. Moreover, viability of the cells was not significantly affected by the microdevices. Figs. B1 and B2 show a single macrophage generated the characteristic Raman spectrum of the carbon nanotubes.

Conclusions: We have developed a novel method for fabricating carbon nanotube-based microdevices for tracking single macrophage by Raman scattering. The microdevices are composed of a polyelectrolyte trilayer, carbon nanotubes and PLGA. The Raman signals generated by carbon nanotubes in the microdevices can be detected at a centimeter-scale sample-lens distance under NIR excitation, suggesting this technique may be used in vivo. Moreover, the microdevices can be engulfed by the primary macrophages without affecting viability and phagocytic ability of the cells. Single macrophages carrying the microdevices can be detected by Raman scattering. The microdevices promise to be useful for non-invasive multiplex tracking of single macrophages *in vivo*.

471195 Study of TL Response of Silver Nanoparticles in Borate Glasses Containing Dy^{3+} and Eu^{3+} Ions for UV and Gamma Dosimetry

Abstract:

Abstract

Dy^{3+} and Eu^{3+} doped the borate glasses samples containing Ag nanoparticles were prepared using a conventional melt-quenching method. The synthesized materials were characterized by scanning electron microscopy (SEM) and their morphology and sized distributions were determined. The photoluminescence (PL) spectra as well as UV-Vis absorbance were also analyzed and show the presence of the rare-earth in the matrix. The thermoluminescence (TL) response of Dy^{3+} doped lithium borate glasses irradiated with ultraviolet (UV) radiation was studied in detail before and after doping with silver.

471313 Effects in Morphology and Thermoluminescent Characteristics of Lif Crystals Synthesized By Using Nonionic and Cationic Surfactants

Abstract:

LiF crystals doped with Ce^{3+} and Mg were prepared using co-precipitation, using as solvent a mixture of ethanol and water in presence of Nonionic (PVP and Pluronic f-127) and cationic (CTAB) surfactants. The LiF crystals were characterized by scanning electron microscopy (SEM) obtaining average sizes of 100 nm, a single cubic crystalline phase was determined by XRD, the PL emission spectra shows two bands characteristics of Ce^{3+} and UV-Vis absorbance shows the presence of the rare-earth in the LiF crystals. The TL response to both X-ray and gamma from a Cs-137 source irradiation was studied.

470030 Thermoluminescent Dosimetric Analysis of Ag and Cu Doped LiB3O5

Abstract:

Methods

Production

Synthesis of Ag and Cu doped LiB_3O_5 crystals was carried out by precipitation, using water as a solvent. Impure crystals were produced at a synthesis temperature of 70 ^oC. A 5g of Ag and Cu doped LiB_3O_5 precipitate was obtained upon evaporation of water in the solution. The precipitates were calcined at a temperature of 750 ^oC for one hour. The calcined materials were grounded finely and subjected to severalcharacterizations.

Characterization

SEM analyses were performed for the doped and undoped samples of LiB_3O_5. The XRD properties of the samples were obtained. The UV-Vis absorbance spectra of the samples were obtained at room temperature. Spectra were recorded between 200 and 1200 nm. The photoluminescence (PL) measurements were carried out. The glow curves of the samples has been obtained using TLD reader.

Results

The XRD patterns obtained for the samples showedthat production of LiB_3O_5 with orthorhombic structure was successful since allof the main peaks of LiB_3O_5 coincide with the JCPDS data. It was seen that the dopant added did

not interfere withthe crystal structure. As it can easily be detected addition of the dopants didnot have any significant effect on the crystal structure of LiB_3O_5 produced.

SEM was carried out to check the morphologies of the samples.According to the SEM images, the samples produced by the co-precipitation method shows the agglomerated surface.

The UV-Vis absorption spectra for the samples showed that the maximum absorbance was observed at 250nm. The result showed that a clear monotonic and increasing dependence of the intensity of the absorbance as a function of dopants.

The dosimetric properties of LiB_3O_5 doped with Ag and Cu were determined by using HarshawTLD Reader Model 3500. The heating rate was chosen as 10°C/s was usedfrom 50 to 350°C. The examined dosimeters were exposed to X-rays at room temperature. The sample amount was 0.5g. The samples have shown the TL response. The pattern of the glow curves changed with the change of dopants.

Conclusions

The TL response of LiB_3O_5 has been compared with the commercial TLD100. The observation concludes that the doped LiB_3O_5 has the TL behavior. The glow curves of the LiB_3O_5 shows different shape with the change of dopants.

469662 Effect of Thermal Treatment on the Characteristics of PES/PVA Nanocomposite Membranes Modified with TiO2 Nanoparticles :a Comparitive Study Between 1-Step and 2-Step Thermal Treatment

Abstract:

PES/PVA thin film composites were modified with TiO2 nanoparticles via dip coating method. Thermal treatment of total modified membranes plays a semi-"catalytic role" in coating process. Hence, 4-temprature levels were selected and investigated, noticing that there is an important restriction due to the stability limitation of polymers under thermal treatment process. This phenomenon also leads to the concept of 2-step treatment. 2-step thermal treatment is a more moderate process which leads to a more stable structure.

Polyethersulfone (PES) ultrafiltration membrane was prepared via phase inversion induced by immersion precipitation. Poly(vinyl-alcohol) (PVA) (1.5 wt%) dissolved in distillated water was coated on the surface of PES barrier layer with dip coating technique. The PVA-coated membranes were then immersed in the cross-linking solution containing 5 wt.% of GA and 0.5 wt.% of H2SO4 for 2min at room temperature in order to reduce the membrane swelling by chemical cross-linking. H2SO4 was used as catalyst of reaction. TiO2 nanoparticles were dispersed in distillated water by sonication for 1hr and stirred vigorously for 30 min. the PVA coated membranes were immersed in 0.2 wt% TiO2 solution for 15 min. the last step is thermal treatment which in this study, four level of temperature (85,95.105.115) were investigated. All modified membranes were thermal treated for 3 min.

The performance of modified membranes was tested with filtration of yeast industrial wastewater(COD=11500 mg/L). Permeate flux and COD of four modified membranes have tested. According to the results, the PES/PVA/TiO2 membrane that was thermal treated in 105 had the best output.

469788 Synthesis, Structural and Morphological Characterization of Cu and Ag-Doped $Li_2B_4O_7$

Abstract:

Method:

In this paper the lithium tetra borate (LTB) was produced by water/solution assisted synthesis method. After the synthesis, metals (doping with Copper (Cu) and Silver (Ag)) were doped into lithium tetra borate ($Li_2B_4O_7$) to enhance its thermo luminescent properties. The heating temperature was 750°C for 2 hr as well as 150°C for 1hr the sample produced by water solution method with 0.1% Cu and 0.1% Ag. The pellets were irradiating by the X-ray.

Result:

The identification and characteristics of undoped and doped $Li_2B_4O_7$ were determined by X-ray diffraction (XRD), Scanning electron microscope (SEM), Photoluminescence (PL), Ultra violet visible spectroscopy (UV Vis). The chemical compositition and its morphologies of the obtained LTB and LTB: Cu, Ag was confirmed by XRD and SEM results. The most intense peak of the XRD pattern of the lithium tetraborate sample is verified by the reference data of the lithium tetra borate with tetragonal structure. The glow curves were obtained by thermo luminescence (TL) reader. The glow curves of the pellets exposed to different doses exhibited a clear response to X-ray irradiation, which is a significant evidence that this material is a candidate for radiation dosimetry. Finally this energy response was also compared with TLD100. The comparison of the results showed that the obtained lithium tetra borate have good thermo luminescence (TL) dose response.

Enhancement of Heat Transfer Coefficient by Using Fe_2O_3 –Water Nanofluids

N.I. Zouli[1], Muthanna Al-Dahhan[1]
[1]Missouri University of Science and Technology, Rolla, MO, 65409
Email: niz8vb@mst.edu

Introduction

Research efforts have been made to increase the heat transfer rate, energy, and efficiency. Some of these efforts include the application of fins, compact heat exchangers, microchannel, and many others. Recently, nanofluids have been introduced to enhance the heat transfer rate of the heat transfer fluids by adding a certain concentration of nanoparticles into the heat transfer media. Most of the work that has been published in the literature focuses on measuring the thermal conductivity of the nanofluid or measuring the overall heat transfer coefficient. Among the efforts for enhancement of heat transfer the application of additives to liquids is noticeable (Bergles, 1973; Ahuja, 1975). Choi (1995) was the first who studied the enhancement of the nanoparticles on the heat transfer coefficient and showed that the nanoparticles could considerably affect the thermal conductivity of the nanofluid to increase. Since Choi is work (Choi,1995), various studies have been conducted on nanofluids and these studies shown that nanofluids increase heat transfer properties and stability while reducing costs, clogging, and energy required to pump the fluid. Heris et al. (2006) found that the heat transfer coefficient increased as the volume concentration of the nanofluids increased when the fluid was composed of water and metal oxides.

In the present study, effects of Fe_2O_3/water on the nanofluid's heat transfer and thermal conductivity will be discussed. The experimental investigations are done to study heat transfer performance of the nanofluid at the entrance region under laminar flow condition varying the particle size and volume concentration of nanoparticles.

Experimental setup and methodology

Schematic of the experimental setup used in the current study is shown in Figure1. It consists of a flow loop made of copper material, a heating unit, a measuring and control unit. A straight copper tube with 190.5 mm length, 25.4 ± 0.02 mm inner diameter, and 31 ± 0.05 mm outer diameter was used as the test section. The heat transfer probe comprises of a heat flux sensor (0.15 in x0.08 in) with a fast response time of about 0.02 sec and a thermal resistance of 0.003 °F/BTU/ft2hr. The sensor was placed in a slit mounted inside the test section wall flashed to the wall without disturbing the flow. The probe measures the heat flux and the surface temperature. In addition, two thermocouples (type T) were mounted inside the test section to measure the bulk temperature. The sensor was placed on the middle distance between the two thermocouples.

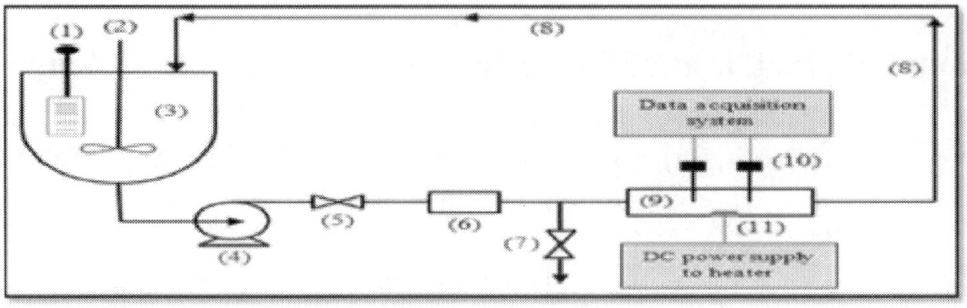

Figure 1. Schematic of experimental setup components: (1) Heater, (2) Agitator, (3) Reservoir tank, (4) Pump, (5) Needle valve, (6) Flow meter, (7) Drainage valve, (8) Copper tube, (9) Test section of copper tube, (10) Thermocouples, (11) Heat flux foil sensor.

Heat flux, temperature of the sensor surface, and the bulk media temperature are directly measured and recorded using Data acquisition system. Further they are used to estimate local instantaneous and averaged heat transfer coefficient using eqn (1) and (2):

$$h_i = \frac{q_i}{T_{si} - T_{bi}} \; ; \; \Delta T_i = T_{si} - T_{bi} \qquad (1)$$

$$h = \frac{1}{n} \sum_{i=1}^{n} \frac{q_i}{T_{si} - T_{bi}} = \frac{1}{n} \sum_{i=1}^{n} h_i \qquad (2)$$

where h_i: local instantaneous heat transfer coefficient (kW/m^2.K)
q_i: instantaneous heat flux measured by the sensor (kW/m^2)
T_{si}: instantaneous surface temperature of the probe (K)
T_{bi}: instantaneous bulk temperature of the media (K)

Results and discussion

The thermal conductivity of Fe_2O_3 water based nanofluids was measured at different concentrations. From Figure 2 an enhancement of 22.35% in the thermal conductivity value has been achieved with 0.09% volume concentration of Fe_2O_3 nanoparticles in water with an average diameter of 20 nm. As can be seen from Figure 2, the thermal conductivity enhancement increases linearly with increased particle volume concentration.

The concentrations for each test run were increased from 0.02-0.09% by volume and run at a constant flow rate of 6 L/min. The experimental results were obtained with 20 nm particles. And it was found that the average increase in the heat transfer coefficient was between 5.95-18.34% with the highest increase being 73.7% at 0.09% volume concentration at 5 mins into the process. It was observed that as the concentration of the nanofluid was increased so did the heat transfer coefficient. The enhancement in heat transfer coefficient was calculated by following;

$$[(h_{nf} - h_{bf}) \, / \, h_{bf}] * 100\% \qquad (3)$$

h_{nf} = heat transfer rate of nanofluid
h_{bf} = heat transfer rate of base fluid

A typical result for 20 nm particles can be seen in Figure 3. Overall the results show that the local heat transfer coefficient increases with increasing the size and the volume fractions of the Fe_2O_3 nanoparticles compared to water. The effect of Reynolds number on Nusselt number is shown in Figure 4. It is seen that the measured Nusselt numbers for nanofluids are higher than those of base fluid, and they increase remarkably with Reynolds number. The observed enhancement of the Nusselt number could be due to the suppression of the boundary layer, the viscosity of nanofluids as well as the dispersion of the nanoparticles.

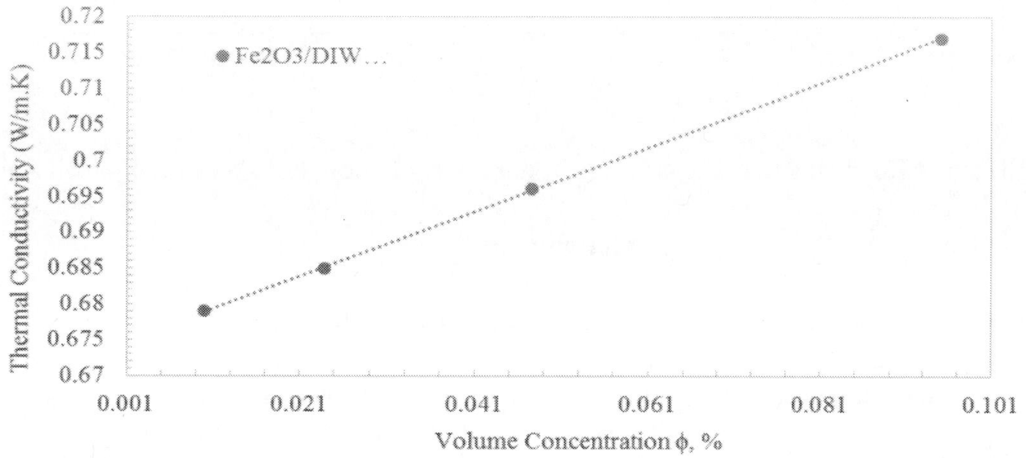

Figure 2. Thermal conductivity of Fe_2O_3 water based nanofluids at different volume fractions (0 and 0.09 %).

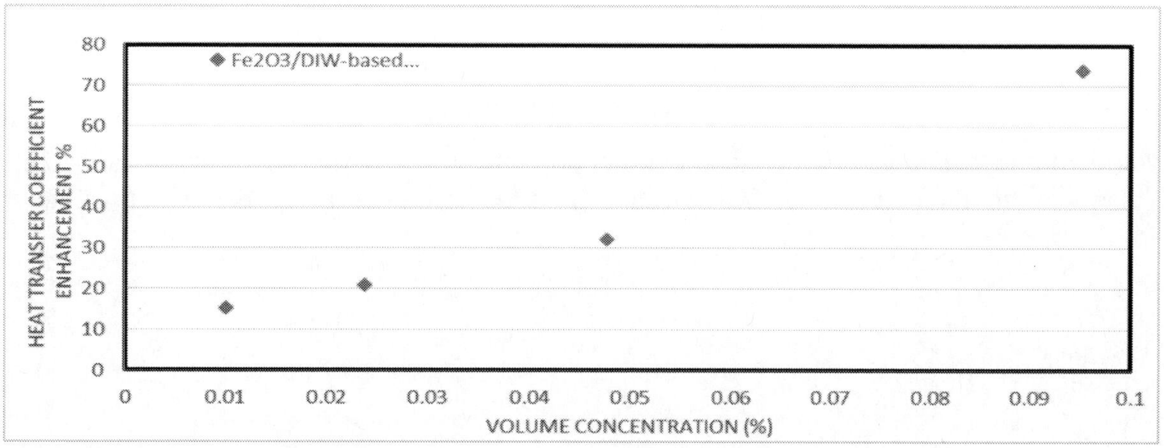

Figure 3. Variation of heat transfer coefficient with different volume concentration of Fe_2O_3 water based nanofluids.

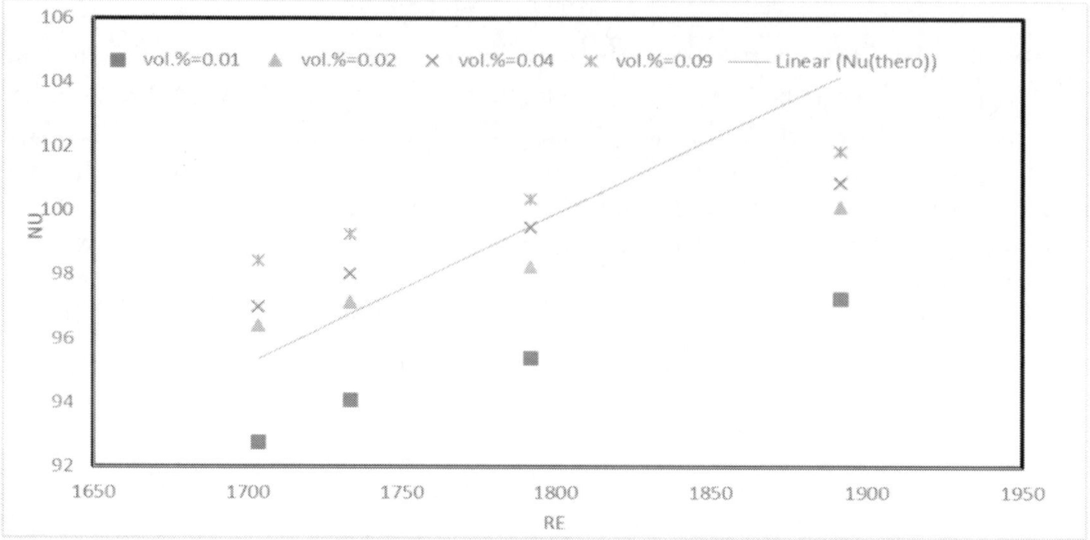

Figure 4. Effect of Reynolds number and Fe_2O_3 nanoparticle concentration on Nusselt number

Conclusions

The paper is mainly concerned with the heat transfer of nanofluids made of water and Fe_2O_3 nanoparticles in the laminar flow regime. It was found that use of aqueous Fe_2O_3 nanoparticles can significantly enhance heat transfer coefficient in laminar flow regime, and the enhancement increases with Reynolds number, as well as particle concentration under the conditions of this work. The enhancement of the heat transfer coefficient could not be only credited to the enhancement of the effective thermal conductivity. A reason for the enhancement, Particle passage which lead to reduce the thermal boundary layer thickness and a non-uniform distribution of thermal conductivity.

Acknowledgments

I would like to acknowledge the financial support provided by the government of Saudi Arabia. I would also like to thank Prof. Al-Dahhan for his supervision and support which made this work possible.

References

1. Brooke, A., Kendrick, D., Meeraus, A. A. (1992). GAMS- A User's Guide (Release 2.25). *The Scientific Press*. San Francisco, CA.

2. Gooding, W. B., Pekny, J. F., McCroskey, P. S. (1994). Enumerative Approaches to Parallel Flowshop Scheduling via Problem Transformation. *Comput. Chem. Eng., 18*, 909.

3. OSL. (1991). *OSL, Guide and Reference (release 2)*, IBM: Kingston, NY.

4. Zhang, X., Sargent, R. W. H. (1994). The Optimal Operation of Mixed Production Facilities - A General Formulation and some Approaches for the Solution. *In Proceedings of the 5th International Symposium on Process Systems Engineering*. Kyongju, Korea, 171.

5. A.E. Bergles, Recent development in convective heat transfer augmentation, Appl. Mech. Rev. 26 (1973) 675–682.

6. A.S. Ahuja, Augmentation of heat transport in laminar flow of polystyrene suspension: I. Experimental and results, J. Appl. Phys. 46 (8) (1975) 3408–3416.

7. Choi, S.U.S. Enhancing thermal conductivity of fluids with nanoparticles. in American Society of Mechanical Engineers, Fluids Engineering Division (Publication) FED. 1995.

8. Heris, S.Z., S.G. Etemad, and M.N. Esfahany, *Experimental investigation of oxide nanofluids laminar flow convective heat transfer.* International Communications in Heat and Mass Transfer, 2006. **33**(4): p. 529-535.

447375 Eco-Friendly Dyeing of Electrospun Cellulose Nanofibers with Reactive Dye Using Ultrasonic Energy

Abstract:

Nanofibers have created enormous opportunities and challenges for the textile and apparel industry, moreover the innovative methods for dyeing of nanofibers have recently gained more attention. Electrospun cellulose nanofibers are increasing significant interest in textile industry in view of its incomparable properties. However, the dyeing of nanofibers faces problem of lower color yield due to the higher surface area of nanofibers. In this study electrospun cellulose nanofiber dyed with using cold pad batch dyeing, a well-known and ecofriendly method of dyeing for cotton base fibers. The dyeing of cellulose nanofibers was carried out with CI Reactive Violet 5 by conventional and ultrasonic (US) method. Scanning electron microscopy and Fourier transform infrared spectroscopy were used to investigate morphology and chemical structure of electrospun cellulose nanofibers before and after dyeing and the color strength (K/S value) and color fastness of dyed nanofibers also assessed.

The results indicates that the use of ultrasonic energy not only shortens the batching time but the color strength (K/S),color fixation and color fastness also considerably increased ,the K/S value of dyed samples enhanced around 5-8% compared with conventional method. Although, washing fastness of sample increased up to 25%, but, color fastness to light was the same as conventional method.

The results proved that, the morphology of dyed cellulose nanofibers with both dyeing method were smooth with structural uniformity and remained unaffected after being dyed.

452910 Combined Quartz Crystal Microbalance with Dissipation and Generalized Ellipsometry to Characterize the Deposition of Titanium Dioxide Nanoparticles on Model Rough Surfaces

Abstract:

Titanium dioxide nanoparticles (TiO_2NPs) are the most extensively manufactured engineered nanomaterials, and understanding of their interactions with soil and sandy aquifer materials is necessary for effective natural resource management [1]. Quartz crystal microbalance with dissipation monitoring (QCM-D) has been widely used to study TiO_2NPs at water-metal oxide interfaces as model systems for soils and aquifers. However, natural sands and sediments exhibit various degrees of surface roughness, and the effects of surface roughness are not accounted for in current QCM-D studies of nanoparticle adsorption, which use the Voigt-Kelvin model [2].

A new mechanical model that considers both viscoelastic and surface roughness effects [3] is described to analyze QCM-D measurements. Obtained parameters include adsorbate areal mass density, adsorbate viscoelastic properties, the vertical roughness length, and the permeability length. Silicon slanted columnar thin films (SCTFs) were deposited onto QCM-D sensors by electron-beam glancing angle deposition and conformably coated with an alumina film via atomic layer deposition to act as model rough surfaces. The adsorption of polyacrylate-stabilized TiO_2NPs onto flat surfaces and the SCTFs was measured by QCM-D and generalized ellipsometry (GE), an optical reference technique.

We report that the surface-roughness-modified mechanical model excludes water that is retained by the SCTF surface roughness and instead considers the surface mass density of TiO_2NPs and water coupled to the TiO_2NPs [4]. GE is also sensitive to the attachment of the TiO_2NPs. When the results of both techniques are combined, new information, the hydration of the polyacrylate-stabilized TiO_2NPs, can be evaluated.

[1] B. Kim et al. J. Environ. Monit., **14**, 1128 (2012).

[2] M.V. Voinova et al. Phys. Scr., **59**, 391 (1999).

[3] L. Daikhin et al. Anal. Chem., **74**, 554 (2002).

[4] N. Kananizadeh et al. J. Hazard. Mater., In press. dx.doi.org/10.1016/j.jhazmat.2016.03.048

461730 Adaptive, Point-to-Point Assembly of DNA Nanotubes Between Molecular Landmarks

Abstract:

In vivo, self-assembly processes often adapt what is assembled to the shape and arrangement of existing cellular nanostructures. This capacity for adaptation makes it possible to build integrated nanosystems consisting of multiple nanostructures elements that work together to perform complex functions. We demonstrate a synthetic self-assembly process in which a DNA nanotube self-assembles between molecular landmarks whose separation distance and orientation are unknown. Connections between DAE-E nanotubes occur when nanotubes grow from molecules at the start and destination locations that act as nanotube nuclei for growth from the two opposing nanotube ends. When the nanotubes grow long enough that their ends can contact one another, these ends can join to produce a stable connection. Time-lapse multicolor microscopy confirms that nanotubes connections generally form by the proposed process and follow the predictions of simple physical models. This process can occur under a variety of physical conditions and may make it possible to build wires between dynamic components through metalization of the nanotubes, or to build complex connected DNA architectures at the micron scale with novel mechanical properties.

472433 Photo-Triggered Self-Assembly and Actuation of DNA Nanostructures and Machines Using Photocaged Nucleotides

Abstract:

DNA nanotechnology has emerged as one of the most powerful methods for constructing complex nanoscale devices with precise addressability. The ability to actuate these devices on-demand would provide a powerful method for creating dynamic nanomachines that can reconfigure nanostructures, apply precisely defined forces, or spatiotemporally control self-assembly. Typically, DNA nanostructures have been dynamically modulated by either incorporating photoswitchable azobenzene nucleotides that, while effective, often suffer from slow kinetics and undesired reversibility. Another approach involves adding displacement strands that can bind to toehold regions and out-compete other interactions to change the state of a nanostructure. However, this approach is limited by the need to add an exogenous strand, which is not feasible for many applications, especially in biology. To circumvent these limitations, we have developed a series of photo-cleavable oligonucleotides that block hybridization until illuminated with UV light. By incorporating caged displacement strands within an existing nanostructure, we can prevent their action until illuminated. These strands serve as effective "internal toeholds" that can be activated on-demand. We have constructed a number of one-, two-, and three-dimensional assemblies that can form only upon irradiation, as well as a nanomechanical tweezer that switches between the closed and open state with UV light. In effect, this device can apply a nanomechanical force within a few seconds of illumination, paving the way for dynamic nanomachines that can exert controlled motion. We will also describe the design principles for more complex nanostructures that can find applications in targeted drug delivery or reconfiguration inside cells to influence biological processes.

461942 Biological Self-Recognition and Self-Assembly for the Next Generation of Hybrid Wires

Abstract:

Biological self-recognition and self-assembly for the next generation of hybrid wires

Xiao Hu[1], Chenbo Dong[1], Rigu Su[2], Quan Xu[2*], and Cerasela Zoica Dinu[1]*

[1]Department of Chemical Engineering, West Virginia University, WV, USA

[2]State Key Laboratory of Heavy Oil Processing, China University of Petroleum (Beijing), Beijing, China

Abstract

The next generation of nanowires that could help advance the integration of functional nanosystems into synthetic applications from photocatalysis to optical devices needs to demonstrate increased ability to promote electron transfer at their interface while ensuring quantum confinement. Current techniques used to synthesize the nanowires such as the interface lithography technology and chemical vapor deposition only allow for a limited control of the nanowireâ€™s structureâ€“property relationships thus limiting their implementation in applications and design of nano-functional devices. Herein we proposed to use the biological recognition and self-assembly properties of tubulin, normally involved in building the microtubule cellular structure, to create stable, free standing and conductive sulfur-doped carbon nanodots-based hybrid bio-nanowires. We evaluated the physico-chemical properties (e.g., composition,

morphology, diameter etc.) of such user-synthesized hybrids using classical atomic and spectroscopic techniques, while the electron transfer rate was estimated using peak currents formed during voltammetry scannings. Such synthesized hybrid bio-nanowires have diameters ranging from 6 to 30 nm and were proved to be more stable than their biological skeletons. Our results demonstrate the ability to create individually addressable hybrid bio-nanowires capable to reduce energy losses at interfaces to be used for the advancement and implementation of nanometer-scale functional devices.

455693 pH Sensing through Silicon Nanoribbon/Carbon Nanotube Porin Hybrid Sensor

Abstract:

pH of local environments is an important parameter for detection of pathological conditions and understanding of cellular mechanisms. Silicon nanoribbon field effect transistor (FET) has been proved to be a high-throughput and sensitive platform for pH sensing. However, the spatial resolution of the silicon nanoribbon is limited by the lithography techniques. Recent studies have demonstrated sub-2nm diameter carbon nanotube (CNT) porins are fast and efficient proton conductors. Furthermore, the CNT porins also can be successfully incorporated into lipid bilayers. In this research, we focus on a bioelectronic approach toward designing pH-sensing devices. We will describe an organic/inorganic hybrid device that combines nanoribbon FET with self-assembled CNT/lipid bilayer for pH sensing. And we will also discuss its potential application in nanometer spatial resolution intracellular biosensing.

463312 Design of Membrane-Embedded Amphiphilic Nanoparticles from Atomistic Molecular Dynamics Simulations

Abstract:

Monolayer-protected nanoparticles (NPs) are a versatile materials platform useful for biological applications because features of the protecting monolayer can be engineered to tailor interactions with the biological milieu. Recently, amphiphilic NPs with surface properties that mimic typical globular proteins were shown to enter cells via a non-endocytic, non-disruptive process that could be of broad interest for applications in drug or gene delivery. Here, we use detailed atomistic molecular dynamics simulations to gain molecular insight into these experimental observations. We find that the amphiphilic NPs insert into the bilayer to obtain a configuration resembling a membrane-embedded protein due to favorable interactions between hydrophobic molecules on the NP surface and the hydrophobic bilayer core. We identify a kinetic pathway for insertion that mimics the early onset of vesicle-vesicle fusion. Our calculations also suggest that charged ligand end groups cross the bilayer on experimentally relevant timescales that are significantly more rapid than expected. Finally, we leverage this mechanistic understanding to design monolayer compositions optimized for bilayer insertion and cellular uptake. This work emphasizes that carefully tuning NP surface coatings to resemble existing biological materials permits favorable interactions with the cell membrane, enabling the creation of new nano-bio hybrid materials with properties similar to those of membrane proteins.

470872 Photo-Electrochemical Characterizations of Photosystem I (PS I) Assembly Under Bio-Mimetic Membrane Confinement

Abstract:

The robust structural and photoactive electrochemical properties of Photosystem I (PSI) make it an ideal candidate for incorporation into bio-hybrid photovoltaic devices. However, the first step towards the rational design of such devices require systematic and oriented assembly of PS I at bio-abiotic interfaces via suitable scaffolding. To this end, our goal is to incorporate and assemble individually oriented PS I complexes into synthetic membrane-bound structures that mimic the natural thylakoid membrane housing of PS I. We investigate the photocurrent generation from such organic-inorganic interfaces in membrane bound PS I systems. Detergent-mediated reconstitution of proteoliposomes is analyzed via absorption and fluorescence spectroscopy measurements. Successful immobilization of membrane bound PS I on gold substrates are mediated by the hydrophillic interfaces of self-assembled monolayers (SAMs). To this end, force-distance spectroscopy during atomic force microscopy (AFM) imaging of lipid bilayer membranes on thiolated Au substrates confirm monolayer formation. Furthermore, topography measurements using liquid-AFM techniques confirm the assembly of membrane-bound PS I. Finally, the current work presents photo-electrochemical (PEC) characterizations demonstrating enhanced photocurrent generation from PS I complexes assembled under membrane confinement as compared to their isolated immobilizations on electrodes. In turn, the aforementioned studies provide valuable insight into future incorporation of bio-mimetic membrane bound protein complexes into novel bio-hybrid devices.

471619 Modeling Protein Folding/Aggregation on Nanoparticle Based Biosensors in Complex Solvent Environments By a Coarse Grained Simulation System

Abstract:

Understand protein behavior in a complex environment is the key in rationalizing the design of many novel techniques such as protein biosensors based on novel nanoparticle (NP) materials. Protein folding and aggregation properties can be largely affected by the environments such as solid abiotic interfaces and complex co-solvents, which would directly affect the expected functionality or performance of the biosensors. Difficulties remain in applying current experimental methods to obtain a detailed picture of protein structures on a solid surface, so our ability is limited by these factors in the design of biosensor surface chemistry a priori for desired protein behavior. To address this point, in this work we present a coarse-grained simulation system that is able to model protein folding, misfolding, and aggregation on various NP surfaces and in the presence of co-solvents based on the Karanicolas and Brooks (KB) Go-like protein model. A flat-surface potential was first included and carefully parameterized based on a large experimental data set. This surface model has been shown in many studies to be able to reproduce and predict protein behavior at different types of surfaces as verified by experimental tests. Further effort in developing the surface potential was applied in describing surface curvature effects on protein behavior as a means of understanding nano-sized biosensor materials. Tanford's transfer free energy model was recently added to this simulation system as a descriptor of co-solvent effects. The energy of each part of the underlying folding/surface/solvation potential has been well balanced so that each element works collaboratively as a single complex environment for the protein. Together, this coarse-grained modeling system is expected to accurately predict the protein adsorption, folding/unfolding, aggregation, and protein-protein interactions in a complex environment as it would experience in differing cellular or solvent environments as well as in the context of a NP-based biosensor.

471963 Aggregation-Induced Emission of Hydrophobically-Modified Metal Clusters in Lipid Nanodiscs

Abstract:

Biocompatible lipid-based nanodiscs (also known as â€œbicellesâ€) showed superior cellular uptake compared with liposomes of a similar length scale [1]. In this presentation we will demonstrate that bicelles can entrap nanometer-sized monolayer protected gold clusters (Au-MPCs) or hydrophobically-modified quantum dot (ZnSe/ZnS QD) through self-assembly, retaining reasonably uniform sizes (< 50 nm) as the bare ones. Aggregation-induced emission is observed in such nanoparticle-in-nanoparticle complexes (NP^2) with nearly two orders of magnitude enhancement in luminescenceÂ in both cases of Au-MPC-NP^2 and QD-NP^2. The outcome of this research will provide the insight into the design of biosensing nanoparticles in a controllable and scalable manner.

[1] W. Aresh,Â Y. Liu, J. Sine, D. Thayer, A. Puri, Y. Huang, Y. Wang, M.-P. Nieh "The Morphology of Self-Assembled Lipid-Based Nanoparticles Affects Their Uptake by Cancer Cells" *J. Biomed. Nanotech.* (in press)

465791 Delta- and Proteorhodopsin-Based Bionanoelectronic Devices for Light-Controlled Conversion of Protonic to Electronic Currents

Abstract:

The burgeoning field of bionanoelectronics aims at interfacing biology with electronics through the use of bio-based or biocompatible nanomaterials. A particular area of interest is the controlled conversion of ubiquitous biological ionic gradients into electrical signals that can be read or processed by conventional electronics. Here, we describe the efficient integration of proton-transporting rhodopsins with palladium (Pd) contacts to achieve light-activated conversion of protonic currents into electronic signals. The process involves engineering archaeal or proteorhodopsins with a Pd-binding peptide in order to bring the proton exit site in close proximity to the electrode; using protein and folding engineering strategies to express high levels of functional rhodopsins that are activated at various wavelengths; optimizing the integration and orientation of these proteins in liposomes; and forming supported bilayers on Pd/PdHx microfluidic devices fabricated by photo- and soft lithography. We demonstrate that addition of a Pd-binding extension leads to a nearly one order of magnitude increase in electronic signal with light-controlled "on" and "off" responses that are complete in seconds, discuss the development of steady state currents and describe devices in which proton transport and the corresponding electronic responses are selectively activated by different wavelengths of light.

462828 Self-Assembly of Protein Nano-Shapes Using Synthetic Coiled Coils

Abstract:

Protein self-assembly is an emerging and powerful approach to create structures at the nanometer scale. Driven by protein-protein interactions, protein subunits can be designed to self-assemble into supramolecular nanostructures with desired dimension and geometry. Recently, computational methods have been used to design complex protein interfaces,[1] while rational protein design approaches using simple protein motifs such as coiled coils have been successfully applied to design artificial structures different from naturally occurring protein folds.[2] Coiled coils are simple, rod-shaped, superhelical protein motifs composed of two or more α-helical peptides. They self-assemble into oligomers stabilized by specific interactions between the α-helical peptide subunits. Because their sequence-to-structure relation is well characterized, coiled coils are potentially useful self-assembling modules for design of artificial nanostructures. For example, an approach of covalent linkage and folding of different coiled-coil segments has been used to create a protein tetrahedron which does not exist in nature.[3] Furthermore, such a rational approach can be extended to design other artificial protein shapes using new sets of coiled coils. In this study, we demonstrate self-assembly of triangular protein nanostructures using synthetic heterodimeric coiled coils. We recently reported interaction profiles of computational designed hetero-associating coiled-coil peptides. Among them, sets of coiled-coiled pairs that make specific binding interactions over weak undesired crosstalk with others are available, and these orthogonal pairs can potentially serve as specific synthetic modules for protein self-assembly.[3] In our design, orthogonal coiled coil pairs are covalently linked and self-assemble into a triangular shape. We investigated mixing and thermal annealing of these protein building blocks, and maximized the yield of self-assembled triangular protein structures by manipulating the annealing temperature pathway. The fraction of self-assembled proteins was purified from the mixture of annealed proteins, using size-exclusion chromatography. Atomic force microscopy revealed the triangular shape of self-assembled proteins, and their hydrodynamic diameter and the supercoiled folding were characterized using dynamic light scattering and circular dichroism, respectively. Our approach is simple and broadly applicable to design of other artificial protein nanostructures, which provide great potential in the area of protein materials development.

1. N. P. King, J. B. Bale, W. Sheffler, D. E. McNamara, S. Gonen, T. Gonen, T. O. Yeates and D. Baker, *Nature*, 2014, **510**, 103-108.
2. H. Gradišar, S. Božič, T. Doles, D. Vengust, I. Hafner-Bratkovič, A. Mertelj, B. Webb, A. Šali, S. Klavžar and R. Jerala, *Nat. Chem. Biol.*, 2013, **9**, 362-366.
3. K. E. Thompson, C. J. Bashor, W. A. Lim and A. E. Keating, *ACS Synth. Biol.*, 2012, **1**, 118-129.

458700 Investigations of the Effect of Overhang and Strand Length on DNA Hybridization at the Solid-Liquid Interface

Abstract:

Many biotechnologies and assays, such as sequence detection and bridge PCR, rely on interactions between mobile DNA analytes or targets and surface chemistries and features to perform their desired functions. Single-molecule techniques allow for investigations of large statistical samples of individual molecular interactions to gain a more fundamental understanding behind what is causing behaviors in a bulk system. This work looks specifically at DNA hybridization at the solid-liquid interface, where Total Internal Reflection Fluorescence (TIRF) microscopy allows us to look at the behavior of single-stranded oligonucleotides. This information elucidates how different degrees of complementarity affect hybridization and searching behaviors as indicated by FÓ§rster Resonance Energy Transfer (FRET) signals, allowing us to observe when two strands are in close proximity to each other. Determining the fundamental processes via this analysis results in better sensing behavior, such as stronger signals on a detector for gene sequencing, mutation detection, and genetic mapping applications.

This study looks specifically at the effect of strand length and the appearance of overhang on hybridization mechanisms and two-dimensional surface diffusion. DNA targets in solution at 100 pM were flown across a solid interface with covalently attached complementary strands. Experiments were performed with systems designed to create short (15bp) duplexes, long (30bp) duplexes, or a partially complementary system where a 30-base target hybridizes with a 15-base probe, introducing a 15-base overhang. Analyses of association lifetimes show that a majority of short associations with long-lived distribution tails, indicative of different melting modes. The longest duplexes maintain longer lifetimes, as expected due to higher stability and more Watson-Crick base pair bonding. The presence of a toehold, or overhang, on the duplex decreases association lifetimes. This suggests exposed and unhybridized DNA nucleobases increase the targetâ€™s attraction to the surface, stabilizing the transition from the hybridized to a surface state. The long-lived events appear to have 1st-order dissociation behavior, indicating that melting is initiated by a nucleation event before undergoing complex unzipping behavior. This gives further insight into the mechanisms within surface-mediated DNA hybridization systems, so that devices relying on these interactions in the future can be designed to give stronger signals and more stable hybridizations to provide their desired functions.

463359 Nitroaromatic Detection and Infrared Communication from Wild-Type Plants Using Plant Nanobionics

Abstract:

Plant nanobionics aims to engineer living plants with non-native functions by interfacing the plants with specifically designed and targeted nanoparticles. Herein, we design and demonstrate living spinach plants (*Spinacia oleracea*) as new materials and functional devices that serve as self-powered auto-samplers and pre-concentrators of nitroaromatics within ambient groundwater, detectors of the nitroaromatic molecules contained therein, and infrared (IR) communication platforms that can send this information to a userâ€™s smart phone. The design employs a pair of near infrared (nIR) fluorescent nanosensors embedded within the mesophyll of the plant leaf, with one engineered through the Corona Phase Molecular Recognition (CoPhMoRe) technique using single walled carbon nanotubes (SWCNTs) conjugated to the peptide Bombolitin II to recognize nitroaromatics via IR fluorescent emission at > 1100 nm with a response time of 5-15 mins after introducing 400 Î¼M of picric acid to the roots.Â The second IR channel is a polyvinyl alcohol (PVA) functionalized SWCNT that acts as an invariant reference signal. As contaminant nitroaromatics in solution are transported up the roots and stem into the leaf tissues, they accumulate in the mesophyll where the pair of SWCNT sensors are embedded. The resulting relative changes in the intensity of SWCNT emission, with a response rate that is mathematically described using a whole plant residence time model. The real-time monitoring of embedded SWCNT sensors also allows residence times in the roots, stems and leaves to be estimated, calculated to be 2 to 5 min, 3 to 6 min and 1.9 min/mm leaf, respectively. We further show that this system is generalizable to the detection of other analytes, such as dopamine which are known to effect physiological changes in plants. These results demonstrate the ability of living, wild-type plants to function as chemical monitors of groundwater and communication devices to external electronics at standoff distances

465082 Multiplexed Detection of Protein Biomarkers Using Recognitive Polymers in a Localized Surface Plasmon Resonance Sensor Array

Abstract:

Synthetic polymers have been designed to differentially bind proteins by incorporating functional groups that form non-covalent interactions with specific residues on protein surfaces. Such recognitive polymers can be synthesized on the surface of nanomaterials to develop receptors with built-in signal transduction. For example, the localized surface plasmon resonance (LSPR) of noble metal nanomaterials is sensitive to changes in local refractive index and can be exploited to detect molecular binding events near the nanomaterial surfaces. In particular, nanomaterials with high aspect ratios, such as gold nanoshells (AuNSs), have enhanced LSPR sensitivity compared to colloidal gold. Here, we synthesized recognitive polymers on AuNSs and take advantage of their enhanced LSPR sensitivity for detecting protein biomarkers of SjÃ¶grenâ€™s syndrome.

Crosslinked copolymers of N-isopropylacrylamide (NIPAM) and methacrylic acid (MAA) (poly(NIPAM-co-MAA)) were synthesized on surface modified AuNSs via precipitation polymerization. After synthesis, additional functional groups were introduced into the polymer shell via post-synthesis modifications to generate a set of polymers with varied recognition properties. Absorbance spectra were measured to detect the LSPR response of each polymer-AuNS composite to increasing concentrations of protein biomarkers. Bare AuNSs showed negligible LSPR shifts upon incubation with any protein. On the other hand, recognitive polymer coated AuNSs exhibited concentration dependent red-shifts in LSPR wavelength upon incubation with proteins for which the polymer had a high binding affinity, but no shifts were observed for low affinity polymer-protein combinations. The polymer-AuNS composites that exhibited the most distinctive recognition profiles were combined in a sensor array that enables multiplexed detection of SjÃ¶grenâ€™s syndrome biomarkers.

471681 Development of a Metal Tungstate-Based Nanoparticle Platform for CT-Guided Chemo/Radio Therapy

Abstract:

X-ray computed tomography (CT) is currently one of the most powerful, non-invasive, clinical in vivo imaging techniques, which has resulted from advances in both X-ray device and contrast enhancement technologies. Our recent work demonstrates that metal tungstates (such as $CaWO_4$) are promising contrast agents for X-ray, radiation and CT imaging, because of the high X-ray mass attenuation of tungsten. We have developed a method of formulation, in which $CaWO_4$ (CWO) nanoparticles (NPs) are encapsulated within a bio-compatible poly(ethylene glycol-*b*-D,L-lactic acid) (PEG-PLA) block copolymer (BCP) capsule. BCP-encapsulated CWO NPs can be used as radio-sensitizers that make cancer cells easier to kill with radiation therapy [1]. The self-assembled nature of BCP-encapsulated CWO NPs is also expected to enable incorporation of additional therapeutic modalities (such as anticancer drugs) inside the hydrophobic domain, thus combining chemo therapeutic and radiological imaging capabilities. As a first step toward realizing this potential, we tested the in vivo safety and radio contrast capability of BCP-encapsulated CWO NPs.

CWO NPs were synthesized by a micro-emulsion method [2]. The PEG-PLA BCP material was synthesized by ring-opening polymerization. CWO NPs were encapsulated with PEG-PLA using a solvent exchange method. This procedure produced fully PEGylated CWO NPs that are devoid of impurities and stable against aggregation under physiological solution conditions for very long periods of time (> several months).

This encapsulated CWO has been confirmed to possess no detectable cytotoxicity against cultured human cancer cells (p53-mutant HN31). A study is in progress to determine the maximum tolerated dose (MTD) of BCP-coated CWO NPs in mice (BALB/C). As of the time of this writing, it has been confirmed that the MTD of this material is, at least, 200 mg per kg body weight following a single i.v. administration; this material is thus, at least, as safe as commercially available dextran-coated iron oxide nanoparticles that are currently used clinically as MRI contrast agents (MTD in mice ≈ 168 mg/kg per dose i.v.). Major excretory organs

(heart, liver, lungs, spleen and kidney) were harvested for histopathologic detection of gross organ damage. No organ damage was detected.

Compared with the conventional radio-contrast agents (iodine and barium), CWO has a significantly higher X-ray absorbing capability because X-ray mass attenuation increases as approximately Z^3 (Z is the atomic number, $Z = 53$ (I), 56 (Ba), and 74 (W)); in the range of X-ray energy used for most diagnostic studies (i.e., 10 – 100 keV) the mass attenuation coefficient of W is three to four-fold higher than that of I or Ba. Our data confirmed this trend. The X-ray attenuation properties of BCP-coated CWO NPs were measured in comparison with a conventional iodine-based contrast agent (Iohexol). Measurements were performed using a clinical CT scanner instrument (140 kV, 700 mA). Results confirm that at identical molar concentrations of I vs. W atoms, PEGylated CWO NPs produce a higher contrast and lower transmission than Iohexol.

Results to date support that PEG-PLA-coated CWO NPs have the potential to be used as theranostic agents for CT-guided chemo/radio therapy.

References

[1] Y.-Y. Won, J. Lee, S. D. Jo, M. K. Joo, B. S. Lee, S. Y. Kim, S. H. Kim, "Radio-Luminescent Particles for Enhancement of Radiation Cancer Therapy", pending *PCT/U.S. Patent*, Application No. 16/12616, filed on January 8, 2016.

[2] J. Lee, N.J. Rancilio, J.M. Poulson, Y.-Y. Won, "Block-Copolymer-Encapsulated $CaWO_4$ Nanoparticles: Synthesis, Formulation, and Characterization", ACS Applied Materials & Interfaces, *8*(13), 8608–8619, 2016 (DOI: 10.1021/acsami.6b00727).

470631 DNA Functionalized, Fluorescent Nanomaterials for the Specific, Ratiometric Detection of Dopamine for *In*Â *Vivo*Â Sensing

Abstract:

Monitoring changes in neurotransmitter concentration is essential for understanding brain function, making the development of new biosensors critical for advances in the neurological imaging of neurotransmitters. Nanomaterials with distinct optical, chemical and material properties show promise for developing such sensitive and specific sensors for temporal and spatial study of biological activity.

First, we evaluate the use of DNA-stabilized fluorescent silver nanoclusters (AgNCs), few-atom fluorescent clusters stabilized using single-stranded DNA, as optical sensors for specifically and sensitively probing neurotransmitter concentration. We identify two particular DNA-AgNC sensors that have broad fluorescence peaks with large Stokeâ €™s shift that respond to exogenous addition of neurotransmitter dopamine. We find that dopamine can specifically quench the fluorescence of one sensor (ex580nm/em650nm) and induces a chromatic shift (-40nm) with a limit of detection of 50 nM. Additionally, when combined with another DNA strand prior to synthesis, dopamine enhances the fluorescence of the resulting AgNCs (ex535nm/em615nm) with a limit of detection of 1 μM. These responses are specific to dopamine and are screened alongside a library of neurotransmitters, interferents and chemical homologues.

Second, we show that AgNC dopamine sensors are readily attached to DNA-dispersed single-walled carbon nanotubes (SWCNTs) through non-covalent coupling, and can be observed via unchanged fluorescence emission. The SWCNT scaffold provides structural stability to more easily perform single-molecule microscopy for analyzing AgNC population variance, as well as colocalizing the fluorescence of the sensors with the intrinsic nIR fluorescence of the nanotubes for ratiometric sensing. The implementation of sensors for *in vivo* work inevitably hinge on invariance of the sensor to anything other than the analyte, for which a colocalized ratiometric signal can be exploited. We extrapolate our results *in vitro* to evaluate their potential for use *in vivo* to correlate external stimuli to neurotransmission within the brain.

469492 Antenna Coupled Plasmonic Nanowire Endoscope for Label-Free Remote Sensing in Single Live-Cells

Abstract:

The ability to perform chemical/biochemical analysis with high spatial and temporal resolution is crucial in describing chemical inhomogeneity, mass transport and reaction dynamics in many areas, such as heterogeneous catalysis, electrochemistry, photochemistry and materials sciences. It is also extremely important in cell biology, where many cellular communication pathways occur within extremely small volumes. Achieving *in vivo* intracellular chemical analysis is one of the fundamental goals of modern biology to provide foundation for complete understanding of complex biological systems and holds the keys to galvanizing the study of stem cell niches, tumor biology, neurosciences and regenerative medicine. To achieve a better understanding of single, live-cell behavior and intracellular communication, physically or optically addressing different areas of the cell with high spatial and temporal resolution is getting more and more important. The volume of a mammalian cell is typically on the order of femtoliter ($1um^3$) to picoliter ($10um^3$). Therefore, to resolve the chemical heterogeneity within a living mammalian cell, the chemical analysis method needs to have an effective sensing volume much smaller than the cell volume, preferably in the sub-femtoliter to attoliter $(100nm)^3$ regime. Meantime, it should also be able to quickly and quantitatively detect specific chemical or biochemical analytes with very high sensitivity and selectivity, and preferably, in a lable-free manner to avoid interference of chemical interference with cellular functions. *Nanowire (NW)-based single cell SERS (Surface enhanced Raman scattering) endoscopes may be uniquely poised to satisfy such stringent requirements.* At the center of this platform lies the integrated plasmonic NW waveguide on the tip of a tapered optical fiber. The plasmonic NW waveguide is coupled to SERS antennas near its tip, which offers not only significant improvement in SERS enhancement, but also allows for surface functionalization with bio/chemical receptors for chemical sensitivity and specificity. NW-SERS endoscope provides (1) sub-femotoliter sensing volume; (2) high signal-to-noise ratio due to high optical transmission that boosts the signal, and the highly localized illumination that excludes background noises; (3) label-free sensing; (4) active position control in all three dimensions with ultra-fine resolution through a micromanipulator (40nm/step) or potentially a motorized piezostage (1nm/step); (5) small insertion volume that avoids damages to cellular membranes, structures and functions; (6) remote sensing to avoid direct photo damage of living cells. We have demonstrated the applicability of such antenna coupled SERS endoscope for in-situ bioanalysis inside single living mammalian cells and label-free remote detection of cellular pH.

458671 Multifunctional Theranostic Silica-Gold Core-Shell Nanoparticles for Breast Cancer Applications

Abstract:

Breast cancer (BrCa) is the second leading cause of cancer death in women, and this year about 40,000 women are predicted to die from BrCa. About 10-20% of BrCa are found to be triple negative, an even more severe type of BrCa that has lower survival rates. Triple negative BrCa (TNBC) does not express the genes for hormone epidermal growth factor receptor 2 (HER-2), estrogen receptors (ER), and progesterone receptors (PR) which means hormone therapy and drugs that target these receptors are ineffective means of treatment. Recently the use of nanoparticles has gained popularity for drug delivery uses, and this research focuses on the synthesis and characterization of multifunctional theranostic silica-gold core-shell nanoparticles (SGCSNs) which will be used for imaging and photothermal therapy of TNBC. The nanoparticles were synthesized by seeding a functionalized silica core with gold nanoparticles, and then forming a complete gold shell around the seeded nanoparticle through gold reduction onto the surface. The silica-gold core-shell nanoparticles were then functionalized with a bifunctional polyethylene glycol (PEG) molecule. One end of the PEG molecule has a thiol group attached which binds to the gold shell of the core-shell nanoparticles by a gold-sulfur bond. The other end of the PEG molecule either has folic acid attached, which binds to the over expressed folate receptor seen in TNBC, or fluorescein isothiocyanate (FITC) attached which serves as a fluorescent imaging agent. A gold shell was chosen for the synthesis of the core-shell nanoparticles because gold serves as a good agent for CT scanning because of its high Z value, making the nanoparticles ideal for imaging. In addition, because of surface plasmon resonance, gold nanoparticles can be used to destroy cells through photothermal therapy. As a result of the core-shell structure, during photothermal therapy, more heat can be generated from the nanoparticle compared to just gold nanoparticles because of the difference in dielectric constants, therefore more efficiently destroying cancerous cells. The characterized and functionalized SGCSNs will be used for applications in imaging and photothermal therapy of TNBC in vitro, using MDA-MB-231 cells. If successful SGCSNs could be useful for both imaging and photothermal therapy, and could greatly improve the quality of life of TNBC patients.

469148 Nanostructured Polymeric Membranes for Overcoming Obstacles Associated with the Use of Exhaled Breath Acetone As a Non-Invasive Biomarker for Diabetes

Abstract:

The detection of volatile organic compounds (VOCs) in human breath as a portable, non-invasive diagnostic technique for human disease has been a long sought after goal. In particular, acetone breath concentrations have been correlated with blood glucose concentrations in diabetics to monitor the patientâ€™s condition allowing for better control of their symptoms compared to blood glucose measurements and could potentially lead to better glucose monitoring compliance.

Unfortunately, practical implementation of acetone breath monitoring utilizing various portable instrumentation methods has been hampered by the interference of water. Accurate VOC detection in the absence of water has been achieved utilizing nanomaterials whose resistance to electrical current is altered by the adsorption of acetone as well as colorimetric approaches wherein acetone reacts chemically with a chromophore. The presence of water vapor can provide false positives by reacting with the detection media thus lowering the selectivity of the sensor, deactivating the catalyst used in the process, or reducing the sensitivity of response in the sensor, which greatly decreases the accuracy and precision necessary to be utilized in a medical device.

A less well-known phenomenon is responsible inconsistent findings in the literature is the significant change in breath acetone concentration that occurs during the brief period of patient exhalation (typically less than 1 minute). VOC content in human breath is governed by a material balance involving exhalation and transpiration from blood vessels lining the lungs. Transpiration, in turn, depends on both the thermodynamic partitioning of the acetone concentration (between the blood, vessel wall, and vapor phases) and dynamic convection through the vessel wall. Both thermodynamic and kinetic processes are governed by the physical make-up of the vessel wall. Consequently, real-time analysis (response times on the order of seconds) are essential to obtain accurate correlations between acetone and blood glucose and approaches based on long-term sample collection are susceptible to error.

We here demonstrate this phenomenon using real-time chemical sensing data obtained from patients utilizing a unique colorimetric approach that mitigates water interference. The method involves the use of nanostructured polymeric membranes that serve as both scaffolds and catalysts to drive highly sensitive and selective colorimetric responses to acetone in humid environments. The fundamental mechanism involves protection of acid catalyst sites within the membranes from de-protonation by altering their local pKa through the incorporation of specific organic acids. Successful clinical trials for the approach for diabetes management and assessment of vascular degeneration will be discussed.

AUTHOR INDEX

AUTHOR INDEX

AUTHOR INDEX

AUTHOR INDEX

AUTHOR INDEX

AUTHOR INDEX

AUTHOR INDEX

AUTHOR INDEX

AIChE
120 Wall Street, FL 23
New York, NY 10005-4020

ISBN 978-1-5108-3441-5

CURRAN ASSOCIATES INC.
proceedings.com